Power Generation
Resources, Hazards, Technology, and Costs

The MIT Press
Cambridge, Massachusetts,
and
London, England

Power Generation

Resources, Hazards, Technology, and Costs

Philip G. Hill

This book was set in Monophoto Times Roman and printed and bound in the United States of America

Library of Congress Cataloging in Publication Data
Hill, Philip Graham, 1932–
Power generation

Includes index.

Includes bibliographical reference and index.

1. Electric power production. 2. Electric power-plants—Environmental aspects. 3. Power resources. 4. Fuel. I. Title.
TK1001.H63 621.312 76–54739
ISBN 0–262–08091–5

Contents

Preface

This book was written to fill the need for a reasoned presentation of the resources, hazards, and costs of power generation in the context of a review of the technology. The great concern in recent years with resources and the environment means that engineering design must be guided and constrained by many more factors than in the past. The engineering profession should be fully aware of widening implications of power generation so that planners can work as intelligently as possible for long-term human benefit. Since there is also a great need to develop an awareness outside the engineering profession of the bases and constraints of power engineering, it is hoped that this book will prove useful to nonengineering students as well.

The book is organized into four more or less independent parts. Chapters 1 to 3 deal with energy resources in light of the demand for power and the possibilities of energy conservation; Chapters 4 to 6 deal with the environmental effects and hazards of power generation; Chapter 7 focuses on financial costs; and Chapters 8, 9, and 10 review the technologies of nuclear, fossil, and solar power.

The problems at the end of each chapter provide the student with occasions to think about the implications of basic factors. The book is intended primarily for upper-level undergraduate or first-year graduate students in engineering, science, or environmental studies.

A first course in thermodynamics would be a help in comprehending the material on power technology.

An awkward and sometimes confusing feature of the energy field is the variety of units in which energy is commonly measured (e.g., joules, British thermal units, barrels per day of oil equivalent, tons of coal equivalent). Where comparisons of energy resources or consumption are made in this text, the S.I. (Système International) unit—the joule—is used as a common base. At various points in the text, however, and in the problems, various non-S.I. units are cited to provide the student with experience in converting from one unit to another. Appendix A provides the necessary conversion factors.

For several reasons, the main focus of attention of this book is on energy use for electricity generation. First, in this century electricity demand growth has far outpaced total energy demand growth in percentage terms. If present trends continue, the all-electric society will approach within a few decades. Although this is unlikely to happen literally, continued rapid growth in electricity demand will necessarily quicken the pace of total energy growth. The trend of doubling electricity demand every 10 years has persisted almost since the turn of the century, though it has attenuated somewhat lately.

Second, electricity generation will be the main site for use of nuclear energy and will therefore be of special significance as fossil fuel resources are depleted.

Third, with its tremendous demand for capital investment and necessary emphasis on efficiency, electricity generation is an area of energy use where technology is stretched extremely hard. This will continue to be true as fuel prices rise and environmental constraints become more difficult to satisfy.

Finally, if the problem of massively expanded, environmentally acceptable, and reasonably low-cost electricity generation can be solved, solutions to many other problems could follow relatively easily, such as electric transportation, synthetic fuels, and integration of power and heating.

The data reviewed in this book indicate that the fossil fuel age will come to an end within a century or less and that world supplies of oil will begin to run out before the year 2000. In principle, coal, uranium, and solar radiation could more than meet all needs, at least up to the middle of the next century, and in a steady-state world for as long as the population can be controlled to a level of perhaps 20 billion or less. There are, however, undoubted perils and costs in using this energy too freely. The use of coal must be severely constrained to prevent excessive pollution of the atmosphere, and nuclear power is also potentially dangerous. Solar power appears at the present time to be an order of magnitude higher in cost than nuclear- or coal-generated electricity. Nuclear and coal power costs are competitive at present (depending upon the local price of coal), but future nuclear costs may be significantly less. Because of the potential hazards of nuclear power, however, it would seem unwise to rely on it solely.

It has often been argued that energy conservation could ease the problem of future energy supply. Modest savings in energy use (without reduction of living standards) have already been justified by rising fuel prices. But beyond this, there may be a need for measures that, in the long term, would reduce comfort and convenience and that may not appear at present to be economically justified. The incentives for such substantial energy conservation may be clear in principle but are unlikely to be effective until the long-term costs of liberal energy use are much more finely appreciated than at present. A rational case for substantial (and costly) energy conservation has hardly begun to be made. This can only be done after energy resources, technologies, costs, and hazards are clearly and quantitatively defined.

There is a rich variety of possible technological solutions to many energy problems. However, many of these solutions are in a quite primitive state of development. This should not be surprising. The abundance of Middle East oil and its low price (both of which were taken so much for granted until recently) effectively discouraged serious innovation. But the world now faces a shortage of petroleum, which prospect has already driven prices up enough to make substitute forms of energy quite attractive. It will, of course, be a long time before major substitutions can be made.

The main concern in the text is with North American power generation, but since the energy market is international, a review has been offered of world resources and power capacity. Also, a comparison of trends in different economies is often illuminating, so data from the United States, Canada, and other countries are compared at various points. In the long run, we do not appear to face an energy famine since there are adequate supplies of energy and enough potential technology to allow continued production of energy at reasonably low cost. However, we are extremely short of time. Major developments that should have been under way 10 to 20 years ago have not yet really begun. The next three decades should bring tremendous opportunites for new engineering in power generation.

Acknowledgments

It was a sabbatical leave during 1973–1974 that enabled me to get this work under way. For this I wish to record my sincere thanks to Queen's University. I am also deeply indebted to the Commonwealth Scholarship Commission of the Association of Commonwealth Universities for a Commonwealth Visiting Professorship and to the Universities of Sheffield and Cambridge for their excellent hospitality and assistance during that year. Professor David E. Newland (then of Sheffield, now of Cambridge) initiated and planned all my arrangements; to him I owe thanks for his intense interest, kindness, and encouragement. I would also like to thank Professor J. K. Royle for making available to me the resources of his department at Sheffield; Professor J. H. Horlock (then Director of the S.R.C. Turbomachinery Lab at Cambridge, now Vice Chancellor of the University of Salford) for welcoming me most warmly to his laboratory; Sir William Hawthorne, Master, and the Fellows of Churchill College for an unforgettable welcome as Overseas Fellow in the College during Lent and Summer Terms of 1974. I derived great benefit from stimulating discussions with numerous people concerned with energy problems, including Sir William, Dr. Norman Kendall of Shell, Dr. Richard Eden of the Cavendish Laboratory, Lord Hinton of Bankside, and many others. World events in 1973–1974 made us acutely aware of the need for fundamental reassessment of our energy future.

But long before the crisis arrived, the late Dr. John J. Deutsch, then Principal and Vice Chancellor of Queen's University, gave me much help and encouragement in studying these matters; I am certainly grateful to him, to Dr. R. J. Uffen, Dean of Engineering, and to my close colleagues who share in the teaching of thermodynamics and power courses in the Department of Mechanical Engineering at Queen's University—Professors H. G. Conn, G. F. Marsters, W. D. Gilbert, C. K. Rush, P. H. Oosthuizen, and J. D. McGeachy—as well as Dr. W. B. Rice, Head of the Department, and Mr. Eric D. Neumann, my graduate student.

I am immensely grateful to a number of manuscript readers who responded confidentially to requests from the MIT Press and provided the incentive for many an important revision.

Dr. Peter Barry of the Health Physics Branch of Atomic Energy of Canada Ltd. at Chalk River reacted most generously to my request for an independent assessment of Chapter 5. His detailed critical review led to major improvement in that chapter. Dr. John McGeachy thoughtfully reviewed Chapters 8 and 9 in the light of his extensive field experience in fossil fuel and nuclear plants, and provided many helpful suggestions. Mr. F. F. Ross of the Central Electricity Generating Board, London, kindly expended much effort verifying the value of approximate calculations on the tall stack hypothesis.

Special thanks are due to Miss Isobel Morgan and Ontario Hydro for permission to use Miss Morgan's sketches of power plants, and also to other organizations who have granted permission to use other illustrative material or data: World Energy Conference, Lawrence Livermore Labs, International Atomic Energy Agency, National Academy of Sciences, Royal Society of Canada, Northeast Utilities, Commission of European Communities, Babcock and Willcox, Central Electricity Generating Board, General Electric, Westinghouse, General Atomic, Atomic Energy of Canada Ltd., United Technologies Corp., Electric Power Research Institute, American Society of Mechanical Engineers, Institution of Mechanical Engineers, American Power Conference, McGraw-Hill, Plenum Press, and the publishers of *Electrical World, Power Engineering*, and *Nuclear Engineering International.*

Of all the help received, there is no doubt that the chief contribution has come from my wife, Marguerite, and specifically from her decision early on that the work was sufficiently important to deserve the enthusiastic support of the family. Marguerite did almost all of the typing of many drafts (helped initially by Mrs. Betty Rowe, Mrs. Rosemary Cox, and Miss Barbara Pearson), checked the English, and took a keen interest in the whole project. Anne, Tricia, and Graham were also greathearted and understanding in every phase and are just as pleased as I now that the job is done.

1
World Energy Demand and Supply

1.1 INTRODUCTION

The energy supply problem is clearly international. Though individual countries may prefer, in principle, to be self-sufficient in energy, they have proved quite willing to import energy when it is cheap and apparently abundant, as oil seemed to be until autumn 1973. This has led to a complex international network of supply in which many of the industrialized countries are now, or are rapidly becoming, heavily dependent on insecure foreign sources of energy. Oil depletion and price will be very much dependent on total world demand. Many of the other serious energy problems of the future will also be worldwide.

1.2 WORLD DEMAND

Figure 1.1 shows the rapid increase in recent decades of world energy consumption.[1] Such a trend cannot continue indefinitely, and it is hard to believe that recent rates of growth will continue even to the end of the century. The dashed lines in Figure 1.1 reflect the belief of the World Energy Conference that growth rates in this decade will be much lower than in the last. Figure 1.2 shows that world energy consumption has grown exponentially over a 20-year period at an average rate of 5 percent per year. This means a doubling of the rate

[1]See Appendix A for energy units and conversion factors.

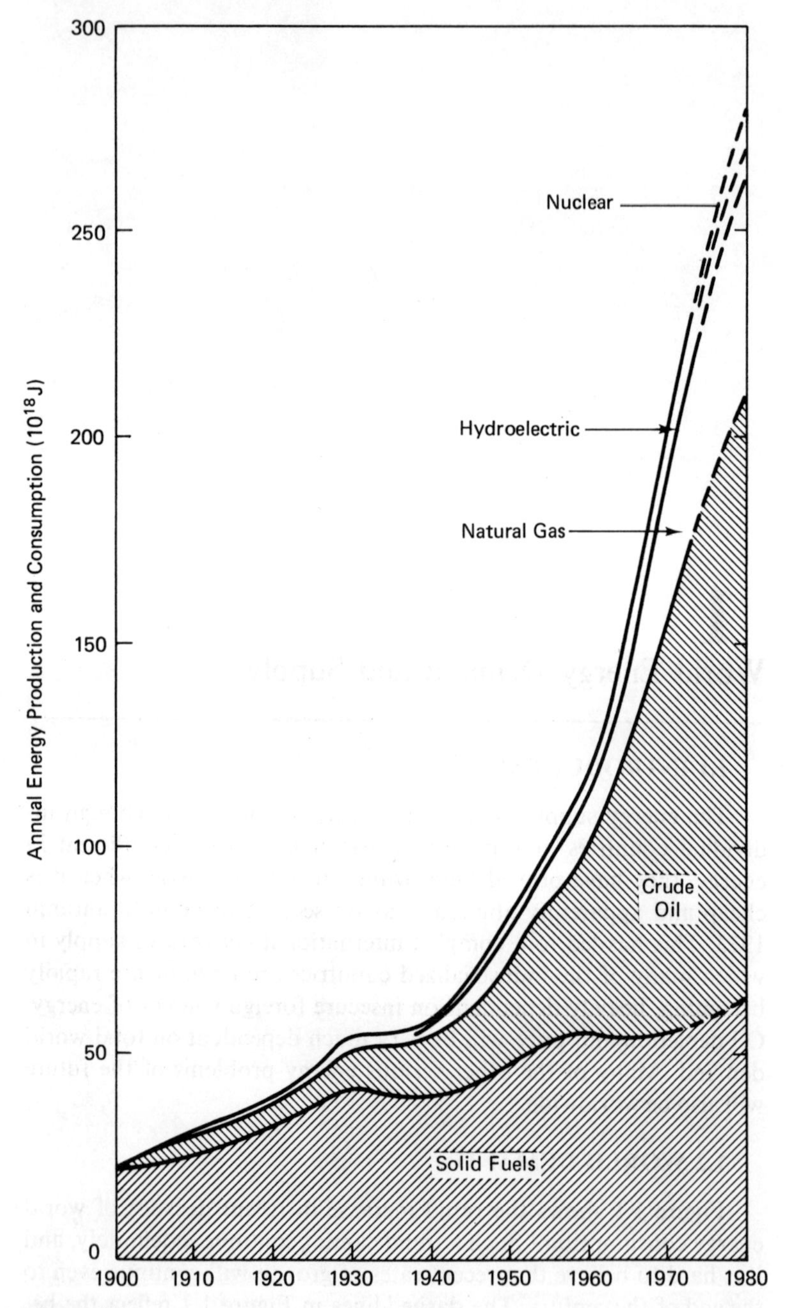

Figure 1.1
Changing use of energy resources in the twentieth century. From *Survey of World Energy Resources 1974* (1), courtesy of the World Energy Conference (London).

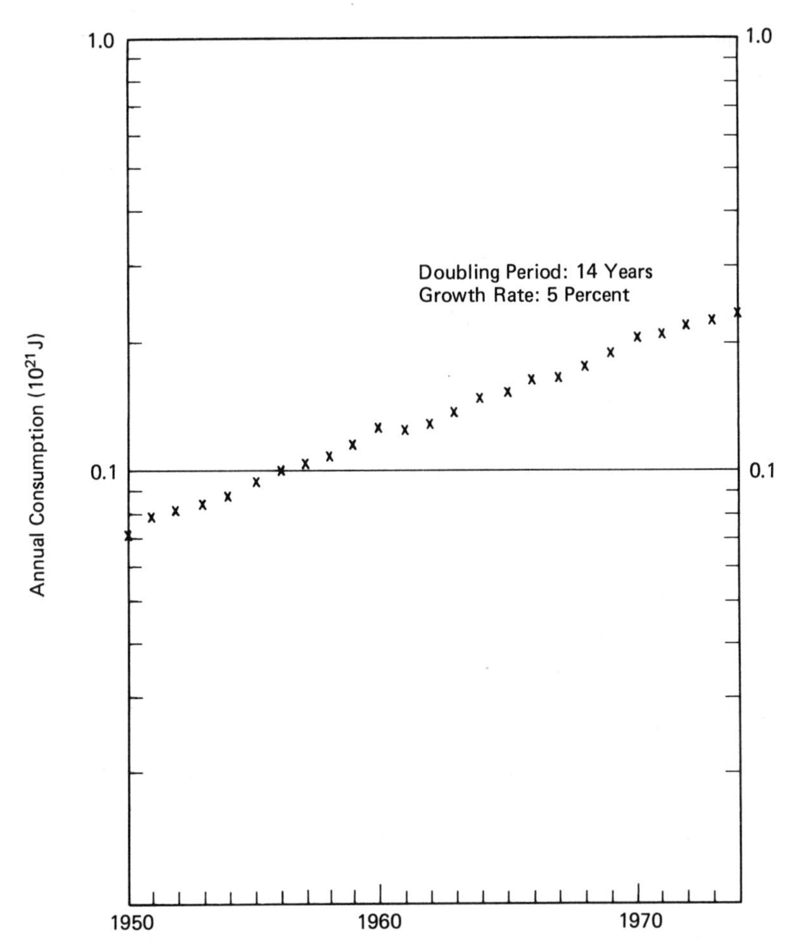

Figure 1.2
World energy consumption, 1950–1974 (2).

every 14 years. (Another 100 years with an average growth rate of 5 percent would raise total energy consumption by a factor of 132.)

Table 1.1 shows growth rates and per capita consumptions of energy for the entire world in this century. In 70 years the consumption rate increased nearly 10-fold. This implies an average yearly growth rate of 3 percent, so that the growth rate has itself been growing in recent decades. The oil embargo of 1973 undoubtedly slowed growth in the period 1972–1974, but the increase in consumption was nonetheless substantial.

Figure 1.3 shows the rates of increase of energy use in various parts of the world. In the United States the average annual increase has been only slightly less than the 5 percent world average. In Canada it has been nearly 6 percent. In the United Kingdom it has been remarkably low, recently averaging 1.5 percent. In Japan the rate

Table 1.1
World Energy Consumption

Year	Annual Consumption (10^{18} J)	Average Annual Growth Rate (percent)	Annual Consumption per Capita (10^9 J)
1900	22		14
		2.9	
1925	45		23
		2.2	
1950	80		32
		4.0	
1960	118		40
		5.5	
1965	155		47
		5.4	
1970	201		55
		5.3	
1972	218		58
		3.3	
1974	233		60

Source: Refs. (1) (2).

of increase has been nearly 12 percent annually. Despite large differences in rate, the tendency for exponential growth appears well established. Growth rates have been greatly upset by the huge increases in oil prices set by the Organization of Petroleum Exporting Countries (OPEC) since autumn 1973, but there is no firm evidence as yet of a long-term leveling off of either national or world demand for energy. Indeed, if one considers the tremendous energy needs of the developing countries and the strength of electricity demand growth, it would appear that without drastic conservation measures future world energy demand could grow at rates significantly higher than 5 percent per year.

Although per capita demand has been everywhere on the increase, there are great differences in present consumption levels from country to country. Figure 1.4 shows relative per capita energy consumption in 1970 as a function of cumulative fraction of the world population. Per capita consumption in the United States is six times the world average, so that to bring the whole world up to the level of the United States would require a tremendous increase in the total consumption rate.

In the United Kingdom and Japan, population has been increasing relatively slowly during the past decade. The difference between the energy growth rates in those countries is very striking, suggesting that population growth is by no means the most

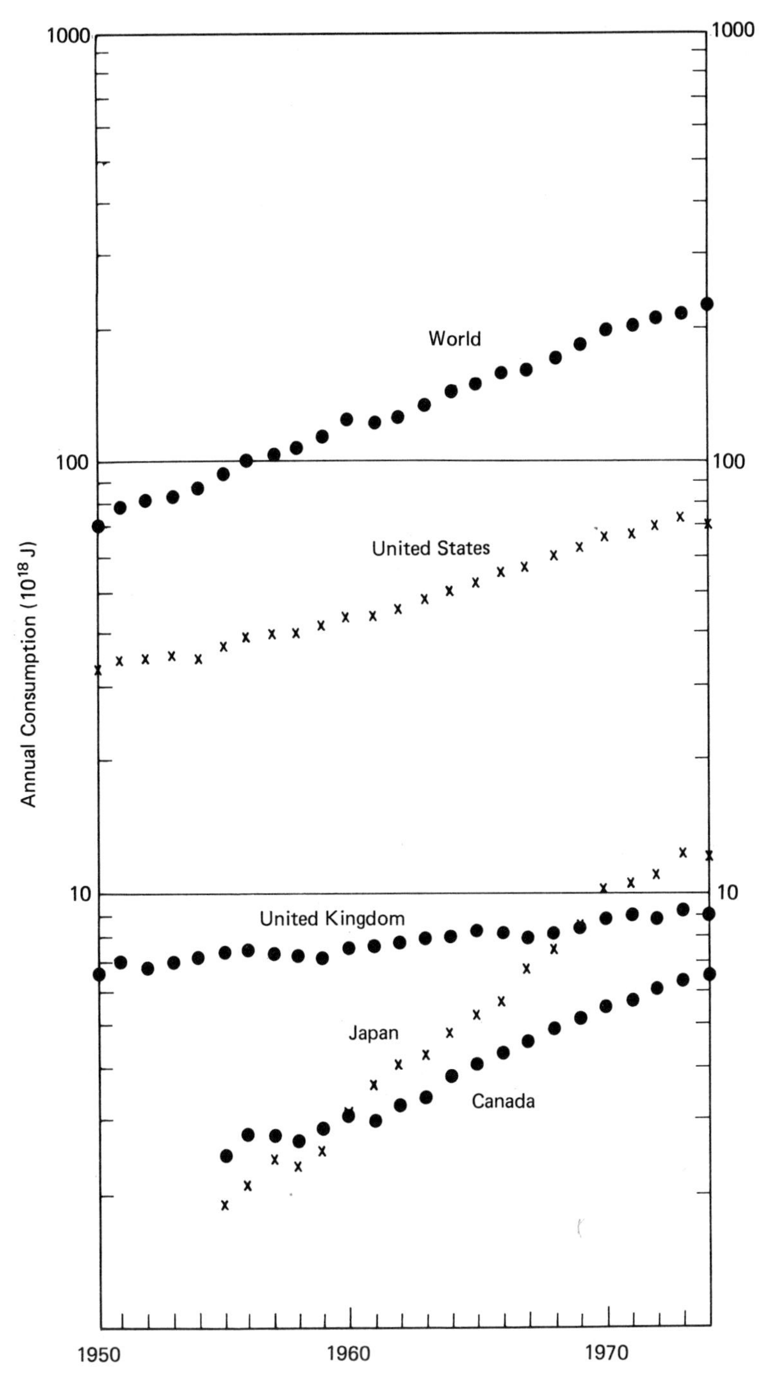

Figure 1.3
Annual energy consumption, selected countries, 1950–1974 (2).

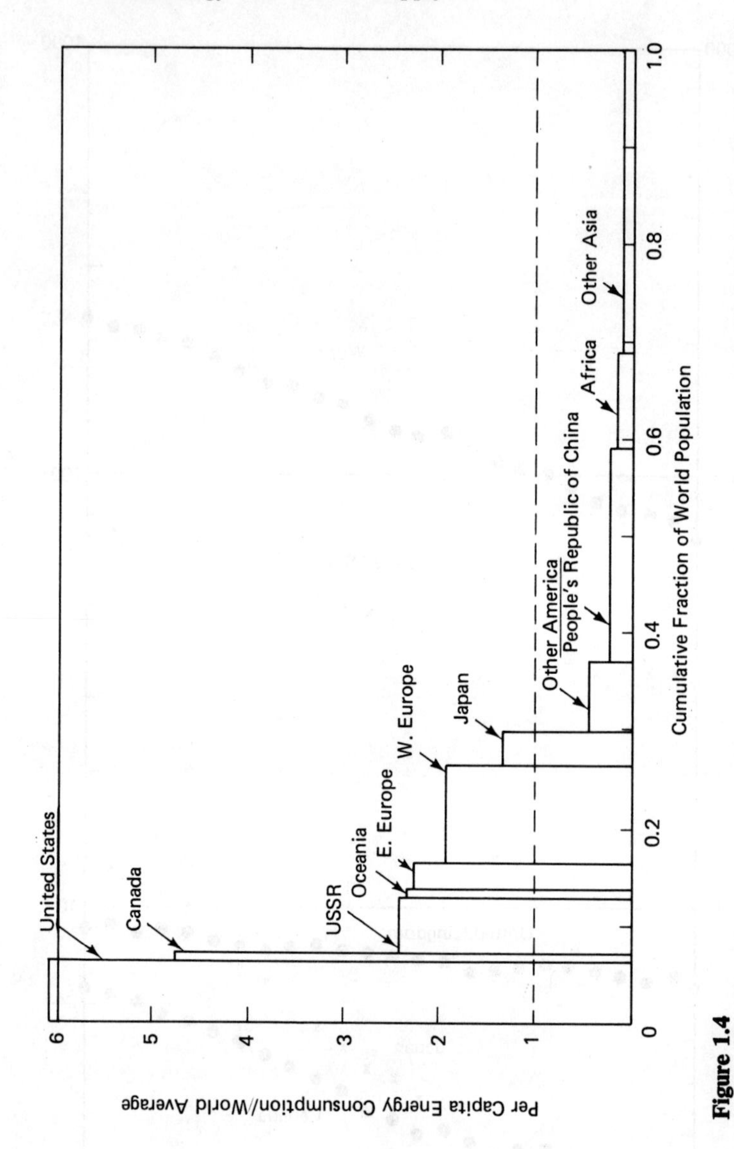

Figure 1.4
Per capita energy consumption as a ratio to the world average vs. cumulative fraction of the world's population. Adapted from Hottel and Howard (3).

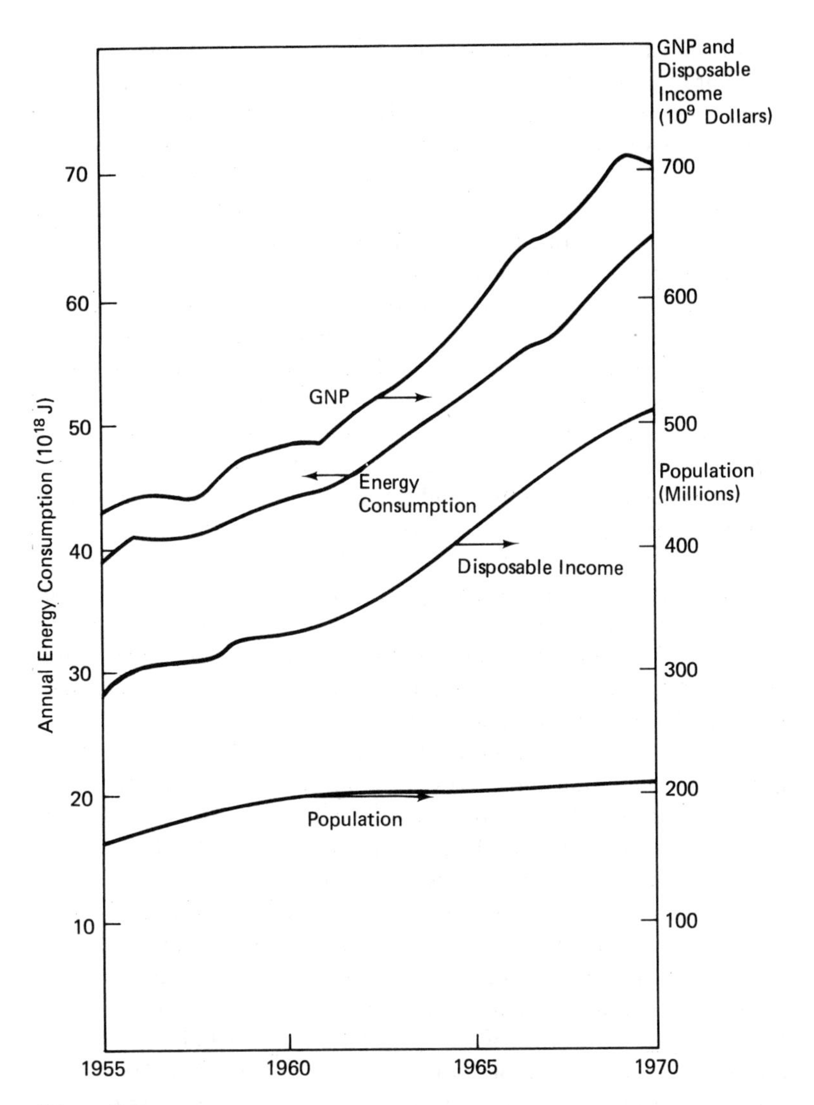

Figure 1.5
Growth of energy consumption and disposable income in the United States, 1955–1970. Adapted from Austin (4), courtesy of Lawrence Livermore Laboratory.

important cause of increase in energy demand. In the United States, as Figure 1.5 indicates, energy consumption has been growing nearly three times as fast as population. Energy growth appears to have kept pace with "disposable income," defined as what the people have to spend on consumables of all kinds. (Disposable income may be calculated as the gross national product minus the sum of all taxes, undistributed corporate profits, transfer payments, etc.)

The apparent close relationship between energy consumption and disposable income suggests that as poor countries become wealthy their demand for energy will grow enormously, whether or not their populations are stabilized. It also raises a serious question about the possibility of long-term energy conservation.

If they are to make any real impact on overall consumption, conservation measures will almost certainly require significant reductions in human comfort and freedom. Unless people know that continued uninhibited use of energy will be unacceptably costly or dangerous (or both), it seems unlikely that attempts at energy conservation will succeed in the long term. In brief crises they may be very effective; the question is whether they can be sustained for a long time. If world energy use could be cut today by 50 percent, but continued thereafter to grow at an annual rate of 5 percent, it would be only 14 years until we were back in the same position.

Although it is quite clear that an indefinite continuation of growth in population and energy consumption is impossible, it is not clear at what point the steady state will be reached. Given the limited land area and resources of the earth, one hopes that the steady state will be reached before conditions become intolerable.

World population growth rate has been variable, from 0.002 percent annually prior to A.D. 1700 to about 2 percent recently. Projection of future population is highly risky, owing to uncertainty in the rate of adoption of birth control on a world scale and to changing life expectancies. With a world population of 3.6 billion in 1970, estimates of ultimate stable population range from 5 to 20 billion (5). Tomas Frejka suggests that the population growth illustrated in Table 1.2 is credible.

Table 1.2
A Projection of World Population Growth

Year	Birth Rate (per 1000 per year)	Annual Growth Rate (percent)	World Population (billion)
1970	33	2.0	3.6
2000	21	1.2	6.4
2050	14	0.3	8.2
2100	13	0.0	8.4

Source: Ref. (6).

In a few countries the birth rate has already declined to the lowest level shown in Table 1.2, but average world rates may be much higher for a long time; it is quite conceivable that world population could rise to two or three times the 8.4 billion estimate cited in Table 1.2.

Another difficult question concerns the energy consumption per capita that would be needed for adequate comfort and convenience everywhere in the world. Weinberg and Hammond estimate that the present per capita rate for the United States (10 kW thermal) might ultimately need to be doubled to allow for large additional energy expenditures for environmental control, provision of food and water, and mining of increasingly dilute ores (7).

Assuming a world population of 20 billion and 20 kW thermal per capita consumption, the total annual energy use becomes 3.5×10^{15} kWh, or 12.6×10^{18} J, or about forty times the present level. With this consumption rate, even the vast coal resources of the world would be used up fairly quickly, possibly within 30 years. Sooner or later, major changes of fuel must be made.

Whether the world can tolerate a waste heat approaching 3.5×10^{15} kWh per year must also be seriously considered, along with the many other environmental problems that accompany high population density and high energy use per capita.

Whether the world actually needs 20 kW thermal per person to assure a high standard of living is also questionable. Serious efforts at energy conservation are only beginning. So far, there has been little economic or other incentive to conserve energy. However, energy use for food production (synthesizing food and fertilizer), water supply (desalination), and air quality control may be much more intensive in the future.

It may also be questioned whether all the developing countries will ever reach reasonable standards of living and whether major wars and other catastrophes will be averted. The prediction of the future is thus quite impossible. Nonetheless, the implications of large-scale future energy consumption deserve serious consideration.

1.3 ELECTRICAL ENERGY DEMAND

An important feature of the growth in demand has been the demand for electrical energy, which has been surprisingly similar in many countries, despite wide differences in total energy consumption. Figure 1.6 shows large and similar rates of growth in electricity demand in the United States, the United Kingdom, and Canada. The world average, at 8.1 percent, is even larger. All of these rates are substantially higher than growth rates for total energy consumption, so that energy used in electricity generation is a rapidly increasing fraction of total consumption.

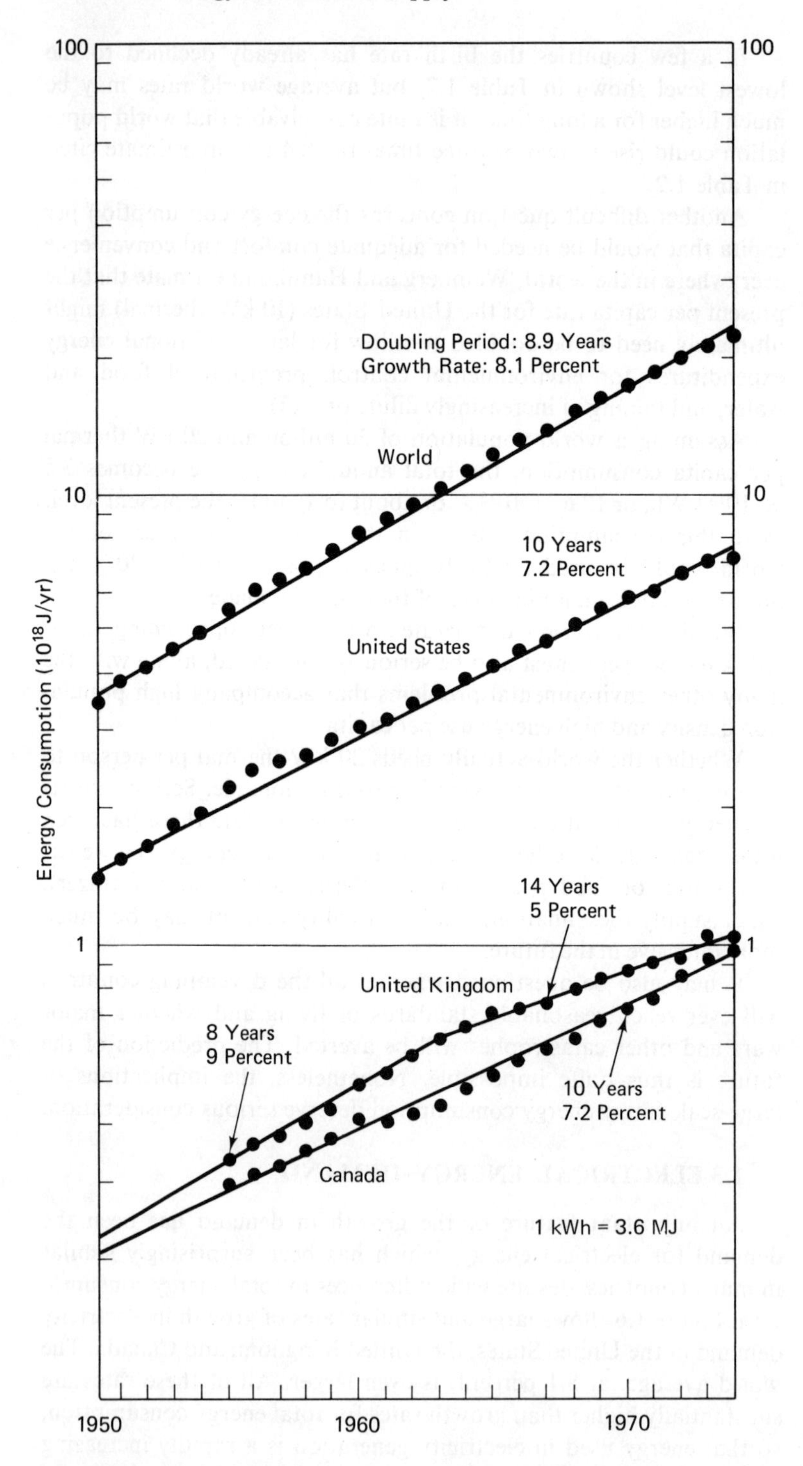

Figure 1.6
Annual electricity consumption, 1951–1974 (2).

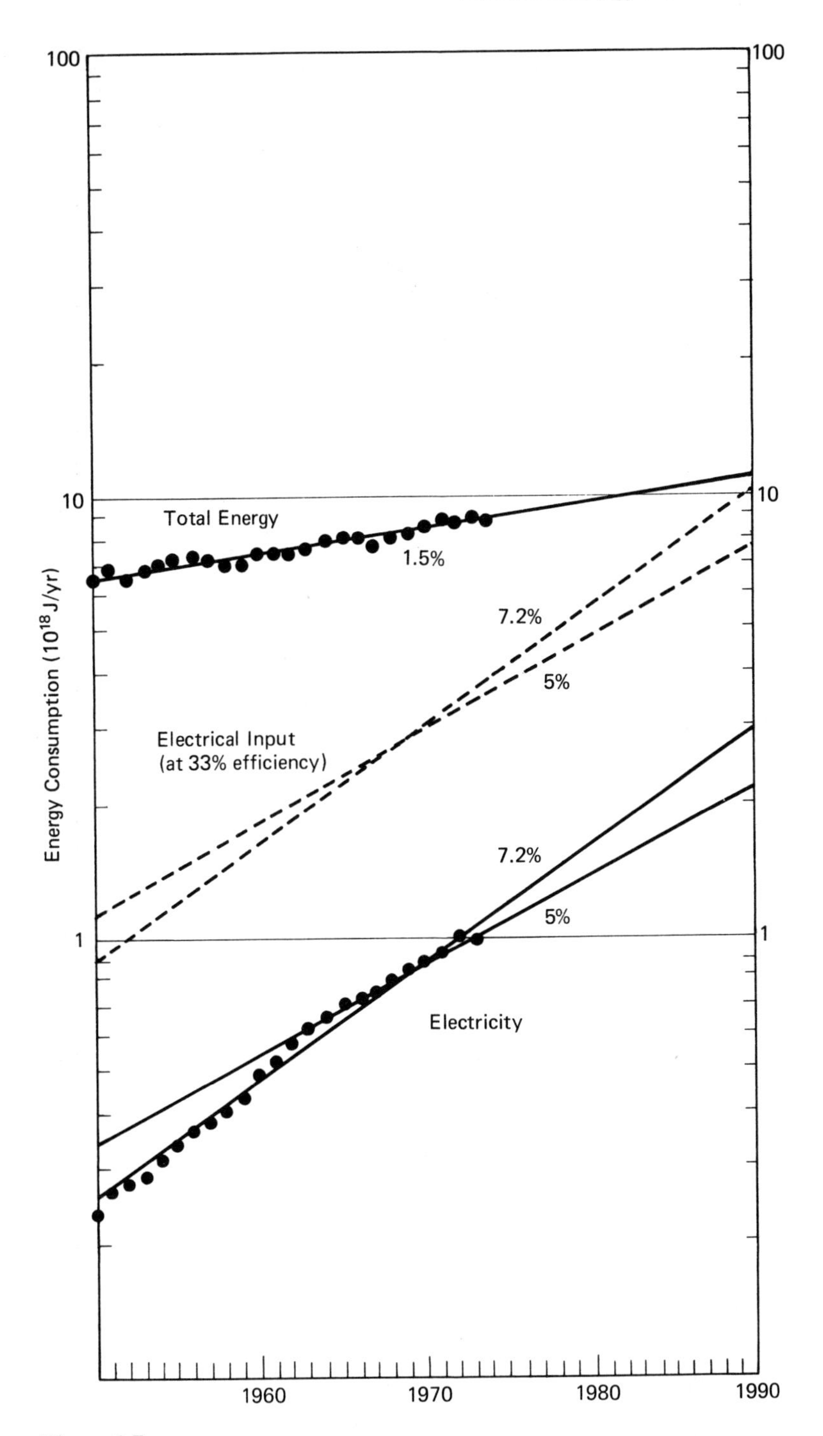

Figure 1.7
Total energy consumption and electricity consumption in the United Kingdom, 1955–1973 and projected to 2000 (2).

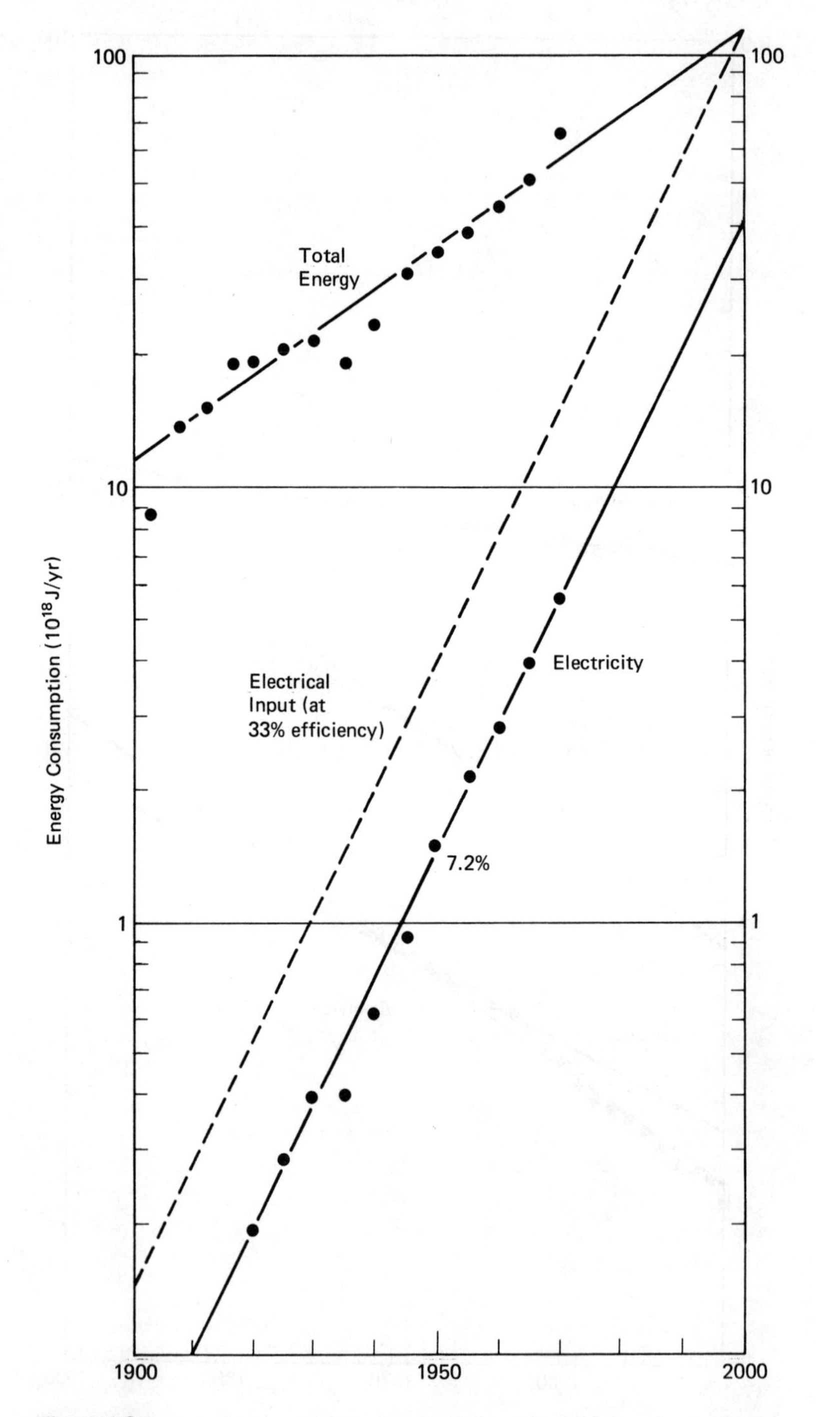

Figure 1.8
Total energy consumption and electricity consumption in the United States, 1920–1970 and projected to 2000 (2, 8).

As shown by Figure 1.7, the growth rate in total energy use in the United Kingdom has been only about 1.5 percent per year over the last decade; in comparison, the growth in electricity demand has been huge. The dashed line in Figure 1.7 illustrates the consumption of energy to generate electricity, assuming an average generation efficiency of 33 percent, somewhat higher than the present 28 percent year-round average efficiencies reported by the U.K. Central Electricity Generating Board. In about 20 years, if recent trends continue, all energy used in the United Kingdom would be devoted to electricity production. Thereafter the growth rate in total energy would jump abruptly; Britain would become an all-electric society.

Figure 1.8 shows the corresponding situation in the United States. Again, the intersection point is reached in a few decades. It is striking not only that the electricity growth rates are so similar in the United Kingdom, the United States, and Canada, but that these growth rates have been nearly constant for so long a time. Figure 1.8 shows 60 years of nearly constant annual growth rate in the United States (the dip associated with the great depression of the 1930s did not affect the long-term trend). Figure 1.8 suggests that due to the rapid growth in electricity demand, the total energy demand growth rate in the United States has already begun to increase.

In 1970 the total world generation of electricity was about 5×10^{12} kWh, which is about 10 percent of the world consumption of energy during that year. If the average efficiency of electricity generation is roughly one-third, then approximately 30 percent of the total world use of energy was devoted to the production of electricity in that year. It can be shown that if the electricity usage rate grows at 8.1 percent per year and total energy use grows at only 5 percent per year, then in less than 25 years the world will have become all-electric. This result seems rather unlikely, but it does suggest that if the current growth in demand for electricity continues, world energy use will grow significantly faster than 5 percent per year.

1.4 FOSSIL FUEL ENERGY

1.4.1 Oil and Gas Depletion

Since nearly 75 percent of present world energy consumption is from oil and gas, the possible imminent depletion of these resources has serious implications. Oil production has shown a steady exponential growth in this century, with a growth rate, shown in Figure 1.9, of around 7 percent per year. This would continue for a long time if sufficient resources remained available. The present level of production is rapidly approaching 20 billion bbl per year. Estimates of proved reserves range from approximately 350 billion to 570 billion bbl; potential reserves are estimated at about 2 trillion bbl.

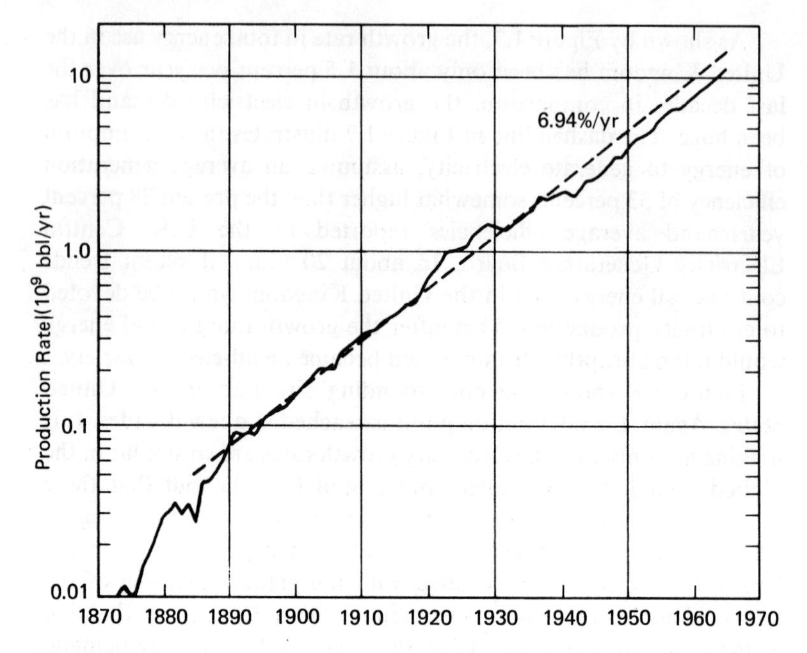

Figure 1.9
World production of crude oil. After Hubbert (9, 10), courtesy of the International Atomic Energy Agency, Vienna, Austria.

At present production rates it would therefore take 100 years to exhaust the potential reserves of oil. At first sight this figure hardly looks alarming, but it must be remembered that this estimate pertains to *potential* oil reserves. Only about one-third of these have actually been located so far—the remainder may not all be discovered or may become unavailable for political reasons. It should also be remembered that the 100-year lifetime refers to the present production rate. If this were to continue to grow at 7 percent per year, then 2 trillion bbl of oil would last only another 29 years, as shown by the mathematics of exponential growth (Appendix B). Reserves of 570 billion bbl would last only 13 years.

Thus it appears unquestionable that exponential growth in oil production will soon be arrested. It is perhaps hard to grasp that depletion will come so quickly when possibly only 15 percent of the world's potential reserves of oil have yet been consumed. It is of course unlikely that exponential growth of production would continue right to the point of exhaustion. Long before that point, in the next 10 to 20 years, major action will have to be taken to substitute new energy sources.

The intensity of the problem will depend strongly on future growth in energy demand, and demand forecasts are highly uncertain. They depend on population growth, a matter of considerable uncertainty, and on other factors such as the effect of rising prices on future energy consumption and the possibility of large-scale energy conservation. Estimates of fuel reserves are also very uncertain. There are often large gaps between "proved" reserves, known to be capable of extraction at reasonable cost, and "potential" reserves, including those deposits not yet discovered but which, extrapolating from previous exploration experience, may reasonably be believed to be available.

Sam H. Schurr has written,

> There is no true measure of the world's endowment of energy resources, nor is there ever likely to be one. Cost alone would prohibit a comprehensive probing of the earth's crust to provide anything approaching a true measure of the resources. More to the point, society's interest is confined to resources that are exploitable now or seem likely to be in the future. As time passes, the standards of exploitability keep changing, mainly as the result of advances in technology and changes in the economic circumstances. Consequently resource supply estimates are subject to at least as much uncertainty as energy demand estimates. (11)

M. K. Hubbert (9, 10) has speculated in a plausible way about future rates of oil production. His results, shown in Figure 1.10, were obtained by assuming what he felt were reasonable upper and lower limits to potential oil resources, and by supposing that the growth and decay of production would follow a Gaussian curve. The curves shown are those with the best fits to the historical variation of oil production, with the total resource quantities indicated. These suggest that, despite our uncertainties about total potential resources, the peak in world oil production will have been reached by the end of the century.

Given that oil is rapidly running out, the question is: How

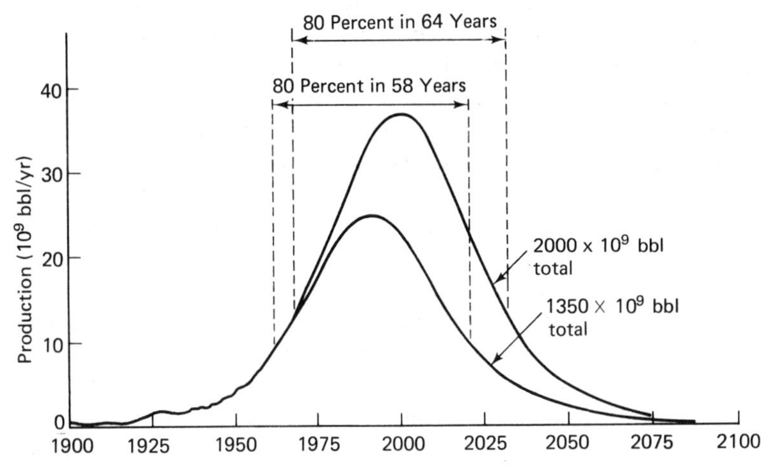

Figure 1.10
Projected world oil production. After Hubbert (10), courtesy of the IAEA.

Table 1.3
World Energy Resources

Resource	Proved Recoverable Resources		Recent Annual Production	
	Quantity	Energy (10^{21} J)	Quantity	Energy (10^{21} J)
Crude Oil	91.5×10^9 tonnes	3.957	2.493×10^9 tonnes	0.1078
Natural Gas	52.5×10^3 km^3	1.932	1.389×10^3 km^3	0.051
Coal[a]	591×10^9 tonnes	13.858	3.00×10^9 tonnes	0.070
Oil Shale (10% kerogen)	230×10^9 tonnes	9.90		
Bituminous Sands (United States)	2.175×10^9 tonnes	0.094		
Bituminous Sands (Canada)	50.25×10^9 tonnes	2.172	3.14×10^6 tonnes	0.136×10^{-3}
Uranium[b]	0.984×10^6 tonnes	0.824 (nonbreeder) 49.4 (breeder)		
Thorium	0.322×10^6 tonnes	16.6 (breeder)		

[a]Total resources in place are estimated at 10.753×10^{12} tonnes.
[b]Lower-cost resources (less than $25/kg). At a price ranging up to $40/kg U, the sum of "reasonably assured" and "estimated additional" resources is about 4×10^6 tonnes (12).
Source: Data from Ref. (1).

abundant are other fuels? Table 1.3 shows proved recoverable reserves of various fuels. These are cited for the entire world, excepting the quantities of bituminous sands, which are cited for North America only. The word "recoverable" signifies that these reserves can be extracted with current technology.

The world quantity of crude oil cited in Table 1.3 is equivalent to 660 billion bbl, only about one-third of Hubbert's upper limit estimate of potential reserves. The ratio of proved reserves to recent annual consumption of crude oil is 37:1. This is small enough, but for the United States and Canada the comparable figures are 13:1 and 15:1.

Table 1.3 indicates that proved world reserves of natural gas are only about half those of crude oil. The ratio of proved recoverable reserves to annual consumption is about 38:1 for the world as a whole. Comparable figures for the United States and Canada are 12:1 and 35:1, respectively. Thus the depletion of natural gas is no less a threat than the exhaustion of crude oil. North American reserves of both will be depleted much sooner than world reserves.

Natural gas has been heavily utilized in North America, where domestic supplies have been tapped and distributed at low cost by pressurized pipelines. In the last two decades the market has expanded rapidly due to the low price of this fuel and its "cleanness" of combustion. The reserves of natural gas in North America are now dwindling rapidly. Sea transport of natural gas is much more expensive than by pipeline; liquefied natural gas tankers do bring natural gas from Algeria, for example, to the United States, but at relatively high cost and in very limited quantity. Natural gas from the North Sea is of course a major resource for the United Kingdom and Europe. Elsewhere in the world natural gas is not used widely, mainly owing to transportation cost relative to that of oil. Much of the natural gas extracted with oil production in the Middle East has been wasted through combustion at the site. Conversion to methanol (liquid) is one method of processing this resource for low-cost shipment to any part of the world.

1.4.2 Coal Reserves

Coal reserves, according to Table 1.3, are much larger than those of gas and oil. The ratio of proved reserves to annual consumption is nearly 200:1, and potential reserves of coal are vast, perhaps 20 times higher than shown in Table 1.3 for proved recoverable reserves. Paul Averitt (13) indicates that the potential reserves of coal may amount to 400×10^{21} J. Possibly only 1 percent of the world's coal reserves have yet been consumed. If the recent average annual growth rate in coal consumption of 3.6 percent (7) were to continue, complete exhaustion of coal could take as little as 130 years (Appendix B), but any projection is quite uncertain.

Owing to problems with mine safety, landscape damage, labor costs, and air pollution, the production of coal has been greatly inhibited in a number of countries in recent years. New technology in mining and coal gasification could change its suitability as a fuel and thus hasten its depletion. In its present main use (electricity generation) coal appears to be increasingly expensive relative to nuclear energy, so that the future for coal in the next few decades will depend strongly on whether the nuclear power program develops with reasonable safety and cost.

1.4.3 Shale Oil Reserves

As Table 1.3 shows, energy reserves in oil shales are vast. Over 95 percent of the commercially recoverable shale oil indicated by this estimate is in North America, but exploration for shale oil in much of the rest of the world is quite incomplete (1). Oil is being extracted from shale in the USSR and China (at about 0.4×10^{18} J per year), but this is less than 0.4 percent of world consumption of crude oil. Shale oil is very expensive to extract. Even high-grade deposits of shale contain only about 10 percent shale oil (kerogen), or about 25 gal/ton of rock. Mining of these shales means handling and disposing of vast amounts of reject material. It also requires crushing, grinding, and heat energy to separate the oil from the rock. The technologies, not to mention the process plants, are far from being available enough to make use of the great energy reserves represented by the shale oil entry. And the environmental problems resulting from strip-mining and waste disposal from shale oil processing could be colossal. If these problems can be solved at reasonable cost, the potential resources are tremendous. World reserves of shale oil from rock containing not less than 4 percent kerogen are 100 times the entry in Table 1.3.

1.4.4 Bituminous Sands

No satisfactory estimate of world reserves of bituminous sands is available (1), although the reserves in North America are substantial (Table 1.3). Again, the costs of mining, processing, and waste disposal are large. Commercial production is under way in Canada, but the economics of future development are uncertain. An investment of $1 billion is required for a plant producing 125,000 bbl of oil per day. As with shale oil extraction, the demand for water and energy during extraction and the associated environmental problems are great.

1.5 NUCLEAR ENERGY

Present-day commercial nuclear reactors are of the "burner" type: they consume the fuel they receive, although not very efficiently. Using natural uranium, U-238, or fuel enriched with the isotope U-235, they transform to heat only a small fraction (1 or 2 percent) of the energy potentially available from the fission process. In contrast to the burner, the breeder reactor breeds or generates additional nuclear fuel while the fission process is under way; the combined process permits extraction of perhaps 75 percent of the fission energy of the natural element.

Total experience with burner reactors is at present on the order of 1000 reactor-years. The breeder reactor is under active development, but because safety, reliability, and cost problems will require much further work, the beginning of widespread commercial application of the breeder could be delayed a decade or two.

As Table 1.3 shows, only a small amount of energy from the proved recoverable reserves of uranium could be used in current commercial nuclear reactors, as compared with the energy from crude oil reserves. It could be made larger by use of the breeder reactor, particularly if supplemented by the use of thorium fuel. But it could also be greatly increased by future uranium exploration and by possible increases in the price level of uranium ore. Exploration for uranium has been active for only a few years; it was rather inactive during the 1960s because supply greatly exceeded demand. Recently, since many nuclear reactors are being built, the pace of exploration and discovery has quickened. An increase in the price of uranium from the present level to \$40/kg could raise reserves by a factor of four, to 4 million tonnes. Since nuclear power cost is a weak function of ore price, however, deposits costing even up to \$200/kg U_3O_8 may be practical to use; this would increase potential world reserves to at least 20 million tonnes. Also, it may be feasible to extract uranium from the oceans, which are estimated to carry a world stock of up to 4 billion tonnes. It is unlikely that more than a small fraction of this total supply could be extracted. Since extraction of uranium from the ocean would be very costly, countries that do not have an abundant local supply of uranium will need the breeder reactor if nuclear fuel is to become their main source of energy.

Figure 1.11 shows data on total and electrical energy consumption for the world up to 1973 and forecasts of future electrical power generation. Two forecasts are indicated for nuclear power. The first, Case 1, was prepared by the International Atomic Energy Agency (IAEA) in early 1973, before the large increases in the price of oil. The second, Case 2, is an estimate of the extent to which increased oil prices will increase the level of nuclear power. Case 2 suggests that a large fraction of electrical power generation will be nuclear by the year 2000. Whether or not nuclear power is introduced as rapidly as Case 2 would indicate, electricity generation will rely heavily on fossil fuels for the next two decades.

The principal difficulty in predicting the future of nuclear power is the safety problem. Nuclear risks have not been clearly shown to be unacceptable; however, insufficient evidence now exists to assure that these risks are satisfactorily small. The development of future reactors, particularly the fast breeder, must proceed slowly enough that adequate assurances of safety can be provided at every stage. Another inhibiting factor is the difficulty in raising capital to build nuclear plants, which are relatively expensive. For these reasons, the pace and the ultimate level of the nuclear power program cannot be predicted with certainty.

Much hope has been expressed that the nuclear fusion process, which performs so effectively in the sun, will one day be shown to be feasible for power production. Fusion typically requires that the reactants meet at a very high density and an extremely high temperature, albeit for a brief space of time. The reaction that appears closest to realization in a scientific experiment is the deuterium–tritium process, which would require a temperature of perhaps 30 million °C. More difficult is the deuterium–deuterium process, which would require temperatures of around 300 million °C.

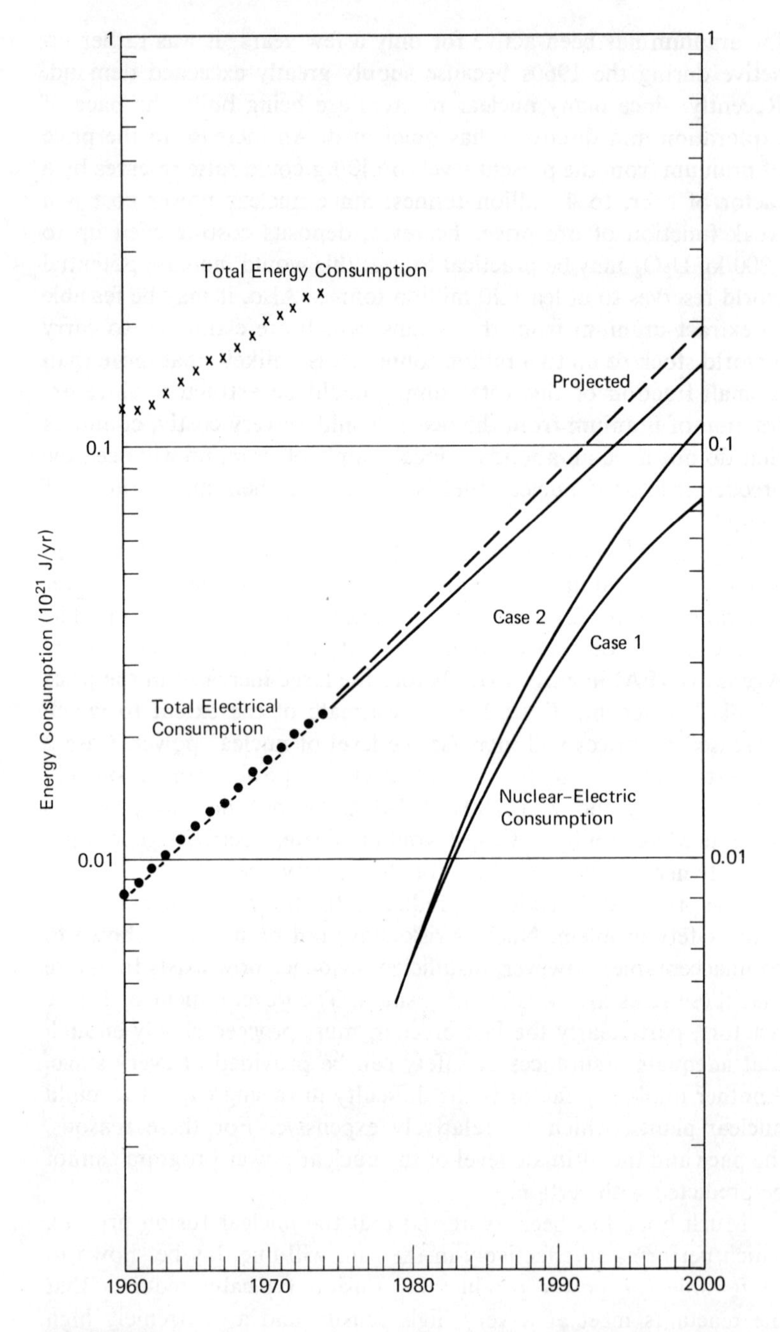

Figure 1.11
Total, total electrical, and nuclear-electric energy consumption for the world, 1960–1973 and projected to 2000. Cases 1 and 2 are IAEA forecasts done before and after the oil price increases of 1973, respectively (14). Other data from (2).

The energy potential of the first process is quite limited, since tritium (an isotope of hydrogen) does not occur naturally to any significant extent and must be produced by the interaction of a neutron with lithium-6, an isotope of natural lithium. Estimates of the world's potential reserves of natural lithium would limit the production of energy in the deuterium–tritium process to about 215×10^{21} J (15), the same magnitude as potential world fossil fuel resources.

In contrast, deuterium is relatively abundant and is readily extracted from the ocean. The exploitation of the deuterium–deuterium fusion reaction could produce virtually unlimited energy for the world. However, the fusion process is as yet so far from practical demonstration that there is no firm basis for predicting when it could actually be put to work.

1.6 CONTINUOUS ENERGY SOURCES

In general, energy sources can be grouped into two classes:

1. Nonrenewable (energy sources): fossil and nuclear
2. Renewable (power sources): hydro, geothermal, solar radiation, tidal, and wind.

The first class includes reserves of finite energy content whose eventual depletion is readily conceivable. The second class includes sources that are essentially continuous (the reserves are effectively infinite); for this class the important question is the possible utilization rate. In accordance with this distinction, we refer to the first class as energy sources and to the second as power sources or continuous energy sources.

By the year 2000, a large portion of world energy will be devoted to electricity generation, which will become the largest single element in energy engineering in the next three decades. The wide array of possible electrical power sources is indicated in Figure 1.12, but many of them are severely constrained by physics, geology, and economics.

Considering first the nonrenewable sources, we have already pointed out the limitations on fossil fuel reserves. A far more stringent limit to coal use could arise from environmental or economic constraints. Use of nuclear reserves could also be limited for the same reasons. Conceivably, both chemical and nuclear energy sources could also be limited in the far future because of their contribution to the heat load at the earth's surface and subsequent climate alteration, either local or global.

Continuous energy sources have several attractive features: they are not subject to depletion or to the drastic cost jump that follows the threat of fuel shortage; they make little or no contribu-

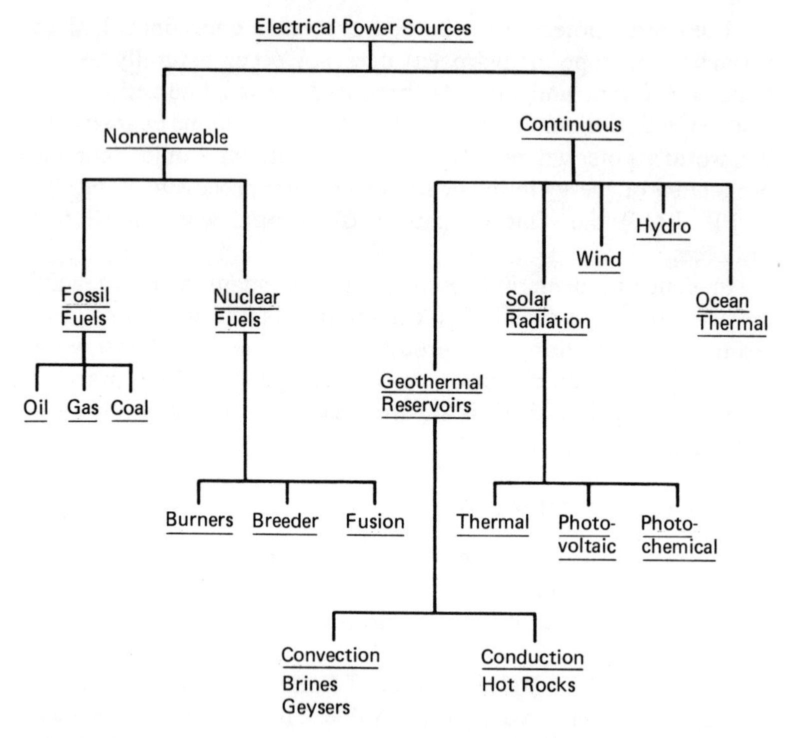

Figure 1.12
Possible sources of future electrical energy production.

tion to air and water pollution; and they have minimal effect on the earth's climate.

1.6.1 Solar Radiation

As shown by Table 1.4, the rate at which solar energy flows to the land mass of the earth is enormous. The total solar radiation intercepted by the earth is about 173,000 × 10^{12} W. However, about 30 percent of that is lost in direct reflection, and another 23 percent is lost in evaporation, leaving only 81,000 × 10^{12} W for direct heating. The figure of 27,000 × 10^{12} W represents a rough average of the solar energy incident on the land mass of the earth. It is a large quantity, perhaps 3000 times the present world consumption of energy by artificial processes. (Present world consumption of energy from all sources is about 10 × 10^{12} W.) Solar energy is so abundant and widely available that many wonder why it is not more extensively utilized in power generation and heating. Unfortunately, it has serious limitations.

First, it is exceedingly dilute relative to other energy sources. A 4000-MW electric plant (not extremely large by today's standards) would require a solar collection area of 400 km^2, even if it

Table 1.4
Potential Contributions of Continuous Energy Sources

Source	Present or Potential Contribution (10^{12} W)
Solar Radiation	
Average on land	27,000
Hydro	
Present installed capacity	0.25
Estimated capacity in A.D. 2000	0.95
Estimated maximum potential	2.86
Tidal	
Present installed capacity	0.00064
Total potential average power of specific world sites	0.064
Total tidal energy flux	3.0
Wind	
Maximum potential	0.1
Geothermal	
Present installed capacity	0.005 (thermal)
Estimated world potential	0.060
Total World Consumption, 1975	10
Possible Steady-State World Consumption	400

Source: Refs. (9) (10).

were located in one of the best solar climates of the United States. Elsewhere, the required land area might be twice as large. A possible future world demand of 3.5×10^{15} kWh/yr would require about 4×10^{7} km^2 of area devoted to solar energy collection.

Second, though solar energy is freely received, it is relatively expensive to use. The cost of the collector, given the area required, is enormous: current estimates of plant cost per kilowatt are an order of magnitude higher than for a fossil fuel or nuclear plant. Because of plant cost and output variation from day to night and season to season, large energy storage capacity would be required to permit effective use of a plant. In limited areas, hydraulic energy storage would be possible using elevated reservoirs, but these and other methods are costly.

If there were no alternative, solar power could certainly be developed. However, the cost differential over nuclear power is a great obstacle. Solar energy is now used effectively in some parts of the world, such as Israel and Japan, for small-scale water heating and even home heating; however, the energies devoted to these functions are a small part of total national need.

Deliberate use of photosynthesis to produce fuel, as in fuel wood or farm waste, has been estimated to have a potential of 2 or 3 $\times$ 10^{12} W. Other photosynthesis-produced fuels are conceivable. One examination of the possibility of forests cultivated for firewood suggests that the land areas required could be in the range of 2000 to 5000 km^2 to provide fuel continuously for 4000 MW of electricity generation (16).

1.6.2 Hydraulic Energy

Developed hydroelectric potential represents less than 10 percent of the ultimate potential; however, the remainder will be increasingly costly to exploit, and many of the best sites are far from where the power is needed. In many parts of the world there will be increasing resistance to the land damage sometimes associated with hydroelectric dams. By the year 2000, it is expected that about one-third of the world's potential hydroelectric power generation will have been developed.

1.6.3 Tidal Energy

Although there is tremendous energy associated with world tidal motions, perhaps 3×10^{12} W, there are very few sites where reasonably large-scale plants can be built to utilize this energy. Sites under serious consideration number nine in North America, nine in France, one in South America, one or two in England, and four in Russia. The total average tidal electric potential is some 64,000 MW; this is comparable to estimates of the potential for geothermal electrical power. However, only two large tidal plants are in operation: a 240-MW unit in the La Rance Estuary in France and a 400-MW unit at Kislaya Bay in Russia. Experience with these plants has not been altogether promising. Tidal power may well be locally advantageous but cannot be expected to make much impact, relatively speaking, on the world energy market.

1.6.4 Geothermal Energy

The thermal capacity of the earth's core is so large (perhaps 10^{30}–10^{31} J) that the idea of drawing a little of it for human use is intriguing. The normal average rate at which the earth cools, 0.064 W/m^2, is very small, however, about 20,000 times weaker than the maximum solar energy flux density at the earth's surface. Donald White (17) has estimated the thermal energy stored in the earth's crust to a depth of about 3.5 km as around 500×10^{21} J, which is still a large entry in Table 1.4, but the problem is extraction. At the present time, a total of over 1000 MW of electrical power is generated from natural hot water and steam sources in Italy, California, Mexico, Japan, and other sites. This quantity is small compared with the present total world electricity generation capacity of around 900,000 MW, but one estimate suggests that it may be practical to produce 60,000 MW from various geothermal sources around the world for both geothermal heating and electricity generation. The total world thermal utilization of geothermal energy is now over 7000 MW.

However, even 60,000 MWe per year is equivalent to only around 5×10^{18} J thermal/yr and does not really become a significant contribution to world energy use, however valuable it may be in certain localities. Hubbert suggests that major geothermal sources would be nearly depleted in less than a century of use.

Reports that the pressure of dry steam used in geothermal plants in Larderello, Italy, and Geysers, California, has been dropping over time (possibly owing to the buildup of deposits in the underground conduits) emphasize that this energy source cannot be regarded as indefinitely continuous. Major efforts may be needed to maintain satisfactory underground plumbing. Even greater effort would be needed to tap the thermal reservoirs of hot, dry rock deposits. Flow passages would have to be created, most likely with explosives, and possibly with the help of thermal cracking that uses pressurized cold water forced down into the rock to create new flow passages through which energy can be brought back to the surface.

1.6.5 Wind Energy

Of the total flux rate of $173,000 \times 10^{12}$ W of solar energy intercepted by the earth, about 370×10^{12} W are thought to be used in driving winds and waves (10). This energy is very large, but is so erratic and diffuse that the wind power electricity generation potential is thought to be only about 0.1×10^{12} W. This would be the power generated by spacing 1000-kW units every 2000 m completely around 25 meridians of the earth, assuming a 20 percent load factor.

Wind power has long been used for electricity generation. Denmark, for example, was reported to have had a total of 100,000 kW from many small wind power units before the Second World War. However, wind turbines have been found to be very costly compared with fossil fuel plants, the major factor being the erratic variation of the energy supply. Suitable large-scale energy storage would be required to make wind power continuously available.

1.7 SUMMARY

World energy demand has been growing exponentially in this century and shows no sign of leveling off. Energy demand in the developed countries is growing much faster than population and appears to be closely related to the growth of disposable income. For the developing countries to catch up to the per capita energy expenditure of the United States, world energy demand would have to increase by a factor of six. Ultimate steady-state world energy demand could be two orders of magnitude larger than the present level.

Electrical energy demand in the United States, the United Kingdom, and Canada has been very strong in this century, doubling every 10 years. It has been growing so much faster than total energy demand that it could force a strong acceleration of total demand growth in the next decade or two. The same is true for the world as a whole.

Oil, which presently supplies about 70 percent of world energy needs, could be depleted early in the next century. Well before then, production rates may fall off and prices may rise to very high levels. The slowness with which substitution of other fuels (coal and nuclear) can be made suggests the possibility of shortages in the next decade or so.

Tar sand and oil shale deposits, which may be of great local benefit to certain countries and may be economical to use as fuel prices rise drastically, will do little to forestall worldwide oil depletion. Prices of oil and gas, rising as depletion threatens, will promote the development of alternatives, but lead times are very long and major effort toward change is long overdue.

The most promising long-term alternatives are nuclear and coal energy. Large-scale use of solar energy offers vast potential but appears very costly. Hydro, geothermal, tidal, and wind power can meet only a small fraction of future total energy needs.

Coal reserves contain enough energy to last the world a century or more, but problems with air pollution, environmental damage, mine safety, and mining costs could mean that many of the world's coal reserves will never be used. Successful demonstration of clean, large-scale, low-cost coal gasification could make a great difference to future projections of coal usage.

Nuclear energy offers tremendous potential, though with present-day nuclear reactors of the burner type, world reserves of uranium would contribute an energy quantity considerably less than that in coal reserves. A nearly 100-fold amplification of the energy to be developed could be achieved with the breeder reactors now under development in several countries. Questions of nuclear reactor safety, integrated long-term effects of low-level radiation, and the problems of nuclear waste disposal continue to raise concern about the future safety and security of nuclear power, particularly with breeder reactors. The scientific feasibility of the fusion reaction has not yet been adequately demonstrated, so there is still no certainty that fusion reactions will ever be exploited in power generation.

References

1. World Energy Conference, U.S. National Committee. *Survey of Energy Resources, 1974*. 345 E. 4th St., New York, New York.

2. United Nations. *World Energy Supplies, 1929 to 1973.* Statistical Papers, Series J, Nos. 1, 2, 3, 7, 8, 11, 14, 18.

3. Hoyt C. Hottel and J. Howard. *New Energy Technology.* Cambridge, Mass.: The MIT Press, 1971.

4. A. L. Austin, B. Rubin, and G. C. Werth. *Energy: Uses, Sources, Issues.* Report UCRL-51221. Lawrence Livermore Laboratory, University of California, May 30, 1972.

5. Nathan Keyfitz. "United States and world population." Chapter 3 in *Resources and Man* (San Francisco: W. H. Freeman, 1969).

6. Tomas Frejka. "The prospects of a stationary world population." *Scientific American,* March 1973, pp. 15–23.

7. A. M. Weinberg and R. P. Hammond. "Global effects of the use of energy." Paper A/Conf 49/P/033, *Proceedings of the Fourth United Nations International Conference on the Peaceful Uses of Atomic Energy,* Geneva, September 6–16, 1971.

8. Joel Darmstadter. *Energy in the World Economy: A Statistical Review of Trends in Output, Trade and Consumption since 1925.* Baltimore: Johns Hopkins Press, 1971.

9. M. K. Hubbert. "Energy resources." Chapter 8 in *Resources and Man* (San Francisco: W. H. Freeman, 1969).

10. M. K. Hubbert. "Energy resources for power production." Paper IAEA-SM-146/1, *Proceedings of the International Atomic Energy Agency Symposium on Environmental Aspects of Nuclear Power Stations,* New York, August 10–14, 1970, pp. 13–43.

11. Sam H. Schurr. "Energy." *Scientific American,* September 1963, pp. 110–127.

12. OECD Nuclear Energy Agency and the IAEA. *Uranium: Resources, Production and Demand.* August 1973.

13. Paul Averitt. "Coal." In *United States Mineral Resources,* D. A. Brobst and W. P. Pratt, eds., U.S. Geological Survey Professional Paper No. 820, 1973, pp. 133–142.

14. J. A. Lane, R. Krym, N. Raisic, and J. T. Roberts. "The role of nuclear power in the future energy supply of the world." Paper 4.1–22, *Transactions of the 9th World Energy Conference,* Detroit, September 22–27, 1974.

15. J. Mandel. "Resources of primary energy." Paper 49/P/359, *Proceedings of the Fourth United Nations International Conference on the Peaceful Uses of Atomic Energy,* Geneva, September 6–16, 1971.

16. G. C. Szego. "The energy plantation." Statement to the Subcommittee on Science Research and Development of the Committee on Science and Astronautics, U.S. House of Representatives, May 1972.

17. Donald E. White. *Geothermal Energy.* U.S. Geological Survey Circular No. 519, 1965.

Supplementary References

N. B. Guyol. *The World Electric Power Industry.* Berkeley: University of California Press, 1969.

M. K. Hubbert. "The energy resources of the earth." *Scientific American,* September 1971, pp. 61–70.

Problems

1.1 The sum of n terms in the series $1 + (1 + r)^1 + (1 + r)^2 + \ldots$ for $r < 1$ is

$$S = \frac{(1 + r)^n - 1}{r}.$$

a. If world oil production in 1975 was 22×10^9 bbl and production grows at a rate of 3 percent per year over the next 10 years (as compared to 7 percent until recently), show that production in 1985 will be 29.6×10^9 bbl and that cumulative production between 1975 and 1985 will be 252×10^9 bbl.

b. Assume that the total world production up to 1975 was 300×10^9 bbl and that total world reserves of crude oil, initially in place, were 2000×10^9 bbl (the upper limit estimated by M. K. Hubbert). If annual production grows by an average of 3 percent per year until 1985 and remains static thereafter, when will 50 percent of the resources initially in place have been consumed?

1.2 Estimate the need for expansion in world coal consumption between the years 1973 and 2000 under the following annual growth rates:

World Energy Consumption: 5 percent per year
Oil (average growth rate): 3 percent per year between 1973 and 2000
Natural Gas: same as oil
Nuclear Power: 20 percent per year
Hydroelectric Power: 3 percent per year

Use Table 1.3 for 1973 oil, gas, and coal consumption, Figure 1.11 to estimate nuclear energy, and Table 1.4 for hydroelectric capacity. Suppose that average hydroelectric power output is 60 percent of installed capacity.

Neglect the possible contributions of oil shales, bituminous sands, and solar power to energy consumption between now and 2000.

Determine the average increase in world coal production rate needed to satisfy the above constraints for the period 1973–2000. Consider the proportions of coal and other energy forms produced by the year 2000.

1.3 Suppose that the promise of abundant nuclear power indicated in Problem 1.2 cannot be realized, and that the annual production rates of oil and gas do not increase during the period 1973–2000. Also suppose that it is not physically possible to increase coal production rates by more than 6 percent per year (over a long period coal production has increased by about 3.6 percent per year) and that hydroelectric and nuclear generation rates each rise by 5 percent per year. Again neglecting the possible contributions of solar energy, estimate the maximum average annual growth in world energy consumption in the period 1973–2000.

1.4 Compare recent global rates of production of oil, gas, and coal energy sources (Table 1.3) with total solar radiation at global land surfaces. A rough year-round average of solar radiation intensity over the earth's land surface is 200 W/m^2. Assume that the global surface area is 30 percent land. (Check the total solar radiation energy with the figure quoted in Table 1.4.)

If world energy consumption were to continue to increase at 5 percent per year, and present total consumption is $\sim 0.26 \times 10^{21}$ J/yr, in how many years would annual energy consumption equal 1 percent of the solar radiation quantity?

World population is presently around 3.6×10^9; if it were to grow to 15×10^9, with a total energy consumption equal to 1 percent of the solar radiation quantity, by how much would world average per capita consumption of energy have increased from the present value?

1.5 Consider potential world hydroelectric resources. It is thought that 330,000 km^3 of water evaporates from the oceans each year. The surface of the globe is about 70 percent water, so only about 100,000 km^3 falls as rain or snow on land areas, and possibly two-thirds of that is directly evaporated, with the remainder flowing in rivers to the oceans.

a. Estimate poential electricity generation (in J/yr and W-hr/yr) from world hydraulic resources under the two sets of assumptions in Table 1.A.

Table 1.A
Two Sets of Assumptions about World Hydraulic Resources

	Case A	Case B
Average Elevation (m)	100	300
Fraction of Water Flow Available for Power	0.10	0.30
Generation Efficiency	0.85	0.85

b. Show that the potential energy generation capacity E (in J/yr) is

$$E = \eta \rho g Q H,$$

where H is the available elevation difference (in m), η is the efficiency, Q is the flow volume (in m^3/yr), ρ is the density of water (in kg/m^3), and g is the acceleration due to gravity (in m/sec^2).

For comparison, present installed world hydroelectric capacity is about 300,000 MWe and annual generation is 1300×10^{12} W-hr.

c. What is the ratio of the annual quantity of electrical energy generated in this way to the chemical energy represented by oil consumption given in Table 1.4?

1.6 One recent projection of world nuclear power capacity is shown in Table 1.B.

Table 1.B
Projected World Nuclear Power

Year	Capacity (GWe)	Capacity Factor*
1975	103.5	(0.6)
1980	316	0.7
1985	888	0.7
1990	1900	0.7
1995	3365	0.65
2000	5330	0.6

*Ratio of actual annual generation to generation using full capacity

With the following assumptions, estimate how long it will take to consume potential world resources of natural uranium (Table 1.3) using nonbreeder reactors.

As a rough approximation, the energy equivalent of 1 gram of U-235 equals 1 MW-day of thermal energy, and thermal energy is converted to electrical energy in the reactor at an efficiency of around 30 percent. Natural uranium contains 0.71 percent by weight of U-235. Nonbreeder reactors can be considered to make use of the energy equivalent in the U-235 part of the fuel only, while breeder reactors (utilizing the U-238 in the fuel) can produce about 60 times as much electrical energy per gram of natural uranium.

In considering the above calculation it should be borne in mind that breeder reactor power capacity may still be a small fraction of total nuclear capacity by the year 2000.

1.7 Estimate the tidal power that could be made available from an estuary enclosed by a dam. Show that the potential energy that could be made available each cycle is

$$E_{max} = \frac{\rho g H^2 A}{2},$$

where ρ is the density of water (kg/m³), g is the acceleration due to gravity (m/sec²), H is the tidal variation, and A is the area of the estuary. Of this potential energy, only 16 to 40 percent is reported by Hubbert to be usable in typical tidal installations, though for the 240-MW installation in the La Rance Estuary in France the figure approaches 50 percent. Taking this into account, and assuming an estuary of area 100 km², estimate the average power level with a turbogenerator efficiency of 0.80. The tidal period is 12 hr 23.2 min. Assume the tidal variation H to be 4 m.

The lower Bay of Fundy has 5.5-m tides and is reported to have a tidal power potential of 1800 MWe (average), with an annual generation potential of 15,800 GWe. The upper Bay of Fundy offers the possibility of a 2176-MWe plant delivering 5600 GWh/yr.

1.8 The Salton Sea geothermal resource in California is estimated to contain 80×10^{18} J of energy in superheated water, capped by impermeable strata through which the heat release is small. This energy can be obtained by drilling wells that release steam and hot water. Estimate the power potential from this site if plant efficiency is 25 percent and the energy is drawn off over a 50-year period. Estimate also the mass flow rate of the condensed steam and warm water discharged from such a plant if the specific work is 50 kJ/kg.

For comparison with the above estimate, the Geysers Plant in California (installed in 1969) has a capacity of 82 MW. The planned additional capacity is 318 MWe (10).

1.9 Hubbert estimates that the world's initial supply of recoverable fossil fuels is *potentially* as shown in Table 1.C.

Table 1.C
Potential World Supply of Fossil Fuels

Fuel	Quantity	Energy (percent)
Coal and Lignite	7.6×10^{12} tonnes	88.4
Petroleum Liquid	2000×10^{9} bbl	5.4
Natural Gas	$10{,}000 \times 10^{12}$ ft^3	4.8
Tar Sand Oil	300×10^{9} bbl	0.8
Shale Oil	190×10^{9} bbl	0.5
Total		100

Source: Ref. (10).

Using the approximate conversion factors in Appendix A, determine the corresponding energy quantities and compare with the estimates of Table 1.3. Compare with Hubbert's estimate of energy percentages for each resource. Note the order-of-magnitude difference between "proved recoverable resources" (Table 1.3) and Hubbert's estimates of potential resources, most of which remain to be discovered.

1.10 Suppose that nuclear power expansion is limited by the following constraints:

1. The maximum rate of growth of nuclear capacity is less than 20 percent per year.

2. The maximum rate of capital investment in reactors is limited to \$150 billion/yr (roughly 5 percent of the present gross world product), while reactor capacity costs \$1000/kWe. This would mean a maximum capacity increment of 150 GWe/yr.

With these constraints, when would nuclear electricity be able to supply 10 percent of world energy consumption? Assume that 1975 world energy consumption (not including the energy losses in generating electricity) is 0.2×10^{21} J, and that this continues to grow at 4.5 percent per year.

Problem 1.6 indicates 1975 world nuclear electricity generation capacity and a projection of capacity growth to the year 2000, with which the results of this calculation can be compared. Suppose for simplicity that from 1975 onward the capacity factor is 0.7 and the growth rate is constant over each 5-year increment.

2
North American Energy Resources and Demand

2.1 FUEL RESOURCES

2.1.1 Proved Recoverable Resources

Oil Table 2.1 supplies estimates of proved recoverable reserves of the major fuels used in North America. Total *proved* reserves of oil in the United States are 5569 Mtonnes. In 1975 the United States Geological Survey (USGS) estimated *potential undiscovered* petroleum resources at 50 to 130 billion bbl of oil (7000 to 18,000 Mtonnes of oil). This estimate of undiscovered reserves is only half of the value given by the USGS during 1974 (which, in turn, was significantly lower than the estimates provided by that same agency in the late 1960s). Thus there is real concern that future undiscovered reserves may not be very much greater than proved recoverable reserves. This is particularly serious since the ratio of proved reserves to annual production rate is only about 13 years.

With rising oil prices, previously uneconomical reserves could be added to proved recoverable reserve estimates. Also, new technology for extraction of oil from reservoirs could increase the total recoverable reserves. At the present time, with water and gas injection techniques, possibly only one-third of total reserves are considered recoverable. With better technology this fraction might be increased to half or more. However, it does not appear that there

Table 2.1
Proved Recoverable Reserves in North America

	Recoverable Reserves (10^6 tonnes)	Energy (10^{18} J)	Recent Annual Production (10^{18} J)	Reserves-to-Production Ratio (years)
Crude Oil				
United States	5,569	241	18.9	13
Canada	1,075	46.5	3.0	15
Natural Gas				
United States	(7,556 km^3)	287	24.2	12
Canada	(2,576 km^3)	96	2.7	35
Shale Oil				
United States	145,000	6,270		
Canada	24,860	1,075		
Bituminous Sands				
United States	2,175	94		
Canada	50,250	2,172	0.136	
Coal				
United States				
anthracite	5,755	193	0.265	
bituminous	118,688	3,463	15.8	
subbituminous	35,282	975	0.03	
lignite	23,056	463	0.14	
total	181,171	5,095	16.2	
Canada				
bituminous	4,195	110	0.298	
subbituminous	885	13.7	0.069	
lignite	457	7	0.046	
total	5,537	131	0.41	
Uranium				
United States	0.329	284 (nb)		
		17,069 (b)		
Canada	0.185	160 (nb)		
		9,632 (b)		
Thorium				
United States	0.051	2,678 (b)		
Canada	0.079	4,128 (b)		

(b) = breeder; (nb) = nonbreeder.
Source: Ref. (1).

is much room for expansion of proved recoverable reserves of crude oil within the United States or Canada.

Gas Natural gas is expected to have a similarly short life in the United States. The reserves-to-production ratio is much larger for Canada, but many of these reserves are remote from centers of consumption. Averaging these reserves for North American consumption would still only make the reserves-to-production ratio about 14 or 15 years.

As Table 2.1 indicates, reserves of oil that could be extracted from the oil shales and bituminous sands of the United States and Canada are about 20 times as large as the energy in proved recoverable reserves of crude oil and natural gas.

Shale Oil Estimates of total reserves of shale oil in the United States have been frequently quoted as 1.8 trillion bbl (250,000 Mtonnes). This figure is not radically different from the estimate of 145,000 Mtonnes of proved recoverable reserves in Table 2.1. It is an enormous quantity, as high as Hubbert's estimates of total initial reserves of crude oil for the entire world (see Figure 1.10).

Opinions vary widely on how much of these resources are economically extractable at the present time. One estimate suggests that only about 6 percent of the oil shale reserves are accessible in seams containing 30 gal of oil per ton of rock (2). These would be the richest deposits, and may still be only marginally economical to extract. In total, these reserves would amount to about 15,000 Mtonnes of oil.

Bituminous Sands The bituminous or tar sands of Canada are another huge oil resource. Commercial exploitation is already under way, though current annual production is still less than 1 percent of North American crude oil production. Of the 50,250 Mtonnes of proved recoverable reserves, less than 9000 Mtonnes may be recoverable by surface mining, at costs of $10/bbl. Capital and environmental problems are a major concern and will strongly restrain the rate at which oil can be recovered from the tar sands.

Coal Total proved reserves of coal in North America are vast—an order of magnitude greater than proved reserves for crude oil and natural gas. Potential reserves are greater still; total potential resources in the United States are estimated to be 2,925,000 Mtonnes. Of these, only 182,000 Mtonnes are expected to be economically recoverable. About 66 percent of the U.S. resources of coal are found in the western Great Plains and the Rockies. These average about 23 percent bituminous coal, 44 percent subbituminous, and 33 percent lignite. About 26 percent of U.S. coal resources are found in the Appalachian eastern and western interior regions. These are mostly bituminous coals, with the exception of anthracite in eastern Pennsylvania. About 1 percent of U.S. coal is found in the Pacific Coast region and another 8 percent in Alaska.

In Canada, potential coal resources are at least 8000 Mtonnes, but only 5540 Mtonnes are considered recoverable. About 99 percent of these resources are located in British Columbia, Alberta, and Saskatchewan; these reserves average 80 percent bituminous, 8 percent subbituminous, and about 12 percent lignite.

Uranium and Thorium The economically recoverable resources of uranium indicated in Table 2.1 for the United States refer to prices of $26/kg or less. Potential reserves are about six times as high, or just above 2 million tonnes of uranium. As will be shown later, this is not large compared with the expected cumulative consumption of uranium in nonbreeder reactors up to the end of the century. The advent of the breeder reactor, however, means that uranium energy use could be amplified greatly. The breeder reactor also permits exploitation of thorium resources.

Another estimate of uranium resources in Canada indicates proved recoverable resources of 0.375 Mtonnes at costs of up to $26/kg. Additional resources recoverable at $26 to $39/kg amount to 0.341 Mtonnes. The thorium resources indicated in Table 2.1 also pertain to extraction costs of about $26/kg.

2.1.2 Potential Reserves

Estimates of potential resources are of interest in considering the far future. For certain resources such as oil and gas, for which exploration has been very intensive for many decades, geologists can estimate upper limits to potential reserves with some degree of confidence. They predict future oil discovery by extrapolating from the past trend in the rate of oil discovery (in barrels per foot of exploratory drilling). Since this rate has been decreasing fairly rapidly with "cumulative exploratory footage," a reasonable upper limit for future discovery may be obtained.

Coal reserves are so abundant that exploration has not been intensive (compared with oil and gas), and estimates of potential resources are probably much less certain. For uranium, exploration has been intense only for brief periods, and again, potential reserves may be quite uncertain.

Table 2.2 indicates the kind of information available on potential reserves. It must be regarded only as an intelligent guess of the resources that will be available in North America in the future. Exploration, new technology, and world energy economics could radically change the picture in 50 years.

For the United States, the estimate for potential oil reserves includes an estimate of the petroleum liquids available from natural

Table 2.2
Potential Reserves in North America

	Potential Reserves	Energy (10^{18} J)	Reference
Crude Oil			
United States	270×10^9 bbl	1,570	(3)
Canada	95×10^9 bbl	550	(4)
Natural Gas			
United States	45.3×10^3 km^3	1,600	(1)
Canada	19.8×10^3 km^3	700	(4)
Shale Oil			
United States	190×10^9 bbl	1,100	(3)
Bituminous Sands			
Canada	300×10^9 bbl	1,740	(3)
Coal			
United States	2.925×10^{12} tonnes	72,000	(1)
Canada	0.118×10^{12} tonnes	3,000	(4)
Uranium			
United States	2.0×10^6 tonnes	960 (nb) 60,000 (b)	(1)
Canada	0.8×10^6 tonnes	480 (nb) 30,000 (b)	(1)

(b) = breeder; (nb) = nonbreeder.

gas. The estimate for shale oil refers to the Green River shales of Colorado, Wyoming, and Utah. This shale resource potential may actually exceed the total world resources of crude oil, but much of the shale may not be economical to process, with yields as low as 10 gal per ton of shale. The estimate of natural gas is based on the oil estimate and past experience of the ratio of gas to oil discoveries. The estimate in Table 2.2 includes a large quantity (10.5×10^3 km^3) of gas already consumed.

Only a small part (16×10^9 bbl) of Canada's potential oil reserves have actually been discovered. Of this figure a large fraction (6.3×10^9 bbl) has already been consumed. A major problem with these reserves is that so many are located in frontier and offshore areas where extraction and transportation will be difficult and costly. Of the 300×10^9 bbl of oil potentially available from the tar sands, about 65×10^9 bbl may be recovered by open-pit mining methods at prices below $10/bbl. Figure 2.1 shows how greatly the potential recovery of tar sand oil augments Canadian reserves.

Coal reserves in Table 2.2 include both high- and low-sulfur coals, representing 50 percent of the coal initially in place in beds of more than 5-cm thickness and depths of up to 2 km. Owing to high mining costs, environmental damage, and sulfur content, a large part of these reserves may not be useful unless new technology is applied on a large scale.

Table 2.3
Maximum Potential Lifetime of North American Resources with Continuing Exponential Growth

	Energy Potential (10^{18} J)	Demand Growth Rate (percent)	Doubling Time (years)	Fraction Already Consumed (percent)	Exponential Life (years)
Oil	2,100	7	10.3	0.15	28
Oil and Tar Sands	3,900	7	10.3	0.08	38
Oil, Tar Sands, and Oil Shale	5,000	7	10.3	0.06	42
Natural Gas	2,300	7	10.3	0.15	28
Coal	75,000	3.6	19	0.01	130
Uranium (burner reactor)	1,440	20	3.5	0.0015	33

Tentative though Table 2.2 is, it gives us some idea of the implications of continued exponential growth for consumption of our major energy resources. In Table 2.3 the resources for Canada and the United States are lumped together; reserves of crude oil, tar sand oil, and shale oil are considered both separately and in combination.

The calculated exponential lifetimes in Table 2.3 follow from the results of Appendix B, using only the fraction r_0 of resources already consumed, and the doubling time t_d for exponential growth. The growth rates and doubling times of Table 2.3 are roughly the values for 1970, and do not include the effects of oil imports. It is most unlikely that these growth rates (except for that of coal) could be sustained over three decades. Oil production in the United States has reached the peak rate already. These estimates therefore represent the times in which maximum credible reserves could be used up if production capacity could be very rapidly increased. They show that oil and gas could be used up before the turn of the century, and that the tar sands and oil shales probably offer only a small addition to exponential life.

Low-cost uranium reserves are also quite limited; the postulated growth rate of 20 percent is clearly not sustainable for 33 years (nuclear energy would be the only energy source at the end of that time). The breeder reactor could in principle extend these reserves by a factor of 60 or 70, and a growth rate of 5 to 10 percent after 1990 could extend the lifetime of these reserves for a very long period. Steady growth of coal consumption at 3.6 percent per year would expand the annual use rate over a century by a factor of nearly 40. Even so, potential coal reserves would be adequate for well over a century.

Established supplies of crude oil and gas are running out rapidly. No large-scale commercial plant is yet producing oil from shale; tar sand exploitation is just beginning. Even if a large-scale commitment were made, neither of these sources would last long under the continued exponential growth in demand for oil. Continuing heavy demand for oil in North America would mean substantial imports of high-priced foreign oil, but world reserves are themselves very limited, as shown in Chapter 1. Thus it seems quite clear that the present heavy dependence on gas and oil (which currently supply 75 percent of the total energy in the United States, 63 percent in Canada) must soon come to an end.

The main alternative fuel sources are coal and uranium. Both of these could in principle supply the needs of North America for a long time to come. Assuming a steady-state population of 500 million and energy consumption of 20 kW per person (300×10^{18} J/yr), potential reserves of coal would be adequate by themselves for 250 years. Uranium (with breeder reactors) could last equally long. In addition, there are the possibilities of solar and fusion energy.

2.2 CONSTRAINTS ON FUEL EXTRACTION AND PROCESSING

2.2.1 Coal

To supply from coal the energy we now get from oil would take many years and many billions of dollars of investment. It could also harm the environment intolerably. One of the most immediate problems is in securing capital for new mines.

Coal Mining Investment in new mines has been slow in the United States. Negotiation of long-term coal supply contracts is hindered by uncertainties in mining costs and in the costs of sulfur removal during combustion. Underground mining costs have been rising rapidly with increased wages and the costs of new health and safety requirements. Surface mining costs depend significantly on requirements for land reclamation, and in some areas surface mining may well be prohibited altogether.

Apart from the uncertainties in the costs of mining and shipping coal, there is still much uncertainty about the costs of controlling emissions from coal combustion. The chief problem is SO_2. Reliable technology for SO_2 removal from stack gases has still not been clearly demonstrated commercially. Clean-burning synthetic fuels may be made from coal; the processes are essentially available but costly. It will take years to bring large-scale demonstration plants to successful operation so that these fuels can begin to play an important role in energy supply.

The Task Force on Energy of the National Academy of Engi-

Table 2.4
Coal Mining Capacity in the United States

Source	Possible 1974–1985 Additions (10^6 tons/yr)	1985 Maximum (10^6 tons/yr)
Eastern Underground	280	480
Eastern Surface	60	220
Western Surface	520	560
Mine Depletion	−200	—
Total	660	1260

Source: Ref. (2).

neering has estimated the maximum rates at which coal mining could be expanded in the United States. Their predictions are indicated in Table 2.4.

According to these estimates, coal mining capacity could be doubled in 10 years. It would, however, require a tremendous expansion of manpower and facilities:

140 new 2-MTPY (million tons/yr) eastern underground mines
30 new 2-MTPY eastern surface mines
100 new 5-MTPY western surface mines
80,000 additional eastern coal miners
45,000 additional western coal miners
140 new 100-yd^3 shovels and draglines
2400 new continuous mining machines.

Most of the new mine capacity would come from western surface mining. Eastern surface mining may even decline in total capacity due to lack of suitable reserves and to legal restrictions on surface mining of steep slopes. Western surface coal can be mined relatively cheaply. Also it is generally noncaking and therefore suitable for processing to make gas or liquid fuel. Table 2.4 implies a 10-fold increase in western surface mining; the requirements and costs of land reclamation for such a huge surface operation are far from being settled.

This large expansion of coal mining capacity would require about $21 billion of investment. The Task Force on Energy believes this investment can only be realized with higher prices for coal and stable regulations for mine operation, land reclamation, and SO_2 emissions from coal-burning power plants. They also believe that this expansion of coal supply is the maximum that could be reasonably expected by 1985.

Unfortunately, even if fully realized, the increase would provide only 28 percent of the 1973 oil and gas energy consumption in the United States.

Table 2.5
Estimate of Coal Energy Transport Capacity Required in the United States by 1985

System	Capacity Required (10^6 tons/yr)
Eastern Rail and Barge Systems	650
Western Rail and Barge Systems	200
Western Coal Slurry Pipelines	100
Synthetic Gas Pipelines	100
Total	1050

Source: Ref. (2).

Coal Transportation Transportation is another possible bottleneck. Moving western coal to eastern markets would require substantial expansion of rail and barge transport. It might mean the building of several coal slurry pipelines (in which coal can be transported 1000 miles at perhaps half the cost of rail transport). Energy transport by gas or liquid pipeline is generally much cheaper than by rail, but the cost of coal conversion to gaseous or liquid fuels is high enough that this will not be done merely to save on transportation; it would be done because these fuels are usable, and coal is not, at the point of application. In fact, many electric power stations now operating on oil or gas could not be converted to coal because of lack of facilities for coal storage and handling, ash removal, and stack gas cleaning. They could, however, deal readily with gasified or liquid fuels made from coal.

The Task Force on Energy estimates that by 1985 coal energy transportation will require the capacities shown in Table 2.5. This would require an approximate 50 percent increase in the national rail and barge capacity and a significant expansion of the pipeline network.

Synthetic Fuels from Coal Processes are available now for producing synthetic gas (largely methane) from coal. This has the same energy content as natural gas, around 1000 Btu/scf, and is therefore just as cheap to transport. Conversion is expected to cost upwards of $\$1/10^6$ Btu. A few years ago, when natural gas cost $\$0.16/10^6$ Btu, the cost of synthetic gas was prohibitive. Now, when natural gas may be $\$1.50$ to $\$2/10^6$ Btu, and is virtually unobtainable for some applications, the cost of synthetic gas does not seem so high.

Successful commercial operation of synthetic gas processes requires careful adaptation to each type of coal, and special problems arise in large-scale operation. This means that four to five years will be needed for large-scale demonstration plants to be brought into successful operation. Many years and substantial investments would be required after that for synthetic gas to carry a significant part of total energy supply.

For electrical power stations, gas of low energy content (150 to 500 Btu/scf) can be produced much more cheaply than synthetic gas. Two factors indicate that this low-energy gas should be generated at the site of the power plant. The first is that its transport cost is high; the second is that integration of gas production and power generation in a single plant permits significant waste heat recovery from the gasification process. The technology for the production of low- or medium-energy gas is essentially available. Several processes have been developed and demonstrated, but the market is not yet clear. The possibilities of using low-Btu gas in combined gas and steam cycle power plants are described in Chapter 9.

Coal liquefaction is another development that could provide large supplies of synthetic oil. It could also supply methanol for transportation, or even gasoline if the market price were right. The two principal methods of converting coal to liquid fuel are hydrogenation and gas synthesis.

Hydrogenation of coal, developed in Germany some 30 years ago, may well be the method of choice (2). It is inherently more efficient than synthesis and can be used to produce a wide range of products. It involves adding hydrogen to coal at high pressure ($\sim$200 atm) in the presence of suitable catalysts. The coal, however, generally needs pretreatment to remove ash, and the hydrogen must be produced separately from the steam and the coal. Successful operation of large-scale prototype plants is necessary before commercial introduction of this process can begin.

Gas synthesis, on the other hand, is already in small-scale commercial production in South Africa, where a Fischer Tropsch gas synthesis plant produces 7000 bbl/day of liquid fuels from coal. The technology is therefore available now, but the production costs are high. Another gas synthesis process can lead to methanol, a liquid fuel that may be useful for automotive and other engines and for home heating.

The Task Force on Energy estimates that even by 1985 the production of synthetic fuels from coal will be very limited, as indicated in Table 2.6. The total will be perhaps 3 percent of the total energy demand.

Canadian Coal In Canada coal production has been expected to rise from 20 million to 37 million tons/yr between 1972 and 1980 (4). But 20 million of the 37 million tons are expected to be exports of Alberta and British Columbia coal, mainly to Japan. The balance is to be distributed among electric utilities in Alberta

Table 2.6
Estimated Maximum Production of Coal-Based Synthetic Fuels by 1985

Fuel	Coal Input[a] (10^6 tons/yr)	Production (10^6 bbl/yr)
High-Btu gas	150	0.8
Methanol	60	0.3
Syncrude	50	0.3
Total	260	1.4

[a]based on use of 8500 Btu per ton of western coal (19,800 kJ/kg).
Source: Ref. (2).

and Saskatchewan (14.5 million tons), electric utilities in New Brunswick and Nova Scotia (0.9 million tons), and metallurgical sales of Nova Scotia coal (1.3 million tons). Because of the cost of transportation, the coal supply for Ontario has come from the United States. Imports of coal to Canada in 1972 totaled 19.3 million tons, almost equaling domestic production. Of this, about 10 million tons were used for thermal-electric power in Ontario, and 7 million tons went to the steel industry.

Thus, aside from exports, little expansion of Canada's coal industry between 1973 and 1980 was foreseen. By the year 2000, however, it was thought that electric utility consumption of coal could rise from about 20 million tons (in 1972) to 79 million tons annually. Such a figure can only be regarded as conjecture in view of all the uncertainties in coal mining and transportation costs and in the extent to which power plant emissions can be controlled.

Canada faces possible loss of self-sufficiency in oil production by the mid-1980s (5), but the main hopes for remedy appear to be frontier oil (Arctic and Atlantic regions) and use of the tar sands of Alberta. There are no coal gasification or liquefaction plants in operation, nor is any large-scale developmental work underway. It is thought that low-cost subbituminous and lignite coal from surface mining in Saskatchewan and Alberta could provide suitable feedstock for synthetic gas production when the price becomes right.

Early estimates suggest that land reclamation after surface mining in western Canada should not be too expensive. The cost of reclaiming strip-mined prairie land appears to be $300 to $600/acre; with 15,000 tons of coal recovery this would add only about 0.2 percent to the coal selling price. In the Rockies costs would be much higher, but still only about 15 percent of mining costs (4). If these estimates are realistic, one could hardly say that reclamation costs would prevent coal from supplying a much larger share of the national energy market.

As in the United States, the big problem is the capital investment required for new mines, transportation, and possibly for synthetic fuel processing. The capital is not likely to be invested until the market is clear. Recent increases in the price of coal suggest that even if emissions control were no longer a serious problem, coal would not compete with nuclear energy for electricity production. For transportation and other fuels, the time may not be far off when the development of synthetic fuels from coal will be seen as urgently required.

2.2.2 Shale Oil and Tar Sands

To understand the problem of large-scale exploitation of the oil shales and tar sands, one has to visualize the size of the solid materials handling problem. Total oil consumption in the United States in 1973 was approximately 16.2 million bbl/day. To obtain equal energy per year, 8 billion tons per year of oil shale would have to be mined; this is about 16 times the tonnage handled by the U.S. coal industry in 1973. Disposal of waste rock would amount to 6000 to 8000 acre-ft (7.4 to 9.8×10^6 m^3) per day. Perhaps only half of this could be returned to the mine, owing to the greater volume of the reject rock. The problems of safe and aesthetic waste disposal would be immense.

The materials handling problem for the tar sands are of the same order of magnitude as for oil shales, since both contain about 10 percent oil in the richest large deposits.

There are methods by which oil can be recovered underground from both oil shales and tar sands; these are frequently referred to as "in situ" methods. They could greatly reduce the materials handling problem, but none has yet demonstrated commerical feasibility.

The tar sand and shale oil recovery methods are both large users of water. Waste water disposal could be a serious environmental problem. Also, it is believed that one of the ultimate limitations on shale oil production may be the availability of water for processing (2).

2.2.3 Nuclear Fuels

Since nuclear fuels have so much energy per unit volume (more than 10,000 times as much as solid fossil fuels), the bulk of materials handled during fuel shipment is not a major problem. Even the quantities of rock handled at the mine are negligible compared with solid fossil fuel mining. Environmental problems around the mine are also relatively small in scale. The great constraints on the rate of growth of nuclear power are not in resources, extraction, land damage, or transportation; they are primarily in safety and capital.

The need to be sure of the safety of fuel processing, reactor operation, and waste disposal has strongly constrained the development rate of nuclear technology and the rate at which regulatory agencies with limited staffs (and encountering strong public concern and intervention) could authorize construction of new plants. The second great constraint, capital, appears to be increasingly important. The capital required to double electrical power plant capacity every 10 years is already so colossal that utilities could find it extremely difficult to raise the money required. Nuclear power is relatively capital-intensive, and so could be hit harder in this way than fossil fuel power.

2.3 CONTINUOUS ENERGY SOURCES

Table 2.7 shows the potential electrical power that could be developed from hydroelectric and tidal power sources in North America. The potential for geothermal power is quite speculative, since it is not yet known to what extent underground drilling and explosions could be used to extract energy from wet and dry deep thermal reservoirs. The estimate in Table 2.7 refers to use of present technology.

There are two serious disadvantages to the potential power sources in Table 2.7. First, most of them are expensive to exploit. The low-cost hydroelectric sites have been used up in most of the continent. Second, they are relatively limited in total magnitude. For the United States, at the end of the century total electrical power generation will likely be 1 million to 2 million MW; tidal power could make little contribution to this even if it were low in cost.

A 2176-MW-capacity tidal power plant is being considered for the upper Bay of Fundy in the period 1980–1990, if the economics are satisfactory at that time. The plant would have an annual output of 6500 GWh. Another possible site is in the lower Bay of Fundy, at

Table 2.7
Potential Continuous Power Sources in North America

	Potential (MW)	Developed Capacity (MW)
Hydroelectric[a]		
United States	187,000	53,404
Canada	95,000	32,501
Tidal Power		
Bay of Fundy	30,000	0
Geothermal Power		
Geysers, California	1,000 (1) 10,000 (8)	400

[a]Average flow conditions.
Source: Refs. (1) (3) (6) (7) (8).

Passamaquoddy Bay in eastern Maine, where the tides are 5.5 m; it would have a capacity of 1800 MW (average), with an annual production of 15,800 GWh. The economics of this scheme were studied extensively in the 1930s and again in the 1960s. At specific sites on the Bay of Fundy, 29,000 MW of average power are potentially available from the tides. So far, however, the costs of building tidal power facilities in North America have been prohibitively high. Prototype plants are in operation in France and Russia.

General studies are still being done on the use of geothermal energy in the United States. Geothermal energy might come from local zones of molten rock fairly high in the crust of the earth. Groundwater coming near these zones is heated rapidly to several hundred degrees Centigrade. Since the technology for using hot, dry rock is not yet available, only the so-called hydrogeothermal resources are now usable. Geothermal energy is used in several parts of the world to produce hot water for space heating and other uses of low-grade heat.

For the most part, geothermal steam will be used to produce electricity. Studies are under way on the possibility of drilling very deep holes in hot, dry rock, fracturing it into many small passages, and then injecting water for the production of steam. One difficulty is that many of the rock elements dissolve in the hot water (as is well known around geysers), so that deposits on pipes carrying wet steam can be a severe problem. Also, noncondensables, including sulfur and CO_2, can be problems. If geothermally heated water were not returned to the underground source and were discharged to nearby streams, it could be a source of unacceptable pollution. Though present knowledge is incomplete, the economic potential of geothermal power in North America appears small, unless an entirely new technology can be developed for extracting heat from hot rock.

In contrast to hydraulic, tidal, and geothermal sources, solar power is sufficiently abundant to supply all of the electricity needed by North America for the foreseeable future. The question with solar power is not the power potential level, but the cost of land, collectors, converters, and energy storage.

Solar energy is receiving significant attention at the present time. In the United States, from July 1973 to July 1975, $61 million was spent on solar energy research and development. The total consists of about $20 million for the development of flat-plate collectors for space heating and cooling of buildings, and $13.3 million for the development of high-temperature collectors for thermal conversion of solar energy to produce electricity and steam. Of the remainder, $9.4 million was devoted to photovoltaic cells and $7.7 million to research on energy extraction from biological material and wastes.

One form of solar energy is wind energy, widely used throughout history. The largest windmill ever built is the 1250-kW generator at Grandpa's Nob, Vermont, operated between 1941 and 1945. It has been estimated that about 500,000 GWh/yr could be generated in the United States by use of 300,000 3200-kW wind power generators. These would be spaced about 1.6 km apart over 900,000 km^2 in the Great Plains. Each generator would have twenty 15-m-diameter turbine blades on a 300-m tower (or one 65-m-diameter turbine). A critical problem for wind power generation is energy storage. In mountainous areas this can be provided by the use of hydraulic reservoirs. Most applications for windmills are for small power units in remote areas, and particularly for pumping water, since temporal variations are unimportant in this application.

Use can possibly be made of ocean thermal gradients to produce power. The vertical temperature gradient in the Gulf Stream is about 16 to 22°C over 1000 m. The Gulf Stream off the southern coast of Florida flows at a volume rate of 2200 km^3/day. Full utilization of the thermal gradient, which averages about 20°C, could produce approximately 20 million MW, or 30 times the present world demand for electricity. A plant of 22 kW was designed and operated by George Claude in Cuba in 1929. Two French power plants of 3500-kW capacity were tested in 1956 off the coast of Africa. One basic limitation of this idea is that the available temperatures limit plant efficiencies to 2 or 3 percent, which tends to make capital costs very high per unit of power output.

2.4 ENERGY DEMAND

2.4.1 Fuel Substitution

For the next two or three decades the key energy problem will be coal and nuclear fuel substitution. Figures 2.1 and 2.2 show the size of the problem. In the 1960s, oil and gas demand grew very rapidly both in the United States and Canada. Coal use was static between 1965 and 1970; nuclear energy contributed very little to total energy supply.

Looking back 100 years, Figure 2.3 shows that major fuel substitution has taken place in the past, though not quickly. Coal took five or six decades to grow to a dominant position early in the century, but has since been largely displaced by oil and gas. Oil and gas, in turn, will probably be supplanted by nuclear fuel and coal within a few decades.

The rate at which this last substitution will actually happen depends on many factors in addition to the decline of oil and gas reserves. One of the key questions is whether the future demand for energy will be substantially reduced by rising prices and conservation actions (see Chapter 3).

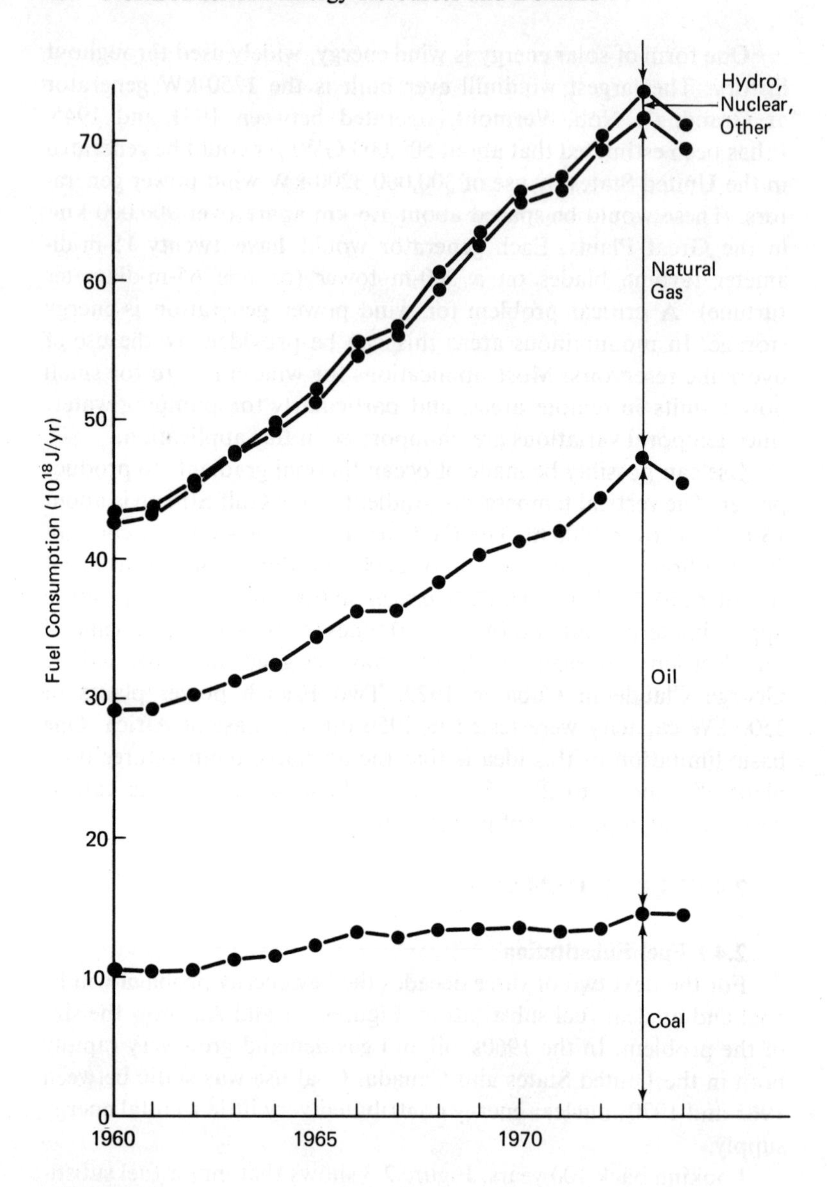

Figure 2.1
Fuel consumption in the United States (9).

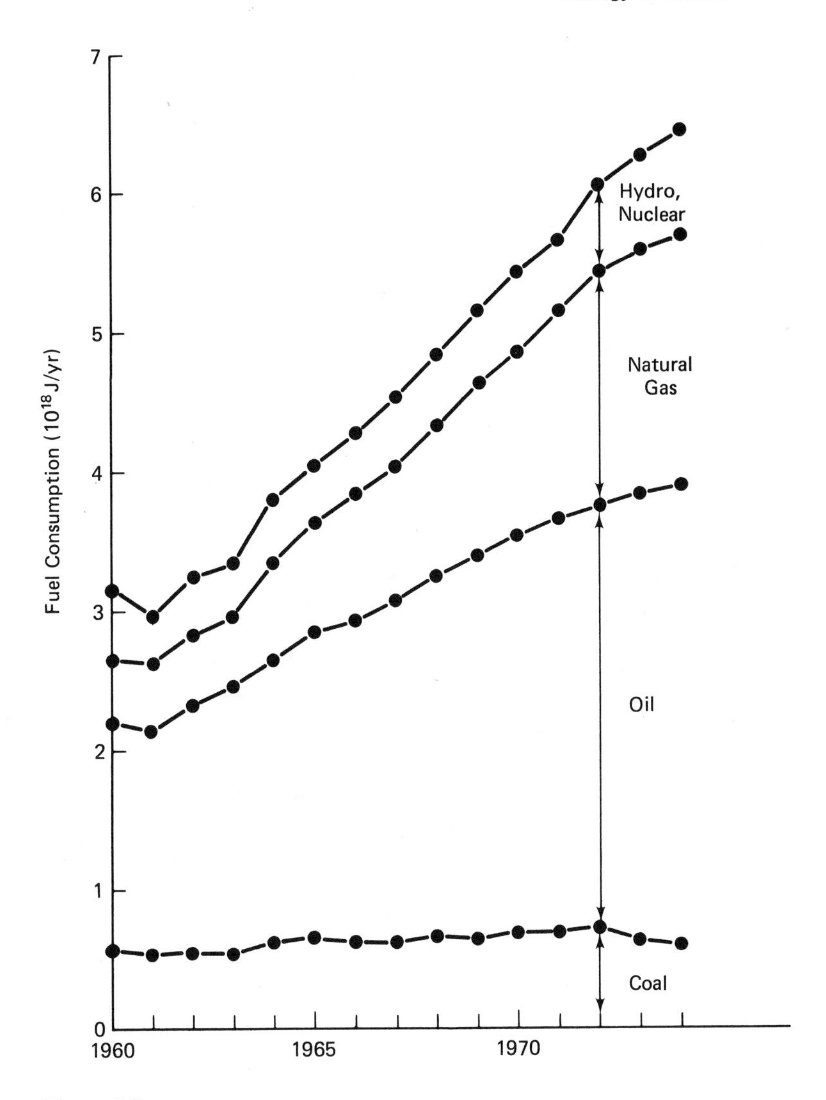

Figure 2.2
Fuel consumption in Canada (9).

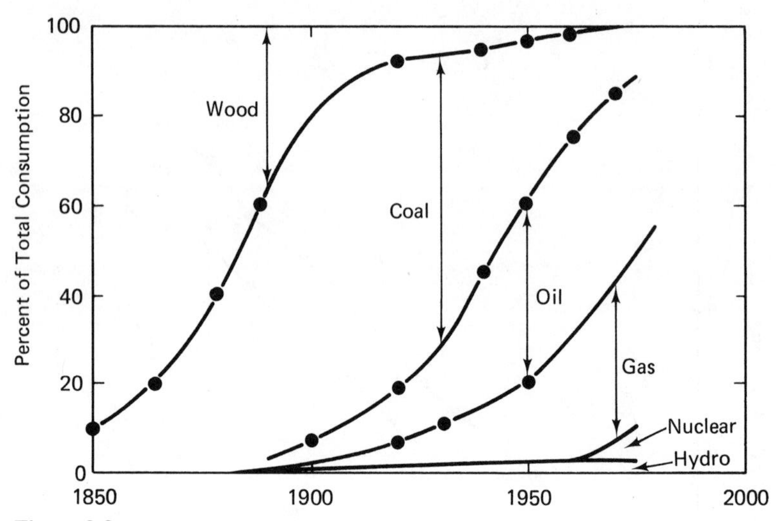

Figure 2.3
Relative contributions of various fuels to total consumption in the United States. Adapted from Ref. (10).

Another question is the extent to which energy use can or should be shifted from oil and gas to electricity. In the far future, nuclear energy may be used to synthesize a variety of clean fuels from coal or other materials. In the near future it will be restricted mainly to the production of electricity. As Figures 1.7 and 1.8 show, electrical energy use has grown much faster than total energy consumption. One advantage of this trend is that it can progressively lessen our dependence on fossil fuels. It is possible that nuclear-generated electricity could become cheaper, even for space heating, than fossil fuels. Fossil fuels, particularly oil and gas, may need to be allocated to special uses, including transportation and chemical feedstocks.

Another key question is the feasibility of coal conversion to clean-burning liquid and gaseous fuels. Many technical possibilities exist. The immediate question is whether coal can be heavily used for electrical power generation. Present costs of coal, and of meeting emissions standards, indicate that in many regions coal-generated electricity will be considerably more expensive than nuclear power. The incentive for coal conversion may come from other fuel needs, but the price of oil and natural gas is not yet high enough.

2.4.2 Demand and Supply Projections

Figure 2.4 shows one recent projection of energy demand in the United States up to the end of the century. Other estimates of total energy demand differ within a range of approximately 20 percent of this projection for the year 2000. Such a projection cannot anticipate all of the future effects of antipollution legislation and changing fuel prices. Either factor may cause radical shifts in the proportions of

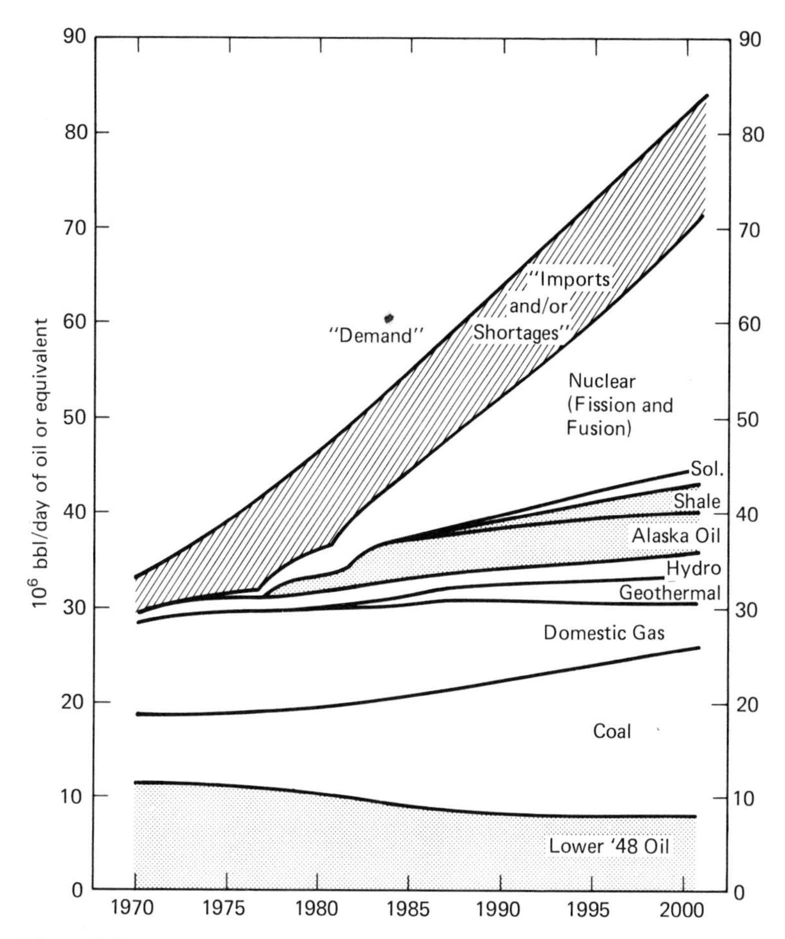

Figure 2.4
Projected energy consumption for the United States (3).

various fuels used, so that Figure 2.4 can only project trends that are now recognized; it cannot predict future rates of fuel supply.

Another great uncertainty is the effect of energy conservation and rising prices on energy demand. Figure 2.4 implies that by the end of the century total demand will be 25 percent less than a simple logarithmic extrapolation of recent trends would indicate.

The most important feature of Figure 2.4 is the indication that major imports of oil will be required up to the end of the century. The import level (12 million bbl/day) would be equivalent to about $60 billion/yr, a significant fraction of the U.S. Gross National Product. To keep import costs even as low as this would require strenuous energy conservation efforts.

Figure 2.4 suggests that coal use will nearly double in the next 10 years and that Alaskan oil could become significant in that time. Shale oil will not make a large contribution even by the end of the

Table 2.8
U.S. Energy Supply, 1973

Fuel	Amount (10^6 bbl/day or equivalent)
Crude Oil and Natural Gas Liquids	10.1
Natural Gas	11.9
Coal and Lignite	6.8
Hydroelectric	1.4
Nuclear	0.4
Total	30.6

century. Nuclear energy is expected to be the dominant new contribution. The projected contributions of solar and geothermal energy should be regarded as early guesses.

The breakdown of 1973 energy supply in the United States is shown in Table 2.8.

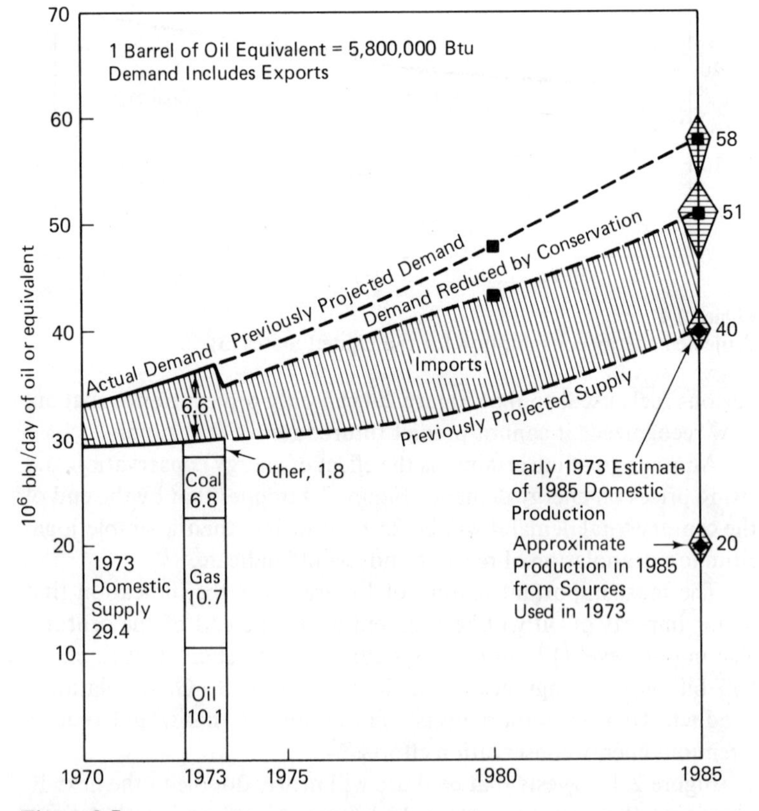

Figure 2.5
U.S. energy supply and demand to 1985 (2). Reproduced with permission of the National Academy of Sciences.

The 1973 domestic oil and gas production of 22 million bbl/day cannot be maintained or increased without intense application of recovery technologies. It is believed that oil and gas production from domestic sources (including Alaska and the outer continental shelf) could possibly be as high as 27 million bbl/day by 1985. This would require $180 billion in capital investment.

Figure 2.5 shows another projection for the United States for a shorter time period. It is reasonably consistent with Figure 2.4 in its projection of the possibilities of energy conservation and the imports of oil required to meet reduced energy demand. It assumes that energy supply in 1985 to the United States will be composed as shown in Table 2.9. Figure 2.5 also implies that as early as 1985 conservation measures and price increases will reduce demand by about 16 percent from what it would otherwise have been.

Figure 2.6 shows energy supply–demand projections for Canada, indicating a large effect of domestic price level on the mismatch between availability and demand. It is now recognized that Canada faces loss of self-sufficiency in energy supply by the mid-1980s, particularly if conservation measures and rising prices do not reduce demand very much. Recovery of oil from tar sands is expected to be only a few hundred thousand barrels per day. There is hope that intensive exploration in Arctic and offshore regions will yield new oil reserves, but these will take a long time to bring to market.

Figures 2.7, 2.8, and 2.9 show energy flows in the United States for 1960, 1970, and 1980 (projected), respectively. They are plotted to the same vertical scale, in units of millions of barrels per day of oil equivalent (10^6 bbl/day of oil = 2×10^{15} Btu/yr). Immediately evident is the tremendous increase (130 percent) in total energy consumption over the 20-year period.

Also apparent is the huge increase (120 percent) in oil consumed over the same period. The projection for 1980 shows that, despite

Table 2.9
Assumed U.S. Energy Supply, 1985

Fuel	Amount (10^6 bbl/day or equivalent)
Oil	13.1
Shale Oil	0.5
Gas	15.6
Coal	10.0
Nuclear	8.3
Continuous Sources	1.7
Total	49.2

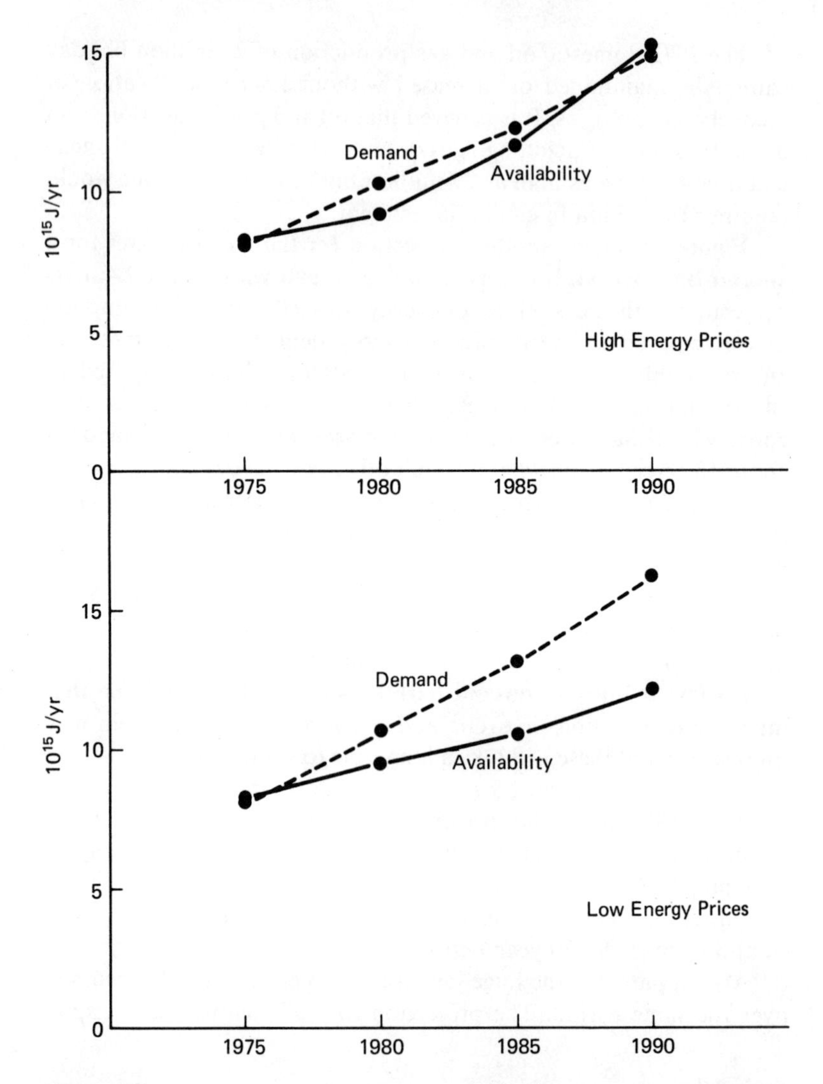

Figure 2.6
1976 supply–demand projections for Canada (4). The "low energy prices" curves assume that energy prices rise only with the general rate of inflation; the "high energy prices" curves assume that domestic oil prices reach international levels by the late 1970s.

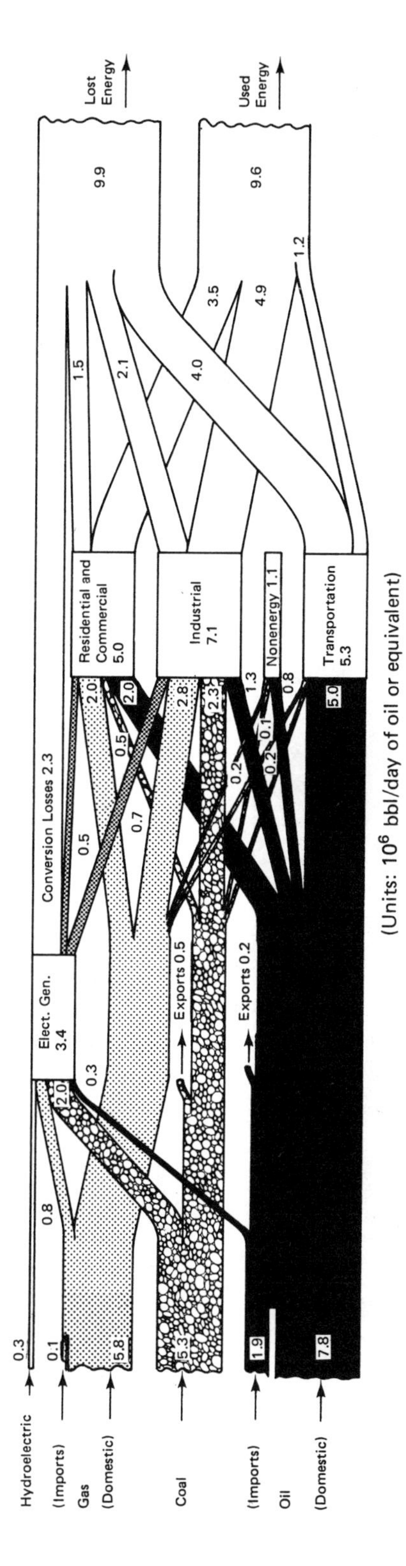

Figure 2.7
Energy flow, United States, 1960 (3).

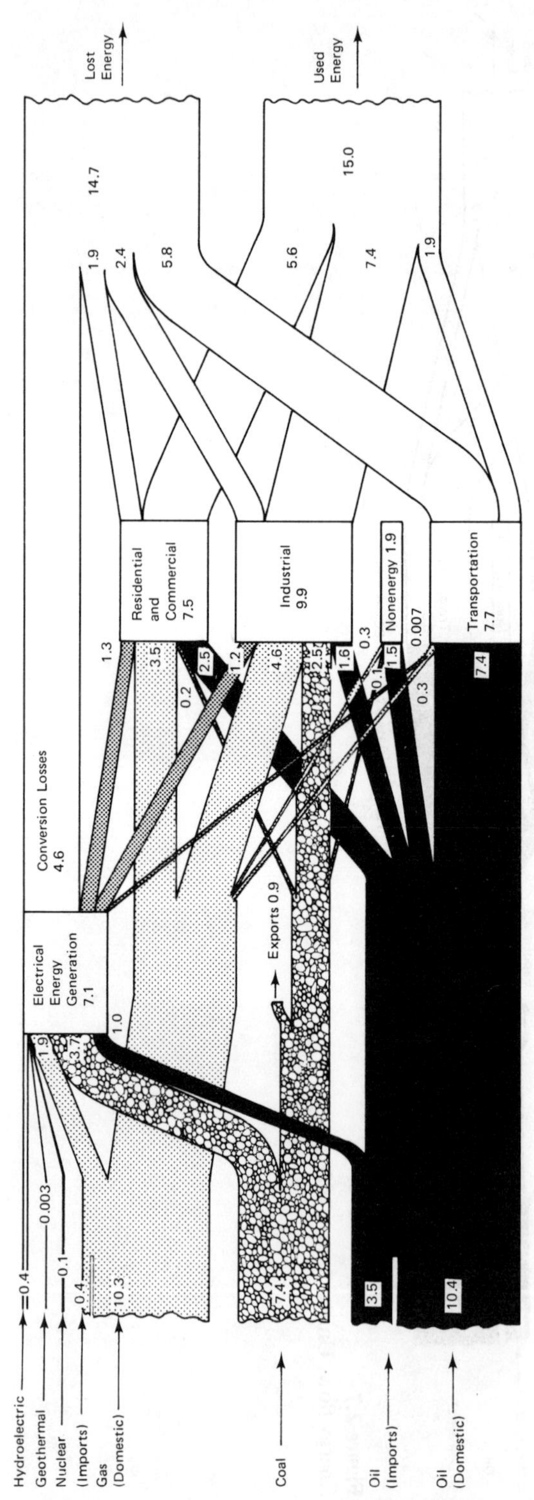

Figure 2.8
Energy flow, United States, 1970 (3).

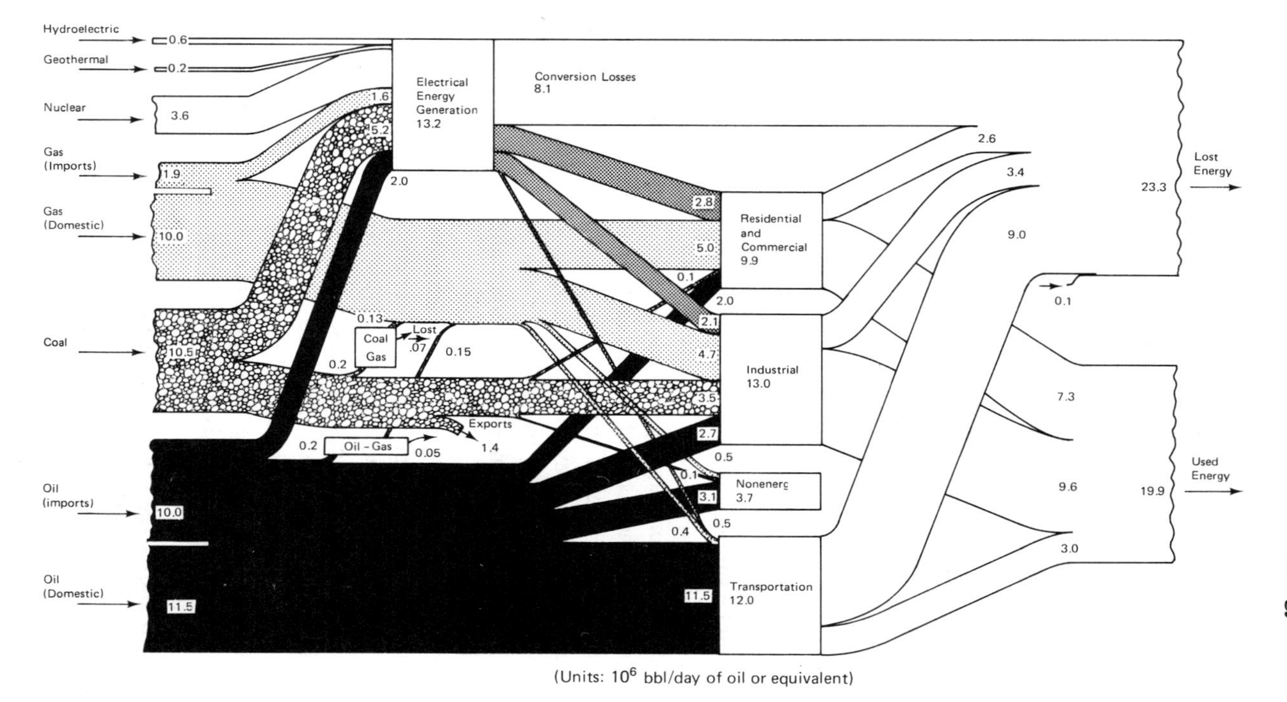

Figure 2.9
Projected energy flow, United States, 1980 (3).

Table 2.10
Overall Generating Capacity in the United States (GWe)

Source	1973	1985
Nuclear	21	325
Coal	143	330
Oil and Gas	175	130
Gas Turbine and Internal Combustion	38	50
Medium-Btu Synthetic Gas	—	20
Hydro	54	75
Geothermal	—	7
Pumped Storage	8	45
Total	439	982

Source: Ref. (2).

the availability of Alaskan oil, domestic supplies of oil will have increased very little over 1970, and nearly half of the total oil consumption will be supplied by imports. It may happen that not all of these imports will be available. Petroleum-exporting countries may not expand production rapidly enough to take care of this huge additional need and the needs of Europe and Japan. Steps should be taken to reduce the rate of growth of consumption, locate new sources of gas and oil, and develop other energy sources as quickly as possible. Nineteen eighty is too soon for major shifts in power plants and fuels or even for many of the important energy conservation measures to take effect.

Between 1960 and 1970, electricity generation increased 110 percent. The projection of Figure 2.9 is conservative; it assumes only an 86 percent increase in electricity generation from 1970 to 1980. Perhaps rising electricity prices will discourage consumption; this remains to be seen. Between 1960 and 1970, 60 percent of the electricity use was absorbed by the residential and commercial sectors. Use of electricity for space heating and appliances has been growing rapidly. A very small fraction is now used in transportation, as in electric railroads.

Table 2.10 shows the relative contributions of various fuels to electricity production in the United States in 1973. It also shows the projections for 1985 of the National Academy of Engineering Task Force on Energy.

In this 12-year period, nuclear power is expected to increase by a factor of 15. The expected doubling of coal plant capacity reflects the belief of the Task Force on Energy that stack gas sulfur removal will have been shown within a few years to be commercially feasible and reliable. Oil and gas plant capacity is expected to decrease considerably by 1985. Pumped storage capacity is projected to increase by a factor of five or six; this is related to the increasing use of capital-intensive nuclear plants.

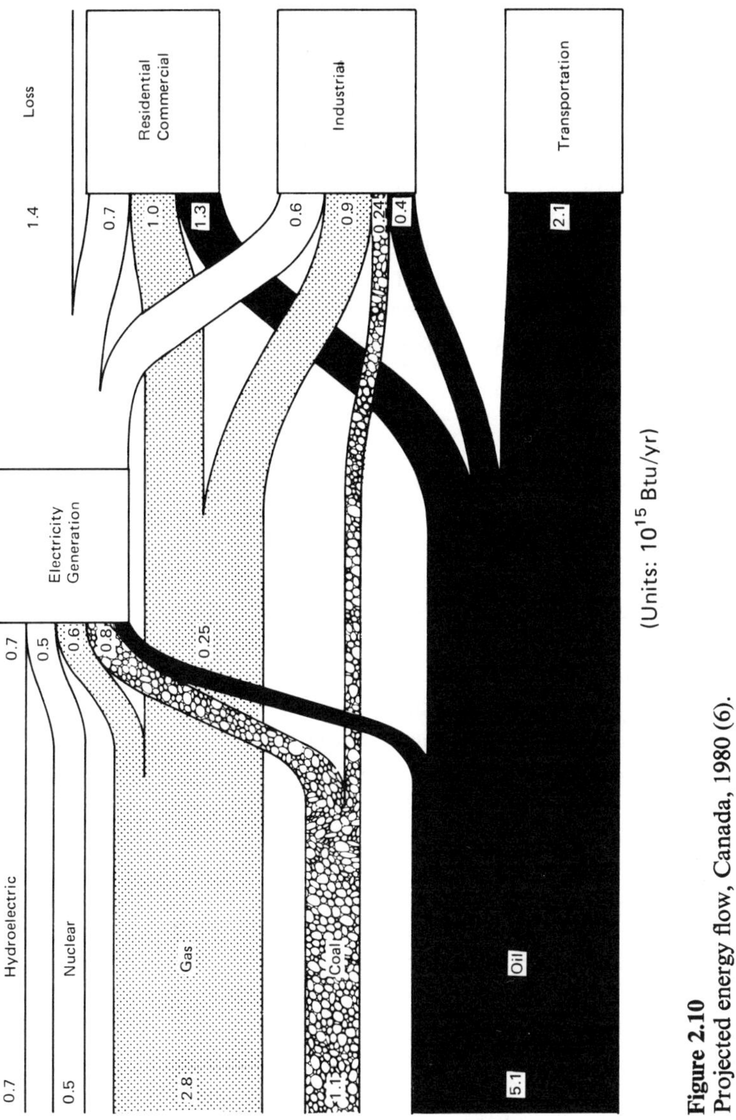

Figure 2.10
Projected energy flow, Canada, 1980 (6).

Figure 2.10 shows the approximate energy flow projected for Canada for 1980. Exports of oil and gas have been roughly equal to domestic consumption; substantial exports of these commodities are expected to continue, reserves permitting. The energy flow picture in Canada is quite similar to that in the United States, with a somewhat higher percentage consumption of energy in the residential and commercial sectors, somewhat lower in transportation.

Growth rates for individual sectors are indicated in Table 2.11. The relatively rapid growth rate for electrical output is noteworthy, but the residential and commercial sectors are also growing rapidly, with space heating, appliances, and lighting being major factors in increasing demand.

2.5 SUMMARY

In the short term, the United States must import large volumes of oil to avoid serious energy shortages. The high cost of this imported oil should accelerate the exploitation of offshore oil, shale oil, and tar sands. Self-sufficiency in energy appears very unlikely in the United States in the early 1980s. The peak in domestic oil and gas production has probably already been passed, despite the recent incentive to develop higher-cost reserves. Canada could also become a net importer of energy in the mid-1980s.

The problem lies not in the absence of energy resources, but in the investment of capital and manpower to bring resources to market. At least 10 years would be required to double coal mining capacity; and this would only replace a small part of the energy supplied by oil and gas. Moreover, it could seriously harm the environment with both land damage and air pollution. The future costs of coal and the effects of environmental constraints are quite uncertain. Investment is retarded both in mining and in coal conversion development.

Production of oil from tar sands is now under way, but the capital demands are so high that expansion of production cannot be extremely rapid. Major environmental problems could result from oil shale and tar sand exploitation.

Electricity generation has been doubling every 10 years. Price

Table 2.11
Energy Consumption Growth Rates in the United States and Canada in the 1960s (%/yr)

	United States	Canada
Electrical Output	7.2	7.2
Industrial	3.9	4.3
Residential and Commercial	5.0	4.2
Transportation	4.1	3.2
Total Energy	4.6	4.8

increases and conservation will both decrease demand. Economic recession can also sharply diminish the demand growth rate. This can be seen in Figure 1.8 for the depression period of the 1930s. In 1975 there was little or no growth in demand, but it is too early to be sure that the long-term growth rate in electricity demand has been sharply reduced.

The future cost and scarcity (or unreliability in supply) of fossil fuels will tend to increase the demand for wider use of nuclear energy to produce electricity. Electrical utilities recall that before the 1960s projections of electricity demand growth rates were underestimates, leading to capacity shortages. Since plant construction takes several years, this condition can take a long time to correct. A number of plans for new electrical plants have recently been postponed or canceled.

The capital requirement for new electrical plants is becoming an obstacle to development. Nuclear power had previously been expected to provide 35 percent of the total U.S. electricity generation capacity by 1980; now this is not expected until 1985.

Accentuating the present problem is the effect of environmental legislation on the choice of fuels for electrical power generation and for other purposes.

Although the United States has some hundreds of years of coal reserves, much of this may not be useful for electrical power production owing to its high sulfur content. The need to reduce SO_2 emissions has greatly increased the demand for petroleum, and especially for low-sulfur oil. It has also escalated the shortage of natural gas.

The major difficulty is the time it takes to develop new sources of fuel and to build new power plants. Up to nine years of development may be required from the go-ahead decision to the commissioning of a new nuclear electricity generation plant. And many years may be required before adequate supplies of synthetically produced gas or petroleum gas, or petroleum from tar sands or oil shales, can arrive on the market.

The short-term North American energy problem involves a natural gas shortage with the possibility of curtailment and rationing, large petroleum imports with serious implications for balance of payments, air pollution hazards, and the prospect of at least a decade of discomfort before satisfactory steps can be taken toward assurance of reliable energy supplies for future development.

In the longer term, North America will face depletion of oil and gas, though coal reserves will last at least a century. Methods of using coal cleanly and at a reasonable price are needed urgently. Several possibilities exist, but considerable time will be needed for development and widespread application. Nuclear energy could assume a large share of the total energy need. The main constraints on its development are the requirements for capital and safety.

References

1. World Energy Conference, U.S. National Committee. *Survey of Energy Resources, 1974*. 345 E. 4th St., New York, New York.

2. National Academy of Engineering, Task Force on Energy. *U.S. Energy Prospects: An Engineering Viewpoint*. Washington, D.C., 1974.

3. U.S. Congress, Joint Committee on Atomic Energy. *Understanding the National Energy Dilemma*. Washington, D.C.: Superintendent of Documents, 1973.

4. Energy, Mines and Resources Canada. *An Energy Policy for Canada, Phase I*, 2 volumes. Ottawa, 1973. See also Energy, Mines and Resources Canada, *An Energy Strategy for Canada—Policies for Self-Reliance*. Ottawa, 1976.

5. Science Council of Canada. *Canada's Energy Opportunities*. Report No. 23, Ottawa, March 1975.

6. M. K. Hubbert. "Energy resources for power production." Paper IAEA-SM-146/1, *Proceedings of the International Atomic Energy Agency Symposium on Environmental Aspects of Nuclear Power Stations*, New York, August 10–14, 1970, pp. 13–43.

7. M. K. Hubbert, "The energy resources of the earth." *Scientific American*, September 1971, pp. 61–70.

8. U.S. Congress, House, Committee on Science and Astronautics. *Hearings on Energy Research and Development*, May 1972, p. 351.

9. United Nations, *World Energy Supplies, 1929 to 1973*. Statistical Papers, Series J, Nos. 1, 2, 3, 7, 8, 11, 14, 18.

10. National Science Foundation. *Energy Research Needs*. Washington, D.C., 1971.

11. U.S. Department of the Interior. *United States Energy: A Summary Review*. Washington, D.C., 1972.

12. Ford Foundation. *Exploring Energy Choices: Preliminary Report of the Energy Policy Project*. 1974.

Supplementary References

National Petroleum Council. *Report on United States Energy Outlook*. Washington, D.C., December 11, 1972.

A. J. Surrey and A. J. Bromley. "Energy resources." Chapter 8 in *Thinking about the Future: A Critique of "The Limits to Growth,"* H. S. D. Cole et al., eds. Sussex, England: Sussex University Press, 1973.

Problems

2.1 Table 2.A shows the decline in oil discovery rates in the lower 48 states in the period 1950–1970.

By plotting oil discovery rate (bbl/ft) of exploratory drilling against cumulative feet of drilling since 1950, estimate roughly the ultimate likely discoveries of oil from 1950 onward. Add to this quantity the discovered reserves of 9.6×10^9 bbl in the Alaskan

Table 2.A
Oil Discovery Rates for 48 States, 1950–1970

Year	Exploratory Drilling Footage (10^6 ft/yr)	Oil Discovery Rate (10^6 bbl/yr)
1950	40	2500
1955	70	1500
1960	55	750
1965	48	930
1972	44	200

Source: Ref. (11).

North Slope region. Add also cumulative discovery prior to 1950 of 96×10^9 bbl.

Allowing for cumulative production from United States wells of 89×10^9 bbl up to 1972, estimate remaining reserves in place as of 1972, and compare the figure with "proved recoverable reserves" of United States oil shown in Table 2.1.

2.2 The Ford Foundation Energy Policy Project (12) reports the per capita energy consumption data shown in Table 2.B.

Table 2.B
Per Capita Energy Consumption

Year	United States (10^6 Btu/ person)	World (10^6 Btu/ person)
1925	180.2	23.4
1930	180.6	
1940	180.1	
1950	223.2	30.7
1955	239.3	36.6
1960	246.7	41.5
1965	274.5	49.0
1967	293.2	
1968	307.7	54.5
1969	320.6	
1970	327.7	
1971	331.8	
1972	345.3	
1973	359.1	

a. Plot these data on semilog paper and estimate the percent annual growth rate of per capita consumption for the United States, 1950–1973; for the United States, 1960–1973; and for the world, 1960–1970.

b. If the annual increase in per capita consumption after 1970 were 2 percent for the United States and 3 percent for the world, how long would it take for world consumption levels to become equalized with that of the United States?

c. Assuming that the population of the United States levels off at 300 million by 2050 (from approximately 220 million in 1970) and that per capita consumption increases at 1.5 percent in the interim, estimate the average rate of growth of total energy consumption between 1970 and 2050.

2.3. The Ford Foundation Energy Policy Project (12) reports the major energy resources of the United States shown in Table 2.C.

Table 2.C
Major U.S. Energy Resources (Ford Foundation Estimates)

	Consumption (10^{15} Btu)	Cumulative Production (10^{15} Btu)	Reserves (10^{15} Btu)	Recoverable Resources (10^{15} Btu)
Petroleum	34.7	605	302	2,910
Shale Oil			(465)	
Natural Gas	23.6	405	300	2,470
Coal	13.5	810	4,110	14,600
Strippable Coal			925	2,600
Low-Sulfur Coal			2,390	
Uranium				
For Use in Light-Water Reactors	0.85	2	228	600
For Use in Breeder Reactors			17,700	47,000
Hydro Power	2.9			5.8

a. Compare proved recoverable reserves in Table 2.1 (from the World Energy Conference) with recoverable resources and reserves in Table 2.C.

b. Compare the ratio of reserves to 1973 consumption calculated from Table 2.C with the ratio of reserves to production in Table 2.1. Note that, with oil imports, production is significantly less than consumption.

2.4 The actual distribution of electricity generating capacity in 1973 and that projected for 1985 by the Task Force on Energy are shown in Table 2.10. Suppose that the average annual growth rates assumed in the table continue to the end of the century, except that the growth of nuclear power from 1985 to 2000 is limited to 15 percent per year (average). Assume annual growth rates of 5 percent for medium-

Btu synthetic gas and 3 percent for geothermal sources, from their 1985 values. If all of this electrical capacity could be utilized with an annual capacity factor of 0.7 and an overall efficiency of 33 percent, estimate the following:

a. The total quantity of energy consumed for electricity generation in the United States in the year 2000.

b. Per capita consumption of electricity in 1973, 1985, and 2000 if population growth is 1 percent per year.

2.5 Consider the environmental implications of developing shale resources rapidly enough over the period 1980–1990 to supply the "imports and/or shortages" indicated on Figure 2.6.

It has been estimated (2) that bringing into production a shale oil capacity of 500,000 bbl/day would require capital expenditure of $3 to $5 billion, 80,000 acre-ft/yr of water supply, and dispersal of a volume 5 square miles × 40 ft of solid waste material each year.

Calculate, for the total required shale oil capacity (12 million bbl/day):

a. the annual rate of investment required per capita;

b. the volume of water flow (ft^3/sec) and the number of 4-ft-diameter pipelines to supply this water at a velocity of 10 ft/sec;

c. the volume of solids wasted each year, total and per capita (ft^3).

Note: 1 acre = 43,560 ft^2.

2.6 Nuclear power capacity in the United States and Canada has been projected to grow according to Table 2.D.

Table 2.D
Projected Nuclear Power Capacity (GWe)

	1975	1980	1990	2000
United States	54.2	131.6	508	1200
Canada	2.5	6.6	41	133

Estimate the requirements of U_3O_8 (in metric tons) for the years in Table 2.D if reactors operate with annual capacity factors of 0.7, and the consumption of uranium is 250 g/kWe-yr for the light-water reactors used in the United States, and 168 g/kWe-yr for the heavy-water reactors used in Canada.

Assuming the breeder reactor is not widely used for power production before the year 2000, estimate the cumulative U_3O_8 requirements for the United States and Canada for the period 1975–2000, and compare these with the proved recoverable resources figures of Table 2.1.

2.7 Possible annual solar energy use in the United States has been estimated for the years 1985 and 2000 as shown in Table 2.E.

Table 2.E
Estimated Solar Energy Use, 1985 and 2000

	Savings in Fossil Fuels Due to Solar Energy Use (10^{12} Btu)	
	1985	2000
Thermal Energy for Buildings		2,100
Conversion of Organic Matter to Fuels		
Combustion of Organic Matter		760
Bioconversion to Methane	270	3,100
Pyrolysis to Liquid Fuels		630
Chemical Reduction to Liquid Fuels		630
Electrical Power Generation		
Thermal Conversion		760
Photovoltaic		1,510
Wind Energy		760
Ocean Thermal Plants		760
Total		11,010

a. For the year 2000, compare the estimated total saving in fossil fuels in Table 2.E with "imports and/or shortages" in Figure 2.6.

b. How much total electrical capacity is projected, assuming an average fossil fuel generation officiency of 35 percent and an annual capacity factor 0.7? What fraction is this of the total electrical capacity that might be projected for the year 2000 (see Table 2.8)?

2.8 A 1973 projection of Canadian electricity production is shown in Table 2.F.

Table 2.F
Projected Canadian Electricity Production (10^9 kWh)

	1970	1980	1990	2000
Hydro	157	235	310	344
Coal	34	76	127	151
Oil	7	29	72	118
Natural Gas	5	11	9	17
Nuclear	1	45	180	502
Total	204	396	698	1132

a. Estimate the growth rate in total electrical energy generation projected for each decade.

b. Estimate average per capita power (kWe/person) for 1970 and 2000, assuming a population of 22 million in 1970 and 1 percent per year growth in population between 1970 and 2000.

c. Estimate average growth rates for coal in the decades following 1980 and 1990.

d. Suppose that total electricity demand is as in Table 2.F, but that shortages permit no growth in gas- and oil-fired generation after 1980, and that capital shortages limit nuclear generation in 1990 to 150 × 10^9 kWh and in 2000 to 300 × 10^9 kWh. How would this affect the average growth rate in coal-fired generation in the decades 1980–1989 and 1990–1999? Assume that hydro generation is as in Table 2.F and that contributions from new energy sources are negligible.

2.9 If 10 square miles of land per million of population could be allotted to solar energy collection for power generation, how much electrical power per person could be generated? Assume that the yearly average incidence of solar energy is 200 W/m^2 thermal and that overall generation efficiency is 10 percent.

Compare your answer with present U.S. electrical power consumption per capita, extrapolating to 1977 from Figure 1.6 and assuming a population of 230 million.

2.10 The U.S. National Academy of Sciences Committee on Mineral Resources and the Environment has estimated U.S. energy resources in March 1975 as follows:

Petroleum: 160 × 10^9 bbl
Natural Gas: 750 × 10^{12} ft^3
Coal: 600 × 10^9 tons
Proved Uranium: 7 × 10^5 tons
Potential Uranium: 3 × 10^6 tons

Convert these quantities to common units and compare with the estimates of proved reserves in the United States shown in Table 2.1.

Determine the reserves-to-production ratio of petroleum and natural gas.

3
Energy Utilization

3.1 INTRODUCTION

As Table 2.11 indicates, total energy consumption grew at 4.6 percent per year in the United States in the 1960s. Such an increase may seem modest, yet if it were to continue for 100 years, total consumption would be 81.6 times what it is today. It has always been clear that exponential growth cannot continue in the long run. What has become apparent recently is that it may be difficult, costly, and hazardous to continue it for even a few more decades.

Continuing growth of this kind would likely mean rapid depletion of gas and oil reserves and costly imports, even with development of offshore and frontier reserves. It could also require substantial exploitation of tar sands and oil shales, a great expansion of coal mining, and a rapid buildup in nuclear plant capacity. John Deutsch remarks,

> Such a course of action would entail extremely heavy commitments for capital and research expenditures[1] and would call for carefully articulated long-range plans. This option, designed to support continued high rates of growth, would carry with it costs and risks which could have substantial effects upon the quality of life in the future. Sharply increased prices for energy, which could not be

[1]In the next 10 years alone, capital expenditure for North American energy supply and conversion could be of the order of $1 trillion.

avoided, would tend to have a regressive impact on the distribution of income. The rapidly rising load of pollution resulting from the massive use of fossil fuels over the next several decades could cause a continuing notable deterioration in man's environment. Our knowledge and technological ability to cope with such volumes of fossil fuel pollutants or with the environmental consequences of a vast expansion of coal mining remain woefully inadequate, or the costs would be so enormously high as to be impractical for some time to come. Consequently, the realistic trade-off between growth and the quality of the environment would tend to be unfavorable to the latter. We are already witnessing this, in striking fashion, in some parts of the world. Finally, this particular high growth option would inevitably entail increased political risks in the immediate future. The bargaining and strategic power of those who control the rapidly depleting reserves of fossil fuels would be greatly enhanced. The adjustments and confrontations involved in the settlement of these matters would inevitably entail shifts in national power and large risks and dangers in both the economic and political spheres. (1)

D.C. White comments,

There are many factors pointing to difficulty in continuing our historic 15 to 20 year doubling time for growth of energy consumption. The new fuel resources required from foreign sources, the development of environmentally clean processing, converting and consuming equipment, the depletion of the domestic prime energy fuels, and the enormous magnitude of total energy consumption projected within only a few decades, are all concerns which shrink in terms of immediate and even longer term importance if the energy growth doubling times extend from 15 to 20 to say even 30 or 40 years. Halving the growth rate in energy consumption gives all segments of our energy system much greater flexibility and allows advances in technology, a better chance to keep ahead of foreseeable and unforeseeable difficulties. All exponential growth processes must stop in time, but in dealing with today's problems, a 40 year doubling time is perhaps much more compatible with technology development and societal adjustability than is one of 10 or 20 years. Conservation to slow down growth while satisfying the needs of society has greater societal payoff than any other single factor today, including new energy supply developments and new resource discoveries. (2)

In this chapter we review the typical pattern of energy consumption in North America; this will help to show how the growth in energy consumption could be slowed. One of the chief concerns is electricity, demand for which has been growing rapidly for a long time. As shown earlier, simple extrapolation of past trends would lead us in two or three decades to an all-electric economy. Possible depletion and rising costs of oil and gas could strengthen this trend; more use may need to be made of nuclear energy, whose main practical function at present is in electricity supply. Thus the questions of how electricity is used, whether it may be used more wisely, and whether the energy wasted in producing electricity may be substantially reduced, all urgently require answering.

3.2 ENERGY USE AND CONSERVATION

Table 3.1 indicates that transportation is the single largest end use of energy in the United States. The three elements in this sector are urban passengers, at 46 percent; intercity passengers, at 35 percent; and intercity freight, at 19 percent (5). Private cars account for about 87 percent of passenger miles between cities and 97 percent in urban areas. The corresponding figures for buses are 2 percent and 3 percent, respectively.

The principal methods for improving the energy effectiveness of passenger transportation are as follows:

1. Making cars smaller. Test data published by the U.S. Environmental Protection Agency (EPA) indicate that automotive fuel consumption in urban driving can be halved by halving car weight. However, safety requirements such as heavier bumpers and frames can lead to as much as a 10 percent increase in car weight.

2. Improving engine efficiency. Early attempts to comply with automotive emissions regulations led to markedly reduced engine

Table 3.1
End Uses of Energy in the United States

	Percentage of Total	
	Office of Science and Technology Estimate (3)	National Academy of Engineering Estimate (4)
Transportation[a]	24.9	25
Space Heating (R, C)	17.9	18
Process Steam (I)	16.7	16
Direct Heat (I)	11.5	11
Electric Drive (I)	7.9	8
Feedstocks, Raw Materials (C, I, T)	5.5	4
Water Heating (R, C)	4.0	4
Air Conditioning (R, C)	2.5	3
Refrigeration (R, C)	2.2	2
Lighting (R, C)	1.5	5
Cooking (R, C)	1.3	1
Electrolytic Processes (I)	1.2	1
Other	2.9	2
Total	100	100

R = residential; C = commercial; I = industrial; T = transportation.
[a]Fuel only; excludes lubes and greases.

efficiency. More recent work in methods of carburetion and combustion control suggests that a 10–20 percent improvement in engine efficiency may be feasible. Other reductions can result from use of radial tires (10 percent) and reduction of aerodynamic drag (5 percent).

3. Shifting passengers from cars to buses or trains. Hirst and Moyers (5) have estimated that shifting 50 percent of traffic from cars to buses could reduce urban passenger energy consumption by 25 percent. They also estimate a 7 percent drop in the total cost of urban passenger transportation (if direct costs per urban passenger mile are 8.3 cents for buses and 9.6 cents for cars). However, if leisure time is worth $2/hr and traveling 20 miles to and from work by bus takes 1 extra hour each day, then the cost of bus transport jumps to 18.3 cents per passenger mile. The 25 percent energy saving mentioned earlier could in this way imply a 50 percent increase in the total cost of urban passenger transportation. Evidence of major shifts in passenger transportation mode has not yet appeared, despite large recent increases in the price of gasoline.

The second largest end use of energy shown in Table 3.1 is space heating. If to residential and commercial space heating we add industrial space heating, the total is approximately 20 percent. If we add hot water heating and air conditioning, the total becomes 26.5 percent, making this combined category the largest end use of energy. It has been shown that space heating and air conditioning energy consumption can be markedly reduced by the following measures:

1. Improving thermal insulation of houses and other buildings. In the past there has been little economic incentive to do this since cheap fuel has been available. With gas delivered at $\$1/10^6$ Btu, good thermal insulation might only save a householder $32/yr even though it could cut energy consumption by 30 to 50 percent (5). Higher fuel prices and revised building standards will no doubt help to conserve energy in coming decades.

2. Improving door and window sealing. Adequate house ventilation is normally thought to require an air change every 2 hours. Such infiltration of fresh air will typically account for 15 percent of the space heating energy requirement. Poor sealing can increase heat consumption by 50 percent.

3. Improved efficiency of furnaces and air conditioners. Efficiencies of residential burners in good condition are normally quoted in the ranges 70 to 80 percent for gas and oil and 50 to 70 percent for coal (6). Some data indicate that burners tend to operate at much lower efficiencies (30 to 50 percent) due to lack of adjustment, intermittency of operation, and other factors (7, 8). It has been shown (5) that household air conditioners are mostly quite inefficient;

for commercial units, electrical energy consumption per unit of cooling energy varies from 0.27 to 0.70, with most units in the upper part of the range.

In industry as in the residential and commercial sectors, there has been little apparent incentive to reduce wastage of energy. Industrial energy costs have typically been only about 4 to 5 percent of average value added. Even in the metallurgical industries, energy costs have been only 7 or 8 percent of value added. One estimate suggests that energy costs for industry as a whole have been only 1.5 percent of the total value of goods shipped. There may be little or no advantage to the manufacturer to install costly equipment for energy saving if the cost of the wasted energy is small and easily passed along to the customer. Rising fuel prices may change this situation considerably.

The potential for energy conservation in the United States has been depicted by the Office of Emergency Preparedness (9), which estimated in 1972 that by 1990 energy conservation measures could reduce demand by at least 30 percent without unacceptable deterioration in living comfort (Figure 3.1). This would mean that the effective rise in the demand curve could be delayed by perhaps six or seven years, and the total energy growth rate could perhaps be halved. Another indication of potential energy savings is provided by Table 3.2, which shows possible reduction in United States energy demand by 1985.

The energy saving in transportation would correspond to (1) a change from the present 30:70 ratio of small to large cars to a 50:50 ratio and (2) an expansion of mass transportation in cities. The residential saving would result from better insulation in 20 million homes and increased use of heat pumps. Industrial processes would become 10 percent less energy-intensive, on the average.

In summary, the major fields for energy saving appear to be space heating, transportation, and industrial processes. Measures that could save substantial amounts of energy include the following:

1. Reduced wastage: for example, installation of adequate insulation in buildings, reduction of unnecessary illumination, improvement of insulation standards for industrial equipment, and provision of heat recovery devices.

2. Increased conversion efficiency: for example, by improvement of the efficiency of the electricity generation plant, development of better equipment (such as the heat pump) for electrical heating, and improvement of automotive engine efficiency.

3. Integration of heat and power supplies in order to reduce total energy consumption while meeting the same needs for heat and power.

Many other possible energy savings, small in themselves, could aggregate to a significant quantity. Packaging is one example. Bruce

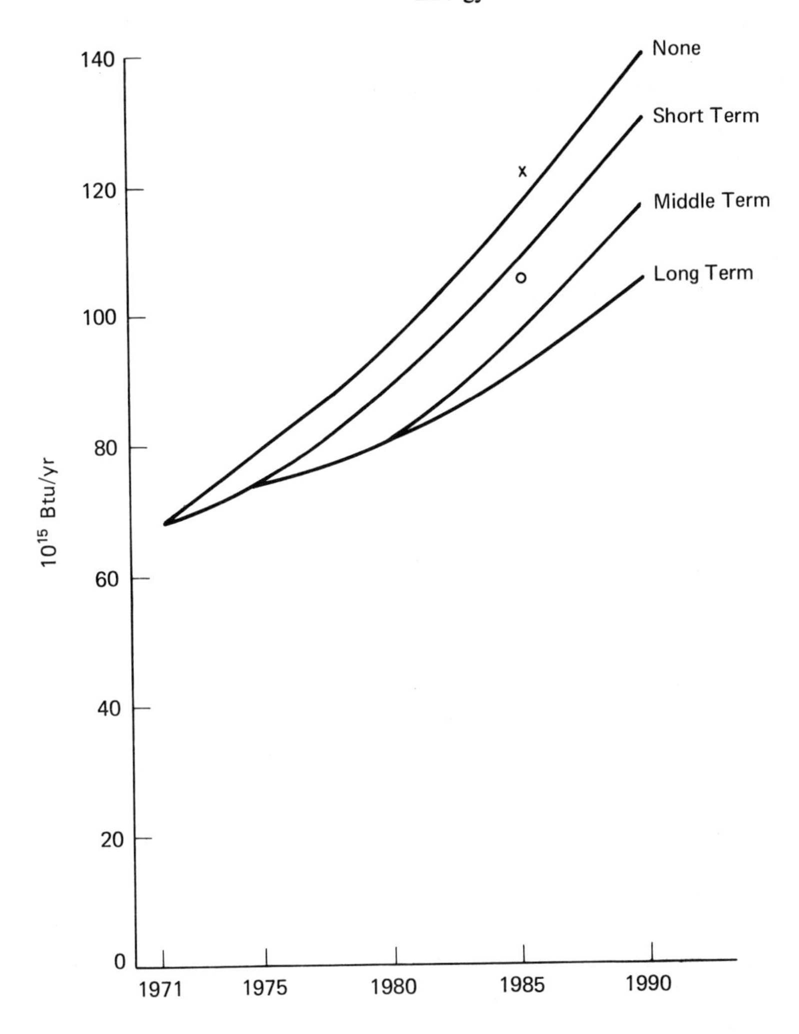

Figure 3.1
Idealized projections of energy consumption based on suggested conservation measures (9). Points marked ○ and × denote estimates for demand with and without conservation measures (4).

Table 3.2
Estimated 1985 U.S. Demand Reduction

	Possible Reduction	
	10^6 bbl/day of oil equivalent	10^{15} Btu/yr
By Conservation		
Industrial conservation measures	1.5	3.2
Transportation		
Lower speeds, car pooling	1.0	2.1
Airplane load factors	0.3	0.6
Space heating efficiency	1.0	2.1
By Use of Energy-Saving Equipment		
Smaller, more efficient cars	2.0	4.2
Other transportation savings	1.1	2.3
Better building insulation standards	1.1	2.3
Residential and commercial equipment	0.4	0.8
Industrial process efficiency	1.0	2.1
Total Conservation Potential[a]	9.4	19.7

[a]A more realistic overall projection might subtract 15 percent for partial overlap, yielding a total of 8.0×10^6 bbl/day or 16.8×10^{15} Btu/yr.

Hannon (10) has estimated savings with various types of packaging and potential (possibly negative) savings due to recycling liquid containers. The total energy consumption in packaging is about 1 percent of the total energy consumption.

To what extent will the savings shown in Figure 3.1 and Table 3.2 materialize? The real problem is human incentive to institute the many corrective measures that would be needed to develop such a large reduction in overall energy use. The authors of Ref. (9) conclude that "little reduction in energy usage will result unless one of two things occurs: (1) energy becomes much more expensive than it is now; (2) some other strong incentive is created."

In the developed countries, uninhibited energy use has been associated with great improvement in individual comfort, mobility, and freedom. Unfortunately, if conservation measures are to be effective in the long run, they will almost surely entail a penalty in human freedom. Legislation requiring high standards of building insulation takes away part of the freedom of the house buyer to spend his capital as he wishes. Measures that force a shift from the private car to public transport could lead to a significant reduction in leisure time. In the short term, and for sound reasons, people will readily submit to restrictions. In the long term, unless the alternatives are obviously too dangerous or costly, they will not. Whether serious conservation measures will succeed in the long term depends on whether vast new supplies of energy can be used at acceptable risk and cost. The first stage in any major move toward conservation in a democratic society must be to define as precisely as possible the resources, hazards, and costs that must be faced with continuing rapid growth of energy use. If these are clearly understood, and the hazards and costs are definitely unacceptable, long-term corrective action may be possible.

Conservation measures to improve the efficiency of electricity use are particularly important because of the rapid rate at which electricity use has been growing; the large capital investment required for new generation capacity; the possibly adverse environmental effects of large power plants and transmission lines; the present shortage of technology for clean burning of coal, and the slowness with which nuclear plants are being developed; and the possible future need (if synthetic fuels from coal are too expensive) to use nuclear electricity on a large scale for space heating and transportation, the two largest end uses of energy.

3.3 ELECTRICAL ENERGY

Figure 3.2 shows the steadiness of exponential growth in electrical demand in the 1960s in the United States and Canada. This growth took place as electricity prices grew very slowly (or declined, if correction is made for inflation), so it may be quite unrepresentative of a future in which electricity (and fossil fuel) prices will be much higher than in the past. The average annual growth rates that one may deduce from Figure 3.2 are shown in Table 3.3.

In each country the greatest rate of growth has been in the residential and commercial sectors. In the United States these sectors could soon be the main customers for electrical energy. In Canada industrial use of electrical energy is still a relatively large part of the total, owing to heavy consumption in certain industries, such as aluminum refining and paper making.

3.3.1 Demand and Price

Prices are rising due to fuel scarcity and environmental restrictions, and should rise even more if the consumer is to pay the full cost to the environment of energy extraction, processing, conversion, and delivery. The EPA estimated, in 1973, that the total annual cost of air pollution in the United States in 1968 was $16.1 billion (12). This was the total of the estimated damages to property, inert materials, vegetation, and human health. Taking the total energy consumption in that year as 62×10^{15} Btu, the unit cost is 26 cents/10^6 Btu, around 10 percent of the average fuel cost to the consumer. A price increase of this size might not do very much to discourage consumption. On the other hand, doubling or trebling of energy prices could quite likely inhibit consumption at least for

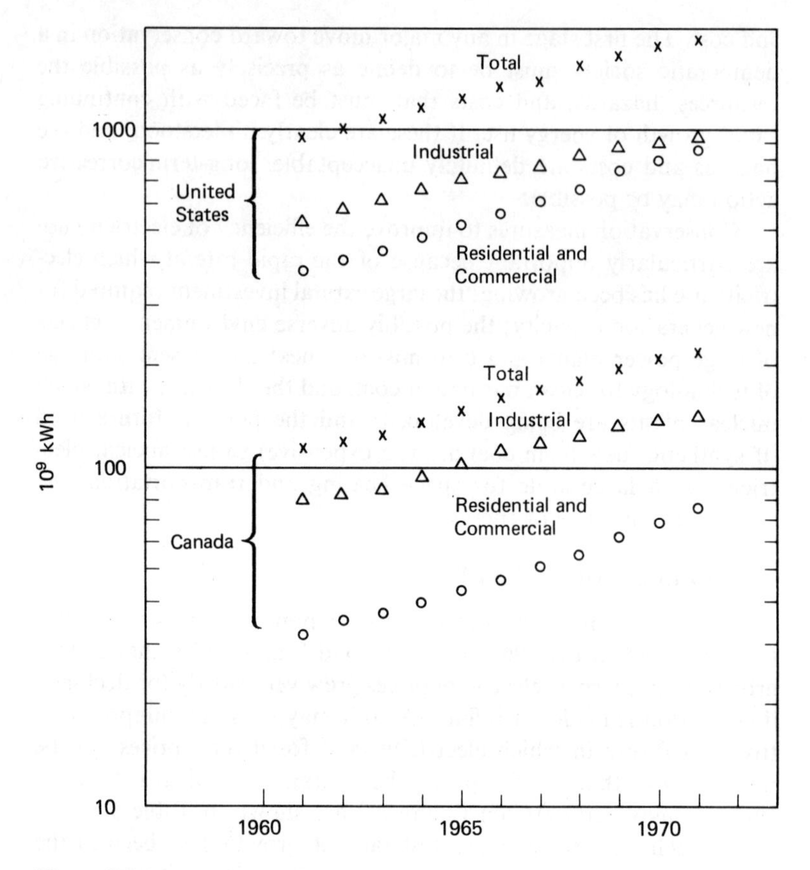

Figure 3.2
Electricity consumption in the United States and Canada (11).

a few years. Whether the effect would be long-lasting would no doubt depend on whether disposable income continues to rise (see Figure 1.5).

The relationship between energy demand and price has been the subject of much study by economists and others, and has been shown

Table 3.3
Average Annual Growth Rate in Electricity Demand, 1961–1971 (%/yr)

Sector	United States	Canada
Residential and Commercial	8.9	8.7
Industrial	6.5	6.4
Total	7.2	6.9

to be complex and uncertain. Apparently the demand elasticity (percent change in demand divided by percent change in price) depends strongly on four factors:

1. Time scale. The short-run domestic consumption of electricity, for example, could be fairly insensitive to price. The consumer has an investment in many electric appliances which he wishes to keep operating. In the long run, however, high electricity prices could discourage him from buying more electrical appliances or encourage him to buy more-efficient appliances.

2. Price level. When energy prices are low, relative to income, demand may be fairly insensitive to price. In the past, the total cost to the consumer of all forms of energy has averaged as little as 5 percent of his income. This could be quite different in the future.

3. Fuel. The elasticity of demand for any one fuel depends, of course, on how easily another fuel may be substituted and on the relative price levels of other fuels.

4. Sector. Demand elasticities for residential, commercial, and industrial sectors can differ considerably.

Table 3.4 provides results from the study of electricity demand made by Mount, Chapman, and Tyrrell (13). They correlated electricity demand in 47 states between 1946 and 1970 with three factors: population, income, and electricity price. Of these factors, the authors concluded that price was the most important and that electricity demand is quite sensitive to price in the long run. In the short run, price elasticities may be an order of magnitude lower. The results shown in Table 3.4 must be regarded as controversial, though other determinations of price elasticity have yielded results between −1 and −2 (14, 15). However, some writers have concluded that the long-run price relationships are relatively inelastic (3). The results of

Table 3.4
Estimated Long-Run Price Elasticity of Electricity Demand

Sector	Electricity Price (mills/kWh)	Long-Run Price Elasticity[a]
Residential	21.39	−1.24
	50	−1.5
Commercial	20.26	−1.45
	50	−1.5
Industrial	10.89	−1.74
	50	−2

[a]The price elasticity is defined as the percent change in the quantity demanded divided by the percent change in the unit price.
Source: Ref. (13).

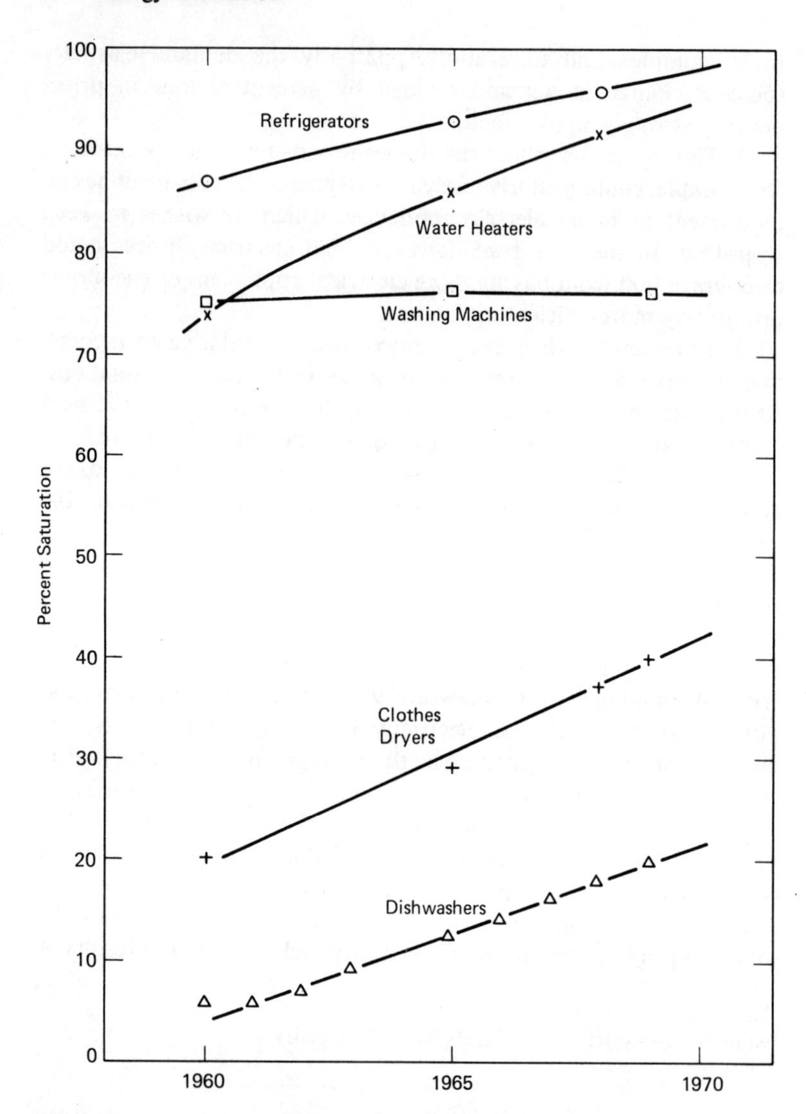

Figure 3.3
Percent saturation of appliances, United States (3).

Chapter 7 suggest that electricity price will rise rapidly in the next decade. If the price elasticities shown in Table 3.1 are realistic, the price mechanism could severely inhibit electricity demand growth; by how much remains to be seen.

Various drastic measures could, in principle, be taken by governments to force major long-term energy conservation, but such actions would have to be taken carefully. It is widely accepted that national energy consumption is closely related to gross national product; drastic reduction in energy use could be accompanied by drastic unemployment. Quite apart from government intervention, rising energy prices could by themselves have a seriously adverse effect on economic growth.

3.3.2 Appliance Saturation

As is evident from Table 3.1, residential space heating, air conditioning, water heating, refrigeration, lighting, cooking, etc., consume a substantial portion of total energy consumption. The growth rate in appliance energy consumption has been so rapid that the issue of appliance saturation in households is of importance. Figure 3.3 indicates that washing machines, water heaters, and refrigerators are apparently approaching a saturation level; however, the apparent saturation of water heaters does not mean a leveling off in energy consumption, for, as Figure 3.3 indicates, the growth of dishwasher installations is only beginning. Clothes dryers are also spreading rapidly. Figure 3.4 illustrates a dramatic increase in the sales of air conditioning units in residences in the past decade and suggests that not only is there no saturation in sight but demand is accelerating. However, there appear to be worthwhile gains coming in improving the efficiency of air conditioners.

3.3.3 Electrical Heating

In the period 1960–1968, electricity use for space heating grew at a rate of 20 percent per year. The number of new households in the United States between 1970 and 1980 is expected to be 13.1 million. Of these, 5.2 million, or 40 percent, are expected to be electrically heated; the remaining 7.9 million will be equipped to burn oil or gas (9).

The advantages of electrical heating to the consumer include low initial cost, compactness, absence of noise, flexibility, avoidance of fuel tank and chimney cost, long life, and low maintenance cost. Despite the disadvantage of higher fuel cost, indicated by Table 3.5, these advantages have been increasingly attractive to builders of residences, institutions, and stores.

Should this trend be forcibly arrested? The factors that weigh against electrical resistance heating are fuel wastage and capital cost. Resistance heating effectively wastes two-thirds of the energy

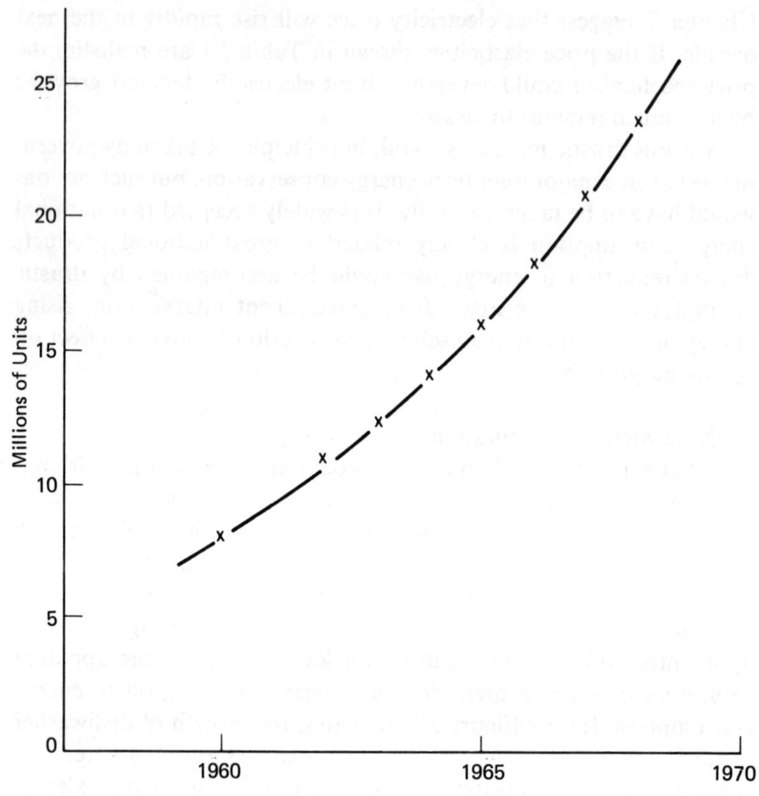

Figure 3.4
Total number of air conditioning units, residential sector, United States (3).

used at the power plant to generate electricity. This is a particularly serious point when oil and gas are used to generate electricity instead of being used directly for heating. However, it is not always appreciated that one-third to one-half of oil and gas energy may

Table 3.5
Relative Cost of Various Fuels to the Consumer

	Unit Cost	Assumed Efficiency	Cost per Million Btu
Gas	$2.00/1000 scf	75	$2.67
Oil	$0.25/gal	65	$3.20
Electrical Resistance	$0.018/kWh	100	$5.27
Heat Pump[a]			
OT = −10°F and COP = 2.0			$3.60
OT = +20°F and COP = 2.6			$2.03
OT = +50°F and COP = 3.2			$1.65

[a]OT = outside temperature.
COP = coefficient of performance = (heat output/electrical input) − 1.

be wasted in inefficient furnaces. Also, this objection does not apply to the heat pump, which can actually raise the overall electrical efficiency above 100 percent (by taking heat from the environment as well as from the fuel). This possibility is illustrated in Figure 3.5.

The capital costs of an electricity generation plant and distribution network add up to over $1000/kWe. The capital cost of gas and oil distribution and heating equipment may be only $100 to $200/kW thermal. Thus electrical resistance heating can be thought of as wasteful of capital resources, though again the objection need not apply to the electrical heat pump.

These two considerations are important, but there are other factors to consider. If gas and oil are in short supply, electrical heating could play a valuable role in shifting part of the heating load to nuclear energy and in improving the security of energy supply. Use of electricity for heating can reduce power generation and distribution costs by raising the ratio of average to peak load (the load factor). Rising fossil fuel prices may, in time, make nuclear–electric heating relatively cheap.

Table 3.5 indicates how the electrical energy cost for space heating can be substantially reduced with a heat pump. The heating performance factor (the ratio of heating energy developed to electrical energy consumed) is typically quite senstitive to the ratio of indoor and outdoor temperatures. The economically optimum electrical heating system would use a combination of resistance heating and heat pumping. Heat pumps have been available for a long time but have not made a big impact on the general market. Problems of initial cost, reliability, and noise have prevented widespread acceptance. Now that energy prices have risen, the heat pump may be

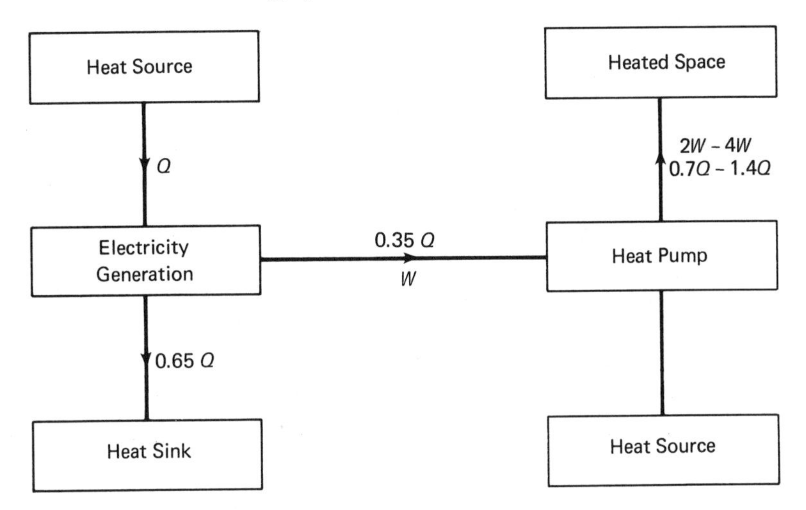

Figure 3.5
Typical energy flow through an electrical heat pump.

entering an era of active development to higher performance, greater reliability, and lower cost (16).

Heat pumps can be used in buildings of any size and have the advantage of providing both cooling and heating. Various heat sources are available, including water from wells, lakes, rivers, ambient air, and coolant discharge from industrial processes and electrical generation plants. In single residences the outside air is commonly used as a heat source or sink, depending on whether cooling or heating is required.

One problem with the air-to-air heat pump has been fan noise, which results from designing heat exchangers for low cost. Another problem may be icing of these heat exchangers in cool, humid weather. Commercial units have been high in initial cost because manufacturing and sales volumes are low. Also, adequate maintenance services are often not available. One increasingly important application of the air-to-air heat pump is in large buildings, where the core may require cooling at the same time that the perimeter needs heating.

The advent of solar heating, which is already closely competitive with electrical resistance heating (17), could be a significant means of conserving fuels. Several studies indicate that solar energy could be a valuable adjunct to a heat pump system.

The future of electricity use for space heating depends very much on how long supplies of low-cost gas and oil can be maintained. Synthetic natural gas from coal could be nearly as expensive as nuclear electricity, so this is not necessarily the answer. If electricity use for space heating is to expand, the heat pump must be further developed to improve efficiency and to lessen the required additional capital investment for electricity generation and distribution.

The future of electricity use for transportation hangs on similar considerations, plus one other—the development of a battery with high power density and high energy density. If this were available, and the supply of refined petroleum fuels were to worsen significantly, there could be a major move toward the electrical car. Allowing for all losses in both propulsion units and the electricity generation loss, the overall efficiency of electrical propulsion could significantly exceed the efficiency of current internal combustion engines. This could permit a valuable substitution of nuclear for fossil fuel, and could also mean the virtual elimination of automotive air pollution. But it could only take place over a long time period since the capital investment for new electrical plants would be colossal—at least equal to the total of present generating capacity. One trend toward electrical transportation—railroad electrification—has been under way for some time.

Table 3.6
Efficiencies of Various Fuels for Electricity Generation[a]

	Extraction	Processing	Transport	Conversion	Transmission	Overall Efficiency
Coal (underground)	56	92	98	38	91	18
Coal (surface)	79	92	98	38	91	25
Oil (offshore)	30	88	98	38	91	10
Oil (onshore)	40	88	98	38	91	13
Natural Gas	73	97	95	38	91	24
Nuclear Fuel	95	57	100	31	91	16

[a]Percentage of reserves not wasted at each step.
Source: Ref. (18).

3.3.4 Generation Losses

In addition to all the losses and inefficiencies in electricity use, there are numerous losses in the extraction, processing, and conversion of various fuels to produce electricity. Table 3.6, which shows estimates of the magnitudes of these losses, reveals that the overall efficiency of conversion of fuel energy (in the ground) to electricity is at best 25 percent. In these rough calculations an important factor is the "efficiency" of resource extraction, defined as the percentage of reserves in place that it is economical to remove. Much effort is being made to raise the recovery fraction of oil resources. Higher fuel prices lead in various ways to expansion of fuel supplies. Except for nuclear fuel, energy losses in fuel processing are moderate, as are transport and transmission losses. On the whole, the greatest loss indicated by Table 3.6 is waste heat loss in energy conversion. This loss would be partially avoidable if uses could be found for power plant waste heat.

3.4 WASTE HEAT OR USEFUL HEAT?

Figure 3.6 shows the trend in heat rejection from electrical power plants in the United States. Waste heat in total represents a colossal expenditure of energy; by 1990 it could account for 25 percent of all energy consumed, a proportion nearly as large as that for the residential and industrial sectors combined. It will be quite large even if electrical energy growth rates fall appreciably.

The prospects for a great increase in the efficiency of large-scale power plants are not very bright. As will be shown in Chapters 8 and 9, certain types of nuclear and fossil fuel plants that now have efficiencies between 30 and 40 percent could evolve toward 50 percent efficiency with major developmental effort. Still higher efficiencies may be possible with fuel cells using gasified coal as fuel, or with

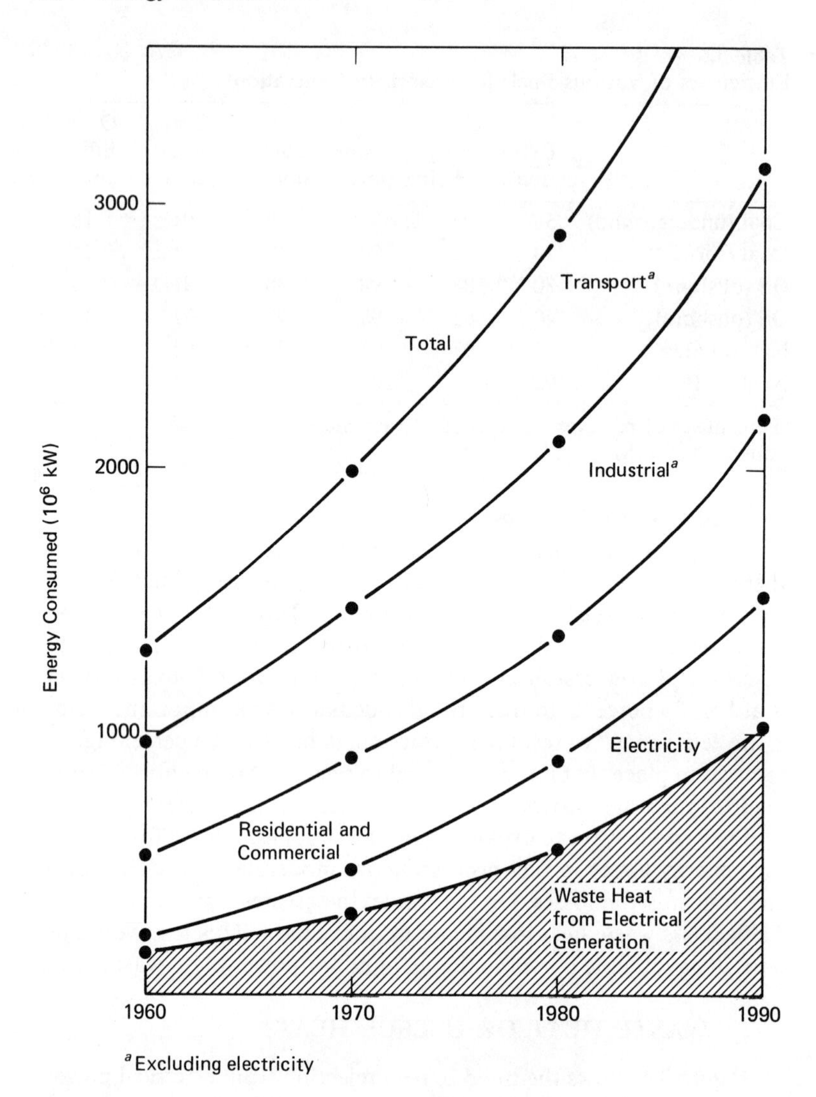

Figure 3.6
Estimated trends in energy consumption, United States (4).

magnetohydrodynamic–steam plants. However, such units are still in the research stage, and it will be a long time before they are shown to be sufficiently durable, inexpensive, and efficient to compete with present plants. Present research and development effort on fuel cells seems too scanty to promise major developments soon. This is even more the case with magnetohydrodynamics. Average system efficiency in electrical power generation is now around 35 percent. For the foreseeable future it appears that electrical output will account for not more than 50 percent of the energy that a plant consumes.

The extent to which waste heat can be put to economical use is an important question. The answer to this question is quite complex, depending strongly on the uses envisaged. Space heating is one possibility, but this depends very much on whether the heat is "warm" enough. The cooling water leaving a large steam power plant is typically only a few degrees above the temperature of the body of water from which it is drawn—certainly not warm enough to heat a house. If the steam plant is modified, its waste heat can be rejected at higher temperature, but at the cost of serious reduction in electrical output. Figure 3.7 shows the sensitivity of the cycle efficiency (defined as work output per unit of heat input) to rising exhaust temperatures for various steam plants. To be useful for space heating in a large network, the cooling water may need to be perhaps 250°F (121°C) in temperature. However, if cooling water were to be heated to this temperature rather than, say, 90°F (32°C), the electrical output of a light-water reactor steam plant would drop by one-third.

Figure 3.8 shows typical energy flows for modern fossil fuel and nuclear plants used to generate electricity only, or to generate electricity and also provide useful heat at 250°F (121°C). The advantage of the latter is a great increase in possible overall fuel energy utilization (from 40 percent to 88 percent for the fossil fuel plant and from 31 percent to nearly 100 percent for the nuclear plant). In an era of diminishing energy resources this must be seen as a strong argument for integrating electricity and heat supply.

An important counterargument is that waste heat utilization decreases the electrical output of the plant. During a period of extremely rapid growth in electricity demand, electricity supply organizations have their hands full raising the necessary capital for new plants. The prospect of dropping peak plant electrical capacity by one-third would hardly be pleasing to an electrical utility, regardless of potential benefits to the nation as a whole.

Other economic and political problems are raised by large-scale use of waste heat. The economic problems are fairly obvious. One is how to balance the extra cost of reduced efficiency of electricity generation and of a large heat distribution system against the savings

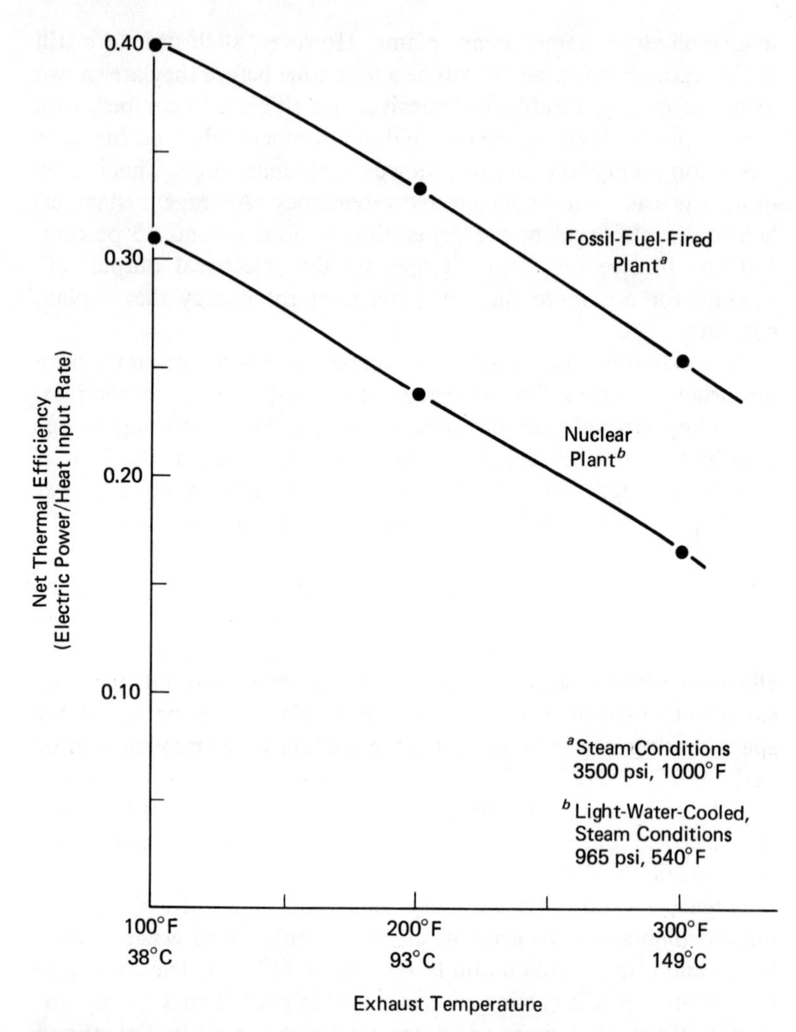

Figure 3.7
Effect of exhaust temperature on power plant efficiency. Adapted from Ref. (19).

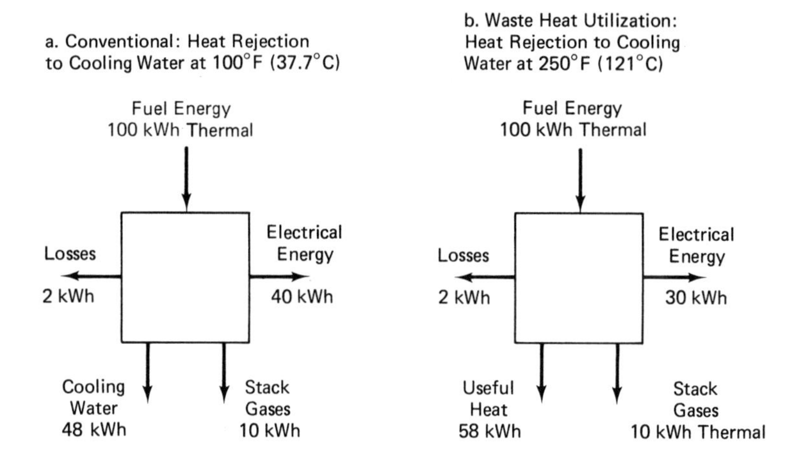

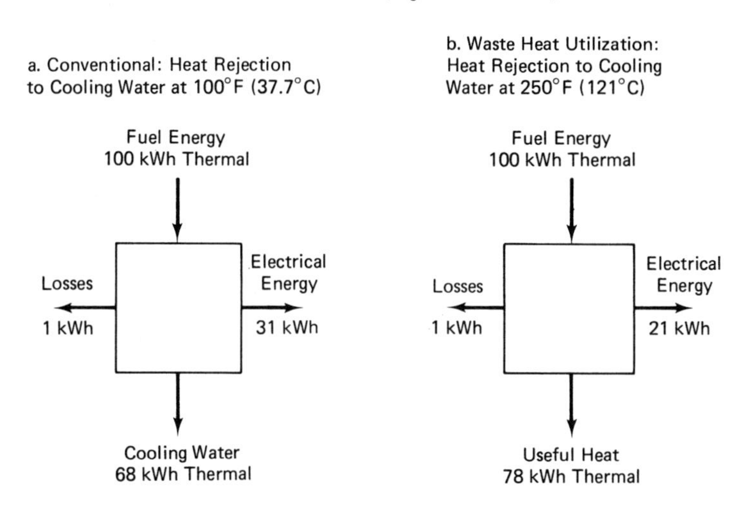

Figure 3.8
Effect of rising heat rejection temperature on power plant output.

due to elimination of many individual heating plants. Another, more difficult problem is to assess the benefit of much-improved overall utilization of fossil fuels and the effects this may have on future energy costs. Still another problem (of special importance in certain areas) is assessing the overall economic benefit of reduced heating of the environment, as well as the benefit of using waste heat to enhance the environment (e.g., to improve food production and recreational facilities).

The political problems are also intricate. First, utility companies, public and private, are primarily charged with the duty of providing cheap electricity (consistent, of course, with reasonable safety, reliability, and return on capital investment). The optimizing of overall energy resource utilization has not been one of their official concerns. Second is the matter of jurisdiction. Many utility organizations simply do not have the legal right or authority to market heat as well as electricity. Third, to be successful, the system would probably need to have a monopoly on heat supply, since a far-flung and fragmented system would be economically unfeasible. Many have concluded for this reason that district heating systems (other than those concentrated in city or industrial core areas) could only be successful in new communities. Even so, major political agreements would be required for system planning and pricing and for setting subsidies and taxes. In the communist countries, integrated supply of power and heat has become quite common. No doubt the concentration of political authority and the particular economic criteria of those countries are significant factors; however, district heating systems are also being successfully used in Sweden, Germany, and the United Kingdom.

As will be shown in Chapters 7, 8, and 9, the provision of cheap electricity requires large plants. Large scale can also lead to low cost in heating supply because fuel processing, pollution control, maintenance and labor costs, and the costs of an efficient plant all tend to decrease significantly with increasing scale. Further, the percentage annual utilization of the plant may be considerably increased by having many users whose daily and monthly demand patterns differ considerably. Large-scale plants can use nuclear energy and a wide range of fossil fuels, with suitable pollution control processes.

For various reasons, the main hope for future use of waste heat appears to lie in very large scale plants. This unfortunately makes the assessment of feasibility exceedingly complex, because supply must then be closely integrated with community development. Optimum pricing cannot be decided by considering only the immediate dollar cost to individual consumers; many long-range economic and environmental effects will become important. The optimum energy supply system will require optimum land use for housing and industry; community subsidy will inevitably be involved, if only in the provision of optimum growth patterns and a monopoly on energy supply. Large-scale development inevitably means long-range planning, and introduces the need to estimate population growth and the costs of fuel and money far into the future. It may also require short-run sacrifices for long-term benefits.

Large-scale plants are so capital-intensive that the annual use of the plant has an extremely important effect on costs. The demand for electricity is generally fairly heavy around the year, but heating demand is highly variable. Though a plant can readily be designed to operate with reduced electrical efficiency while supplying "useful" heat and at maximum efficiency while not, the capital cost of the heat distribution system is large. This provides great incentive to develop alternative uses for heat in industry or possibly in agriculture, to balance heating and electrical demand as far as possible around the year. In addition to the complexities mentioned earlier, this may lead to a closer integration of industrial and residential development than has occurred when energy costs have been only a small percentage of home-owning costs and a small fraction of the value added in manufacturing.

Given the economic and political difficulties of large-scale use of waste heat, is it a worthwhile pursuit? Yes, because it offers the chance to increase overall energy utilization (associated with electricity production) from 30 or 40 percent to 70 or 80 percent. It may provide cost savings to the consumer even in the short term, as well as delaying the depletion of important energy resources. At a future time it may be a valuable method of increasing food production. In areas where the environment may suffer from heat rejection, it can provide the benefit of reducing overall heat rejection. If fossil fuels become very scarce and costly, it may be the best way to make use of nuclear energy for space heating. (Of course, this will not be the only way, since nuclear energy could be used for production of fuels such as hydrogen which could be distributed in pipelines for heating use.) A serious examination of the feasibility of large-scale use of waste heat would require a detailed study of the local situation. In the next section we consider the characteristic costs and efficiencies of such systems.

3.5 THE ECONOMICS OF DISTRICT HEATING FROM POWER PLANTS

3.5.1 Heat Generation Costs

There are clearly many ways in which the costs and merits of waste heat utilization can be viewed. Deciding how the costs should be shared among the products of any multiproduct plant is of course an arbitrary matter; there are various methods by which the costs could be computed for the heat and electricity from a plant that produces both. We use here the idea that the costs of generating useful heat should be identified with the value of the "lost" electricity. This method seems appropriate for considering whether it is worth developing a market for waste heat when the electricity market is already developed and has a fairly stable price. It consists in principle of allocating to the useful waste heat the entire cost of the decrease in electricity output of the plant due to exhausting energy at a higher temperature. The value of the electricity is the cost of its production at maximum efficiency in a plant that has no waste heat utilization.

The total cost of the waste heat has two components: the production cost at the plant, and the distribution cost. The distribution component is calculated from the capital charges and maintenance expense for the pipeline system conveying warm water or steam from the plant to customers.

If the ratio of heating load to electricity load is quite uniform throughout the year (as with industrial process heating), the power and heating plant can be designed to operate most economically for a fixed high back pressure, which corresponds to the required temperature of the waste heat fluid. This is referred to as the "back-pressure turbine" system.

In general, the ratio of heating to electrical power output can vary widely, from zero to as high as four or five, depending on the required temperature and the demand pattern. This kind of flexibility can be obtained by designing the plant in the usual way to expand to minimum pressure, corresponding to the temperature of available cooling water, but providing for the possibility of extracting variable amounts of steam from several points in the turbine. This is called the "extraction turbine" system. Before returning to the boiler, steam from each extraction point condenses while warming the fluid that is to transfer the useful heat to the customer. At maximum useful heat output, all of the turbine steam could be extracted, none flowing to the low-temperature condenser. The extraction system, while more flexible, calls for more capital investment than the backpressure turbine system.

In the nuclear plant example in Figure 3.8, production of 78 kWh of heat at 250°F entailed a loss of 10 kWh of electricity; in this case the unit cost of the heat produced could be set at 10/78, or 0.128, times the unit value of the electricity produced by a plant generating electricity only. The relative costs of heat and electricity for other exit steam condensing temperatures are shown in Figure 3.9, whose derivation is described in Appendix C. The ratio of heat to electricity cost at the plant appears to be somewhat higher for fossil fuel plants than for nuclear plants.

Since the effect of extraction temperature on cost is so large, it pays to use several extraction temperatures. If, for example, district heating water were to be heated from 65°C to 110°C (149 to 230°F), perhaps three extractions could be used, with one-third at 110°C

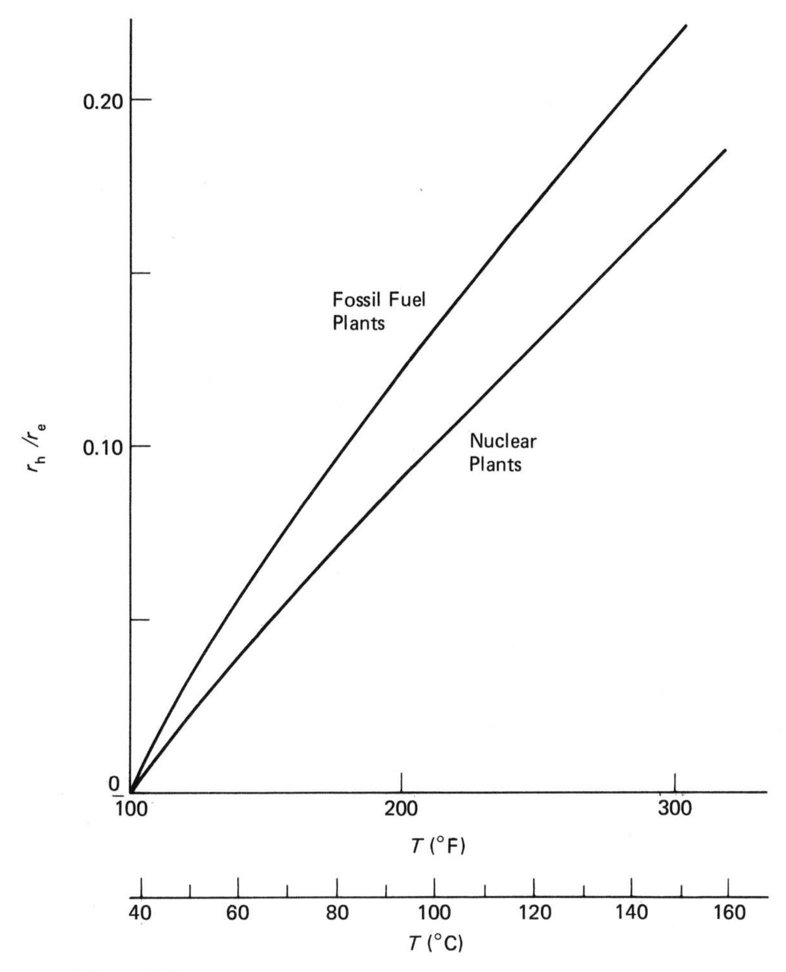

Figure 3.9
Relative costs of useful heat and electricity for the plant described in Appendix C (r_h is in mills/kW thermal, r_e in mills/kWe). No allowance is made for the cost of the district water heating and distribution system.

(230°F), one-third at 95°C (203°F), and one-third at 80°C (176°F). This would give an average cost ratio (for a nuclear plant) of $r_h/r_e = 0.094$, where r_h and r_e are the units costs of heat and electricity, respectively. For a large steam plant it may be reasonable to allow for a capital charge of $40/kWe for a condenser for the extracted steam to heat water for a district heating system. This would be equivalent to about $20 per kilowatt of heat flux in the condenser. With an annual capital charge rate of 15 percent, the annual capital cost of the system would be around $3 per kilowatt of heat generated. If the system operated for 50 percent of the year, the cost per kilowatt-hour of heat generated would be ($3/kW) × (1000 mills / $) / (8760 hr × 0.5), or 0.69 mill/kWh. With this estimate the cost of useful heat generation can be calculated as in Table 3.7.

These numbers are in rough agreement with the results of A. J. Miller and his colleagues (19), who calculated heating costs from some 31 examples of nuclear and fossil fuel plants supplying energy to district heating water (from 150 to 380°F) and also supplying process steam at various pressures. They estimated the capital costs for modifying these plants to extraction systems, and allocated all costs (beyond those necessary for an electricity-only plant of the same electrical output) to the cost of producing useful heat. Their estimated costs for nuclear and fossil fuel plants were in the range 30 to 50¢/10^6 Btu for electricity costs in the range 5 to 6 mills/kWh.

3.5.2 Heat Distribution Costs

District heating is already widely used in central core areas of large cities. Many large buildings can be heated by steam or hot water from a single large central combustion plant. Also, space cooling can be provided by using the hot water or steam to actuate an absorption-type refrigeration plant. The idea has a number of advantages, including lower maintenance, labor, and fuel costs than the total for many individual heating and cooling units. Pollution control and safety measures are provided with less cost, and overall plant utilization may be significantly higher. Against these cost advantages must be weighed the cost of the required district pipeline system. Miller et al. report that in 1968 the average heating cost to the customer for a total of 43 district heating systems in the United States was $1.42 per million Btu. For the same year the overall average cost of heat production in office buildings and apartments was $1.17 and $1.42, respectively, per million Btu. Thus the cost advantage of district heating over individual heating has not been clear. The effect of future large increases in fuel costs would, however, favor district heating since much of its cost is due to invested capital.

At least half the cost of using the waste heat from a large central

Table 3.7
Costs of Useful Heat Generation (Nuclear Plant)[a]

Unit Value of Electricity, r_e (mills/kWh)	Value of Lost Electricity, r_h (mills/kWh)	Capital Charge on Heating System (mills/kWh)	Heat Generation Cost	
			mills/kWh	¢/10^6 Btu
5	0.47	0.69	1.16	34
10	0.94	0.69	1.63	48
20	1.88	0.69	2.57	75
30	2.72	0.69	3.41	100

[a]Three-stage steam extraction heating to 110°C (230°F); $r_h/r_e = 0.094$.

power plant is associated with the pipeline distribution system. This raises the question of whether waste heat recovery could not be better done locally, in small centers. This idea is practical for very large industrial heating needs, where the industrial site and the power plant are side by side, and is often feasible for certain residential or commercial complexes with a reasonable annual balance of electricity and heat loads.

Apartment buildings, shopping complexes, and certain communities have been able to achieve quite high overall energy utilization using diesels, gas turbines, or steam turbines to provide a combined supply of electricity, heat, and possibly refrigeration. In general, the cost savings of these plants are marginal and are dependent on local fuel prices. Indeed, for a variety of reasons the economics of small-scale total energy plants makes their widespread use doubtful:

1. Capital costs per kilowatt for diesels and gas turbines in the 1 to 10-MW size range can easily be a factor of two higher than capital costs for very large power plants, and electricity generation efficiencies are typically lower, particularly for the gas turbine.

2. Small-scale plants are generally restricted to gaseous or distillate fuels.

3. Fuel costs can be considerably higher for small than for large systems.

4. The cost of providing the same reliability of electricity supply as with large grid systems (such as one day of failure in 10 years) has often been seriously underestimated. Small power generators that can run unattended year in and year out at low capital cost and high efficiency have not yet been developed. The fuel cell could conceivably meet this need; it has a high efficiency of electricity generation in home-sized units, and its waste heat could be coupled with a combustion source for space heating. But the initial cost and reliability of fuel cells are still not satisfactory.

5. Maintenance costs are typically much higher. In a very large power plant, the manpower requirement is about 10 to 20 MW/man. For high reliability, smaller plants may require as many as three men for a 1 to 10-MW unit, or 0.3 to 3 MW/man. For this reason manpower and maintenance costs may be an order of magnitude larger for small-scale local electricity generation than for large central power stations.

6. Once installed, conventional small-scale plants may not be readily adaptable to load changes.

7. The space required and the noise emitted by local plants may be objectionable.

In general, there is substantial economic advantage to largeness of scale, in both electricity generation and waste heat utilization. A large plant size typically implies an extensive heat distribution system. This means that markets for waste heat would need to be developed to keep the demand steady, and the unit cost of delivery would have to be as low as possible. But it would also provide an opportunity for nuclear energy to start displacing fossil fuels for space heating and industrial process heating, as well as other uses.

Extensive Swedish experience with district heating from large central power plants suggests that warm water distribution systems may be economical for distances of 50 miles or more (20). The warm water supply temperature will typically be in the range 175 to 250°F (81 to 121°C). Steam appears to be much more expensive than warm water for transmission over distances greater than about 1 mile. Figure 3.10 shows Swedish estimates of the costs of nuclear plant heat generation and distribution as a function of system size, water supply, and return temperature. The unit cost of the heat at the plant is in the range 0.10 to 0.16 of the unit cost of the electricity,

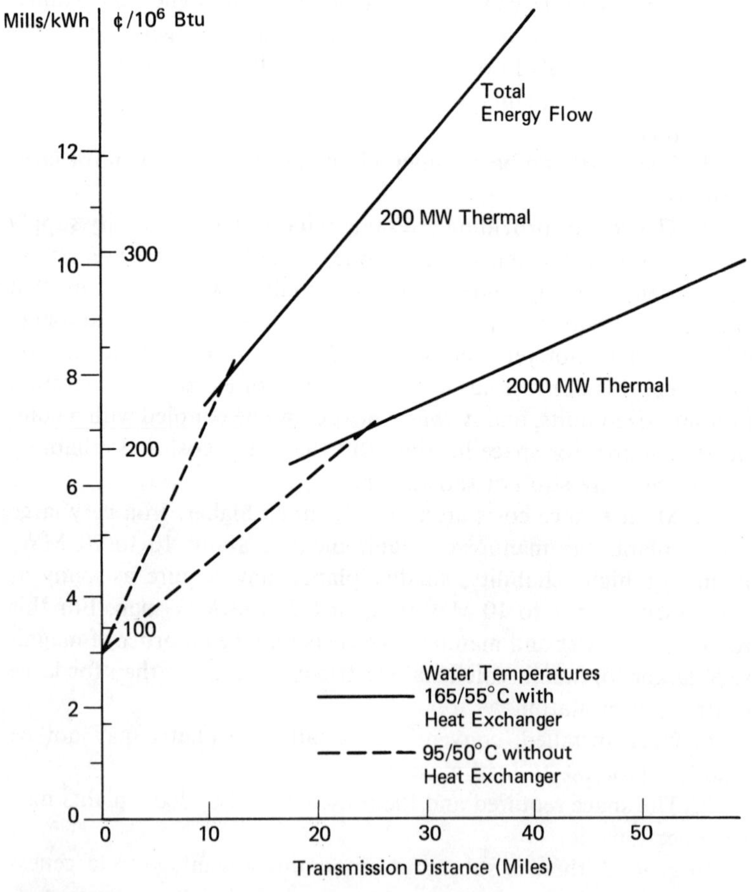

Figure 3.10
Cost of district heating from nuclear plants in Sweden. Pipes protected below ground by concrete culverts. Data from Ref. (20).

depending on the water transmission temperature selected. For short transmission distances, Figure 3.10 suggests that low water temperatures are more economical. Thermal energy in warm water can be transmitted over long distances with relatively low pumping power and relatively low heat loss (2 to 4 percent). Figure 3.10 also indicates that the cost of delivered heat is dependent on the total size of the system and on the heat transmission distance.

In the last decade, with oil prices (delivered to bulk consumers) running at 30 to 40¢/10^6 Btu, the economic incentive for waste heat use for space heating did not appear to be high. Now, with world oil price around \$2/10^6 Btu, the situation is quite different. The total cost of oil heating to the individual consumer (capital plus fuel) may approach \$4/10^6 Btu. Thus it appears that waste heat utilization from nuclear power plants could be relatively cheap and could effect a large reduction in gas and oil demand for space heating. It could considerably extend the usefulness of nuclear energy in the economy, with overall fuel utilization of 60 to 80 percent. The annual energy utilization would depend much on the degree to which the demands for heat and electricity could be balanced.

Miller et al. estimated the cost of hot water heat delivery from a nuclear power plant to a hypothetical city of 389,000 people (19). Two-thirds of the population were assumed to live in apartments. District heating was assumed to supply 172,000 people within a 5- to 10-mile radius, and 86,000 between 10 and 12 miles from the nuclear reactor. The climate assumed was that of Philadelphia. Useful heat was to be supplied for sewage distillation, industrial process heating, and space heating.

The average annual energy flows from the nuclear reactor are shown in Table 3.8. The overall energy utilization from this system was about 67 percent, more than double the energy utilization of a water-cooled reactor used for electricity production only. The flexibility of the system is indicated by its ability to alter the district heating load from 0 to 1144 MW thermal. Cost of the district heat piping system was estimated in 1971 as \$74.8/kW thermal. Currently,

Table 3.8
Annual Energy Flows from a Nuclear Reactor

	Energy Flow	Percent
Reactor Heat Production	2041 MW thermal	100
Electrical Power	463 MWe	23
District Heating	457 MW thermal	22
Industrial Steam Heating	368 MW thermal	18
Sewage Distillation	90 MW thermal	4
Condenser	634 MW thermal	31

Source: Ref. (19).

costs as high as \$200/kW seem justified. With an annual capital charge rate of 15 percent and the district heating system effectively used about 50 percent of the year, \$200/kW would be equivalent to a heat delivery cost of about 7 mills/kWh, or \$2.04/10^6 Btu. Adding the costs of heat generation at the plant, total cost of the delivered heat could be in the range \$2.50 to \$3.00/10^6 Btu.

3.6 OTHER USES FOR HEAT FROM POWER PLANTS

3.6.1 Seawater Distillation

Steam extracted from a turbine at 266°F (130°C) may be used to supply evaporation energy to a seawater distillation unit, consisting of 15 to 40 stages. According to Jorgen Christenson (21), the range of heat requirement would be 40 to 60 kJ/kg (or 17 to 25 Btu/lb) of distilled water. The cost of the required heat (at \$1.00/10^6 Btu) would be 14 to 21¢/1000 gal; the capital and pumping costs of the desalination plant would be extra. J. A. Lane and his colleagues (22) asserted in 1974 that the cost of water from an 1800-MW thermal dual-purpose power and desalination plant lies in the range 70 to 80¢/1000 gal. These costs could be quite acceptable in the absence of natural supplies of fresh water.

3.6.2 Sewage Distillation

Distillation of sewage to obtain drinkable water is somewhat more expensive than seawater distillation and may be a practical solution only where water is in very short supply. Miller et al. (19) and Christenson (21) have made estimates of the relative costs of sewage distillation and water recirculation. Although their estimates differ in details and in magnitudes, they both indicate that sewage distillation and recycling would increase the cost of domestic water supplies by about a factor of two.

3.6.3 Agriculture

Another possible use of heat rejected from power plants is in soil warming and greenhouse heating to increase agricultural output. The heat rejected from a 1000-MWe power plant would be sufficient to warm a greenhouse of 750 to 1500 acres in winter (19). In summer, the heat could in principle be used for evaporative cooling of the same greenhouses. The greenhouse would thus serve as an extended cooling tower for the power plant. Conventional greenhouses cost around \$22,000/acre, and the cost of heat distribution equipment has been estimated as about \$5000/acre. This would be equivalent in total to a capital cost for plant cooling of \$20 to \$40/kWe and compares with \$6 to \$10/kWe for conventional cooling towers. Depending on the market, however, the value of the food production due to greenhouse heating might compensate for the added expense.

Soil warming without cover has been done successfully at lower capital expense in Holland and Germany, with plastic tubes penetrating the soil to a depth of a few feet and spaced a few feet apart. In this way the soil can be kept at temperatures of 25 to 30°C for a long season. Simple plastic covers have been used for particular vegetables. These techniques may not be generally economical now, but they suggest the possibility of greatly expanded food production should the need arise.

3.6.4 Fish Culture

Fish culture is another potential beneficiary from use of power plant waste heat. At Hunterston, in the United Kingdom, the use of waste heat to warm water near a thermal power plant has raised the production of plaice to around 2 kg/m^2/yr. High productivity requires that the water temperature be kept above 12 to 15°C; the heat required to do this will, of course, vary greatly with the situation. This annual catch would be equivalent to about 2000 tonnes per square kilometer of enclosed bay or fish farm. If such a large area could be kept warm by the waste heat output from a 500-MWe plant, the annual fish production could be as high as 4 kg/kWe of power; this does not suggest a very high economic value for the waste heat. However, with ocean fishing resources in danger of depletion in particular areas, artificial fish culture in protected lagoons could become a valuable source of food in future years.

Much thought has been given to other uses of low-grade heat, but apart from space heating, industrial process heat, water or sewage distillation, and possibly food production, none of these uses seem to have any large market that would be economically significant.

3.7 SUMMARY

A review of energy use suggests many ways in which energy consumption could be reduced. These include reduction of waste, improvement of conversion efficiency, and better organization of transportation. This potential reduction in energy use would greatly alleviate the problem of maintaining supply at reasonable cost in future years.

Increasing energy prices may effect a considerable reduction in demand in the long run, but by how much is uncertain. Other mechanisms for demand reduction are unlikely to be attempted or effective until the true costs of expanded energy use are better known than at present.

The wise use of electrical energy is a particularly important issue since demand has been rising so quickly and the capital requirements of new generation capacity are so large. Most electrical appliances have still not reached saturation level, and there has been a strong trend toward electrical heating. Measures to improve overall utilization of energy, as with heat pumps or waste heat systems, offer substantial possible benefits.

Waste heat from large central power stations represents an energy source of great magnitude. To be useful, this energy must be associated with a reasonable temperature, perhaps 100°C or more, rather than 25°C, a typical steam power plant exhaust temperature. Extracting energy from the power plant at a useful temperature will typically reduce its electrical output by perhaps 25 percent and could only be done economically if there were a balanced market for the waste heat. The overall utilization of energy by the plant could double, depletion of both fossil and nuclear fuels could be delayed, and total environmental heating could be markedly reduced if waste heating were widely adopted.

Markets for useful heat from large power plants have not been easy to develop, since cost savings relative to conventional methods have been modest. Rising fossil fuel prices make waste heat use more attractive economically. For a new city, in fact, district heating from a combined heat and power station might well be economical today. Seawater distillation or sewage distillation by waste heating from power plants might double the cost of the domestic water supply, but in areas or times of water shortage these ideas may acquire practical importance. Substantial augmentation of food production by use of waste heat also seems quite feasible.

Use of waste heat recovery on a small scale saves the cost of huge piping distribution systems but introduces other problems, including higher fuel, capital, labor, and maintenance costs, which have often militated against economic feasibility. Given the growing shortage of clean fossil fuels and the increasing availability of nuclear energy, it would seem that the best prospects for waste heat utilization are from large-scale central power plants.

Because of the complexity of large-scale systems and the time delay likely to be encountered in building them, serious studies are needed of the possibilities of large-scale integration of heat and electricity supply in areas where substantial population growth is expected.

References

1. John J. Deutsch. "The effects of energy on life styles." *Proceedings of the Symposium on Energy Resources*, Royal Society of Canada, Ottawa, Ontario, October 15–17, 1973.

2. D. C. White. "Energy supply and demand: The role of conservation." Statement before the Hearing of the Subcommittee on Science Research

and Development of the Committee on Science and Astronautics, U.S. House of Representatives, May 9–11, 23–25, 29, 1972.

3. Office of Science and Technology. *Patterns of Energy Consumption in the United States*. Washington, D.C., January 1972.

4. National Academy of Engineering, Task Force on Energy. *U.S. Energy Prospects: An Engineering Viewpoint*. Washington, D.C., 1974.

5. Eric Hirst and John C. Moyers. "Efficiency of energy use in the United States." *Science*, March 30, 1973, p. 1299.

6. American Society of Heating, Refrigerating, and Air-Conditioning. *Handbook of Fundamentals*. New York, 1972.

7. National Bureau of Standards. "Some technical aspects of energy conservation." Memorandum, June 16, 1972.

8. Matthew C. Cordaro, Bruce C. Netschert, and James R. Mahoney. "Energy utilization and pollution aspects of two space heating alternatives." Paper 6.2–5, *Transactions of the 9th World Energy Conference*, Detroit, Michigan, September 22–27, 1974. See also the "Discussion by R. L. Dunning" in the same *Transactions*.

9. Office of Emergency Preparedness. *Energy Conservation*. Staff Study prepared for the Energy Subcommittee of the Domestic Council, Executive office. Washington, D.C., July 1972.

10. Bruce Hannon. "System energy and recycling: A study of the container industry." Paper 72-WA/Ener-3 presented at the Annual Winter Meeting of the American Society of Mechanical Engineers, New York, New York, November 26–30, 1972.

11. Organization for Economic Cooperation and Development. *Statistics of Energy 1951–71*. Paris, 1973.

12. L. B. Barrett and T. E. Waddell. "Cost of air pollution damage." Status Report prepared for the Environmental Protection Agency. Washington, D.C., February 1973.

13. T. D. Mount, L. D. Chapman, and T. J. Tyrrell. "Electricity demand in the United States." Chapter 24 in *Energy: Demand, Conservation, and Institutional Problems*, Michael S. Macrakis, ed. Cambridge, Mass.: MIT Press, 1974.

14. Charles J. Cicchetti and William J. Gillen. "Electricity growth: Economic incentives and environmental quality." Chapter 25 in *Energy: Demand, Conservation, and Institutional Problems*, Michael S. Macrakis, ed. Cambridge, Mass.: MIT Press, 1974.

15. John Tansil and John C. Moyers. "Residential demand for electricity." Chapter 28 in *Energy: Demand, Conservation, and Institutional Problems*, Michael S. Macrakis, ed. Cambridge, Mass.: MIT Press, 1974.

16. Eugene G. Kovack, ed. *Technology of Efficient Energy Utilization*. Report of a NATO Science Committee Conference held at Les Arcs, France, October 8–12, 1973.

17. National Bureau of Standards. "A discussion of a research program for the application of solar energy to space heating, air-conditioning, and water heating in buildings." In *Solar Energy Research*, Staff Report of the Committee on Science and Astronautics, U.S. House of Representatives, 92nd Congress, December 1972.

18. Council on Environmental Quality. *Energy and the Environment:*

Electric Power. Washington, D.C.: U.S. Government Printing Office, August 1973.

19. A. J. Miller, H. R. Payne, M. E. Lackey, G. Samuels, M. T. Heath, E. W. Hagen, and A. W. Savolainen. *Use of Steam–Electric Power Plants to Provide Thermal Energy to Urban Areas*. Report ORNL-HUD-14, Oak Ridge National Laboratory, January 1971.

20. Peter Margen. "District heating: The philosophy." *Modern Power and Engineering*, July 1975, pp. 56–57.

21. Jorgen Christenson. *Low-Grade Heat from Thermal Energy Production*. Report AE-448, Aktiebolaget Atomenergi, Studsvik, Nyköping, Sweden, 1972.

22. J. A. Lane, R. Krymm, N. Raisic, and J. T. Roberts. "The role of nuclear power in the future energy supply of the world." Paper 4.1–22, *Transactions of the 9th World Energy Conference*, Detroit, Michigan, September 22–27, 1974.

Supplementary References

Joel Darmstadter. "Limiting the demand for energy: Possible? Probable?" *Environmental Affairs* (Boston College Environmental Law Center), vol. 2, no. 4 (spring 1973).

F. M. Fisher and C. Kaysen. *The Demand for Electricity in the United States*. Amsterdam: North-Holland, 1962.

Ford Foundation. *Exploring Energy Choices: Preliminary Report of the Energy Policy Project*. 1974.

Serge Gratch. "Energy consumption by the transportation industry." Paper presented at the Annual Winter Meeting of the American Society of Mechanical Engineers, Detroit, Michigan, November 11–15, 1973.

F. H. Knelman. "Energy conservation." Science Council of Canada Background Study No. 33, July 1975.

A. C. Malliaris and R. C. Strombotne. "Demand for energy by the transportation sector and opportunities for energy conservation." Chapter 32 in *Energy: Demand, Conservation, and Institutional Problems*, Michael S. Macrakis, ed. Cambridge, Mass.: MIT Press, 1974.

John C. Moyers. *The Value of Thermal Insulation in Residential Construction: Economics and the Conservation of Energy*. Report ORNL-NSF-EP-2, Oak Ridge National Laboratory, December 1971.

Problems

3.1 Compare the capital investment consequences of a continued 7 percent per year growth in electricity demand to those of a growth rate of only 3.5 percent per year. For simplicity, assume that all further expansion of power generation is either coal-fired (40 percent) or nuclear (60 percent) and that the 1980 installed capacity is 700 GWe.

The capital costs directly and indirectly associated with new generation facilities are shown in Table 3.A.

Estimate average capital investment per person per year over the

Table 3.A
Capital Costs of New Electricity Generation

	Nuclear ($/kWe)	Coal ($/kWe)
Transmission and Distribution	500	500
Power Plant	850	700
Transport	0	50
Mining and Milling (including land costs)	200	500
Total	1550	1750

period 1980–2000. Assume that the average population in this period is 250 million.

3.2 Infer the price elasticity[2] of per capita demand for electricity from the following approximate data corresponding to conditions between 1960 and 1970:

Increase in total demand:	100 percent
Decrease in electricity price (1960 dollars):	75 percent
Increase in population per year:	2 percent.

In making this calculation assume, for simplicity, that electricity demand depends only on price and population. It seems reasonable to suppose that the elasticity of demand with respect to population is +1.0.

This sort of calculation may be useful if the price of electricity has been changing at a low rate for a long time. It may not be helpful for assessing the effects of the large, fairly sudden increases in electricity price encountered in this decade. In this case the effects of elasticity on price could be long delayed.

Suppose, for example, that the elasticity of demand to price varied linearly from −0.1 to −1.0 over a period of 15 years from the price increase, and that the price of electricity doubled suddenly in 1980, remaining constant thereafter (in 1980 dollars).

Compute relative demand level (1 at 1980) for the years 1985, 1990, and 1995, assuming

a. 10-year doubling, or growth at 7 percent per year;
b. demand elasticity to price and population as above;
c. annual population growth is 1.5 percent.

[2]The price elasticity is defined as the percent change in per capita demand divided by percent change in price (for small changes):

$$dD/D = e\,dP/P,$$

where D is per capita demand, e is price elasticity, and P is price. For large changes, with constant elasticity, this is integrated to

$$D_2/D_1 = (P_2/P_1)^e.$$

Note that this example does not allow for the possibility of different price elasticities in the industrial sector and the commercial or residential sectors. It could also be quite misleading in that it does not account for the effect of increasing personal disposable income.

3.3 Estimate the net economic benefit (\$/yr) of insulating the four walls and the ceiling of a single-story house 30 ft by 60 ft in plan, with an 8-ft ceiling, under the following assumptions.

The rate of heat loss Q can be estimated as $Q = \sum u_i A_i \, \Delta T$, where u and A represent thermal conductance (Btu/ft^2/hr/°F) and area (ft^2), respectively, ΔT is the temperature difference (°F) between the inside and the outside of the house, and the sum includes all four walls plus the ceiling. The conductances with and without insulation are shown in Table 3.B.

Table 3.B
Conductance as a Function of Insulation

	Thickness of Insulation (in.)	Conductance (Btu/ft^2/hr/°F)
Walls	0	0.190
	1	0.155
Ceiling	0	0.120
	6	0.042

Assume a fuel cost of \$3.25/10^6 Btu and a furnace efficiency of 70 percent. The house is located in a climate of 8000 degree-days (i.e., equivalent to $\Delta T = 40$°F for 200 days in the year). The attic and outside temperatures are identical.

Assume a total insulation cost of \$750, which the householder must borrow from the bank and repay over a period of 10 years at an annual interest rate of 12 percent, compounded monthly. For a loan of S dollars repayable at an interest rate i in n equal payments, each payment P is

$$P = S \frac{i}{1 - (1 + i)^{-n}};$$

in this case, $i = 0.12/12$ and $n = 10 \times 12$.

3.4 Hirst and Moyers (5) indicate that transportation accounts for about 25 percent of total national energy expenditure. The distribution of transportation energy expenditure is approximately as follows:

Intercity Passengers:	6 percent (automobiles 5.6 percent)
Urban Passengers:	9 percent (automobiles 8.7 percent)
Intercity Freight:	4 percent
Urban Freight, other:	6 percent.

Approximate energy consumptions by various modes of passenger travel are shown in Table 3.C.

Table 3.C
Approximate Energy Consumed in Passenger Travel

	Load Factor	Energy Consumption (Btu/passenger mile)
Intercity		
Bus	0.45	1600
Railroad	0.35	2900
Automobile	0.48	3400
Airplane	0.50	8400
Urban		
Mass transit	0.20	3800
Automobile	0.28	8100

Estimate the percentage savings in national energy consumption under each of the following assumptions:

a. Automotive engines can be made 10 percent more efficient.

b. Automobiles can be made 30 percent smaller (assuming fuel consumption proportional to vehicle mass).

c. Twenty percent of urban passenger miles can be shifted from automotive to mass transit.

d. Twenty percent of intercity passenger miles can be shifted from automobiles to buses.

3.5 Extrapolate the trends shown in Figure 3.2 for industrial and residential–commercial sectors (using the calculated average growth rates shown in the text) to the year 2000. Suppose that by the year 2000 transportation use of electrical energy is still insignificant, and that total energy consumption has grown at 5 percent per year in the interval 1970–2000.

Estimate the fraction of total energy consumption in the year 2000 accounted for by:

a. electricity output,

b. energy used to produce electricity, assuming a 33 percent average conversion efficiency.

c. Repeat the calculation with the assumptions that due to approach to saturation of electrical appliances, demand growth in the residential–commercial sector drops to 5 percent, while due to insecurity of gas and oil supplies, industrial demand for electricity grows at 8 percent.

3.6 Consider the potential reduction of fossil fuel consumption if waste heat from new nuclear stations could be used for space and process heating. Suppose, for simplicity, that all new electricity generating capacity after 1980 is nuclear, and that half of the new stations can be designed to supply waste heat. Assume also that a market exists for half of the waste heat that the combined power–heating stations could produce year-round. Because of heat supply the electricity generating efficiency drops from 0.3 to 0.25 (annual average).

Take 1980 electrical generation capacity as 700 GWe, and assume that growth in capacity from 1980 onward is 7 percent per year.

a. Compare the total annual energy that could be supplied by waste heat in 1990 and 2000 with the "imports and/or shortages" entry in Figure 2.6.

b. Show that the maximum heat supply from a 1-MWe capacity plant is $c(1 - \eta)/\eta_0$ where c is the capacity factor, η_0 is the efficiency with no waste heat utilization, and η is the efficiency with waste heat utilization. For a combined heat and power plant the capacity factor can be defined as average heat input divided by maximum heat input.

3.7 The Pickering Nuclear Station produces 2000 MWe of electricity at a thermal efficiency of 28 percent. Suppose this plant had been designed to supply heat to a nearby community for 6 months of the year via a circulation of water warmed to 200°F. For simplicity, assume the community absorbs 100 percent of the heat rejection from the plant for one 6-month period and none for the rest, and that with turbine steam extraction the thermal efficiency of the plant (electric power output divided by nuclear thermal input) drops from 28 to 22 percent. The plant operates with 0.8 capacity factor in the nonheating period, and electrical output drops in proportion to efficiency in the heating period.

a. Determine the annual average energy utilization factor (heat plus electricity output divided by heat input).

b. Looking at the problem another way, one might say that waste heat recovery improves the effective electricity generation efficiency (i.e., the electrical output divided by the difference between nuclear heat input and waste heat used). Show how the effective electrical efficiency improves under these conditions.

3.8 Estimate the cost ($\$/10^6$ Btu) of waste heat delivery to residential customers from a 1000-MWe nuclear plant, under the following conditions.

If the plant were designed for electricity generation only, the electricity generation cost is 15 mills/kWh at a capacity factor of 0.7. The heat is taken from the plant via water heated by steam extracted from the turbine at an average temperature of 200°F (93.3°C). The plant efficiency drops with extraction temperature, as shown in Figure 3.7.

The maximum heat output of the plant is required for 7 months in the year. For the remainder of the year the plant generates electricity only, at an efficiency of 0.31 (Figure 3.7).

Capital costs of the heat supply system include a heat exchanger system for warming supply water ($15 million) and a system to distribute hot water for district heating ($200 million), for a total of $215 million. Assume that the capital and maintenance charges on this investment are 17 percent annually.

Note that the heat input to the nuclear plant is the same with and without waste heat utilization.

3.9 In the city of Winnipeg (population 560,000) the annual average space heating load is 4700 Btu/person-hr.

A district heating system is proposed to supply all the heat needed by 70 percent of the population. The estimated cost of heat available at the central plant is 50¢/10^6 Btu, and the capital cost of the district heating system has been estimated as $150 million. Annual maintenance costs for the system are estimated as $1.25 million.

Capital is available at 10 percent interest (compounded annually), and the life of the system is expected to be 30 years. The yearly payment P (principal plus interest) for a loan S repaid in n equal payments at an interest rate i (compounded annually) is

$$P = S \frac{i}{1 - (1 + i)^{-n}}.$$

Estimate the cost of heat delivered to the customer ($/$10^6$ Btu).

3.10 Estimate the cost of seawater distillation (¢/1000 gal) using waste heat from a nuclear power plant under the following conditions.

The nuclear plant can produce electricity only, at an efficiency of 0.31 and a cost of 12 mills/kWh. However, it can also operate with average steam extraction temperature of 100°C, and can supply all of its heat to a distillation plant which absorbs 50 kJ/kg of distilled water produced. The capital cost of the distillation plant is $1.5 million per 10^6 gal/day of distilled water output. Annual capital and operational charges total 17 percent.

4
Power and the Environment

4.1 INTRODUCTION

Any comprehensive assessment of power generation must include consideration of hazards to the environment. These hazards arise because energy conversion and use affect virtually the entire physical world, including plant and animal life, agricultural productivity, air and water purity, landscape beauty, and climate.

Some of these effects may damage not only the environment but also man himself. Human hazards due to power generation include occupational dangers in the mining, transporting, and processing of fuels. They also include public health hazards associated with power plant emissions, particularly in the case of accidents. These will be discussed in Chapters 5 and 6.

The main emphasis of this chapter will be on the local effects of fuel extraction, transportation, and utilization. A discussion of the particular global problem of environmental heating is also included.

4.2 FOSSIL FUELS AND THE ENVIRONMENT

Table 4.1 describes the sources of adverse environmental effects associated with fossil fuel use. These effects can be grouped in several categories:

1. Effects that can be removed at reasonable cost with existing large-scale technology. One example is the removal of particulates

Table 4.1
Sources of Adverse Environmental Effects of Fossil Fuel Use

	Air Pollution	Water Pollution	Land Damage	Limitations on Land Use
Extraction	Fires in abandoned coal mines and in refuse heaps	Acid drainage	Solid wastes Strip mining Underground coal mine subsidence	Strip mining Oil and gas wells
Transport	Dust	Oil spills	Dust Oil spills	Oil and gas pipelines
Processing	Coal dust Coke emissions Oil refinery emissions	Oil refinery emissions Condenser antifouling chemicals		Refineries Coking plants Gas works
Utilization	Particulates SO_2 NO_x CO CO_2 Waste heat	Waste heat		Coal piles Oil and gas tank farms Electric power plants Cooling towers Transmission lines

from combustion gases (by electrostatic and mechanical separators). At least 99.5 percent of the particulate matter can be removed with equipment costing only a few percent of the capital value of the plant. Unless, as some fear, the submicron-sized particles, which are too small for removal by existing methods, are a health hazard due to buildup of deposits in lung tissue, particulate emission is a solved problem.

2. Effects that may be scientifically worrisome but whose implications for human welfare are not yet so clear that they should affect energy planning. An example of this is the thermal effect of the increasing carbon dioxide content of the atmosphere.

3. Effects that change the beauty of the land but do not harm health. An example is high-voltage overhead electrical transmission lines, which, though unsightly, may be unacceptably expensive to replace by underground lines.

4. Effects that consume large land areas otherwise useful for agriculture, recreation, industry, or urban settlement. The feasibility of reclaiming the vast tracts of land that could be used for strip mining of coal, oil shales, and tar sands is still uncertain. The time may come when shortage of agricultural land puts a severe constraint on availability of land for power plants and refineries. Already the problems of siting near densely populated regions are severe. The floating of large power stations offshore is costly but could become a necessity.

5. Effects that damage plant and animal life to some extent, but are not known to have any direct effect on people.

6. Effects that may damage people and for which large-scale preventive technology may or may not be commercially proved. These include emissions from fossil fuel plants, particularly in the vicinity of large population centers.

4.2.1 Air Pollution

One of the most widespread ill-effects has been pollution of the air with sulfur dioxide, oxides of nitrogen, unburned hydrocarbons, and particulate matter, as shown in Table 4.1. In the past, oil refineries and fossil fuel power plants have been major emitters of SO_2, NO_x, and particulate matter. Other gaseous emissions from fossil fuel power plants include heavy metals (such as nickel, molybdenum, manganese, zinc, cadmium, bismuth, mercury) in concentrations of 2 to 10 ppm at the stack. Possible hazards may also arise from emissions of certain carcinogenic hydrocarbon substances. Many of these substances are being monitored regularly. Attention has even been drawn to emissions of minute quantities of radioactivity in coal ash particles. For many of these substances the effects of pollutant concentration on human health are not well understood. There is considerable scientific uncertainty about health effects of

SO_2, even though recommended standard concentration limits have been formulated.

The EPA has estimated that by 1977 the total economic cost of air pollution from all sources in the United States will average about \$100/person-yr (1). This includes damage to health (\$40), residential property (\$30), and materials and vegetation (\$30). About half of the total cost could be ascribed to combustion in fossil fuel power plants.

4.2.2 Water Pollution

As Table 4.1 suggests, bodies of water can be polluted in various ways as a result of fossil fuel energy conversion. Oil spills from ocean-going tankers (or in offshore drilling operations) can cause considerable damage. Large tankers have capacities as high as 500,000 tons, but even 1000 tons can pollute several miles of shoreline and lead to the death of wildfowl and damage to beaches. Oil spills from pipelines can destroy vegêtation in limited areas for long periods.

The term "acid drainage" in Table 4.1 refers to the formation of sulfuric acid in mines in the presence of air and moisture. Water drained or pumped from coal mines may have sufficiently high acidity to kill vegetation. The volume of water flowing through many mines is such that neutralizing the acidity, by using lime for example, could be quite expensive. Other methods exist, but none is wholly satisfactory. According to Harry Perry (2), the total cost of abating acid mine drainage for the nation would be \$6.6 billion. He concludes that "correction of acid mine water pollution will be the most difficult and expensive of all the environmental problems caused by coal mining."

Heating of rivers and lakes by flows of cooling water from power plants can also be regarded as a form of pollution. It may be accompanied by dispersion of chemicals added to the cooling water to prevent deposits of marine growth on the power plant condenser tubes. The warming of river and lake water can have numerous adverse effects on marine life, including the killing of certain species of fish and excessive growth of undesirable plants or algae.

4.2.3 Land Use and Land Damage

Another important environmental effect of fossil fuel consumption concerns land use. Vast tracts of land are used by fossil fuel power plants directly and indirectly. The plant itself may occupy some hundreds of acres. An oil-fired plant may occupy 100 or more acres; a coal plant with fuel storage area may occupy 300 to 400 acres. High-voltage transmission lines also require large rights-of-way—10 to 100 acres per mile of length. Further expenditures of land are required for oil pipelines or coal transport by railways. Mining, and particularly surface mining, can also affect large areas of natural landscape.

Surface mining has provided some notorious examples of land damage. According to Perry (2), about 1.3 million acres in the United States have been disturbed by the strip mining of coal, mostly in Pennsylvania, Ohio, West Virginia, Illinois, and Kentucky. Of this total area, some 800,000 acres need reclamation, possibly including grading, construction of ditches, channels, ponds, and dams, and seeding and planting to stabilize the soil. The cost of reclamation after strip mining operations may be several hundred dollars per acre, but this can be equivalent to as little as 10 cents per ton of coal removed (2). In the absence of proper reclamation, land damage has been tragic. Acid drainage from deep and surface mines in Appalachia is said to pollute thousands of miles of streams (3). Watersheds have been destroyed. Soil erosion losses from freshly strip-mined areas have been 1000 times greater, per unit area, than from undisturbed lands. Landslides, timber damage, and various adverse chemical effects in the soil have been prevalent in the Appalachian region.

In the past, strip mine operators have not generally had to pay for the effects of their operations on people living downslope or downstream from the mining site. In recent years laws have been passed in a number of jurisdictions to control surface mining and require reclamation. There is hope that if these laws are enforced, future environmental damage from strip mining can mostly be prevented, though the rectification of old damage might be too expensive.

In West Germany, a policy of complete reclamation and even improvement of agricultural and other lands used for strip mining of coal has been in force for more than 20 years (3). This has proved worthwhile even with reclamation costs as high as \$3000 to \$4000/acre. Such a program may not be economically justified unless the land value is high. Also, it may not be feasible in mountainous terrain, where the only acceptable measure may be to prohibit strip mining on steep slopes.

There are perhaps 50 million acres in the United States which could still be used for strip mining of coal. Some of the greatest reclamation problems would be in the West. In certain regions, where rainfall is less than 10 inches per year and where soils do not retain moisture, reclamation may not be feasible. Some 60 percent of Western coal underlies the mixed grass region of the Northern Great Plains and the Ponderosa pine and mountain shrub zone of the Rockies. Earth moving costs may be only a few cents per ton of coal, but restoration of natural growth in these areas would take perhaps 20 years or more. The extent of the necessary reclamation work would be unprecedented.

While some effects of energy use (in the past deemed to be inevitable) now appear to be correctable at reasonable cost, others may be unacceptably expensive to remove. High-voltage transmission lines are often considered a blight on the landscape, yet the cost of "corrective" technology (buried power lines) appears to be prohibitively high. Underground lines cost perhaps a factor of ten more than overhead lines. Such a penalty may be considered acceptable only in cities, where the value of land is sufficiently high. Land requirements for rights-of-way are much less for underground than for overhead transmission lines. Much thought has been given to replacing conventional transmission towers in some areas by single steel poles that might look less ugly. Even measures like this could increase transmission cost by a factor of two.

The clearing of rights-of-way and the construction of maintenance roads for transmission lines can also cause soil erosion on steep slopes. The use of herbicides to suppress vegetation under power lines can have adverse side effects because toxic chemicals are transported with runoff. Some of the substances that have been used are persistent, and some concentrate in living organisms. Power lines that run through nesting or concentration areas can disrupt and reduce several forms of wildlife. In some regions, complex environmental cost–benefit studies are advisable before transmission line siting (4).

4.3 NUCLEAR FUELS AND THE ENVIRONMENT

Many of the environmental problems of fossil fuel use mentioned in Table 4.1 are insignificant for nuclear fuels. The volume of the fuel itself and the volume of the gases emitted from a nuclear plant are orders of magnitude smaller than those from a fossil fuel plant. The volume of uranium ore handled at the mine is not necessarily small, however, since it is economical to mine relatively dilute ores (less than 1 percent uranium). Also, more energy is rejected in the cooling water of a nuclear plant than from a fossil fuel plant of the same power, since most nuclear plants operate at low thermal efficiency. Land use for power plants, fuel transport, and waste disposal are typically much less than for coal-fired plants. Transmission line problems are, of course, comparable.

4.3.1 The Fuel Cycle

The most important environmental concern with nuclear power is the dispersion of small quantities of radioactive material as a result of the mining, processing, transporting, consuming, and possibly reprocessing of the fuel, and in the disposal of fuel wastes. Stages in the typical fuel cycle for reactors operating on enriched uranium fuel include:

1. Mining uranium ore
2. Concentration of the ore to U_3O_8
3. Enrichment (except for reactors that use natural uranium fuel)

4. Fabrication of fuel elements
5. Neutron bombardment of the fuel in a reactor
6. Storage of spent fuel
7. Reprocessing of spent fuel to recover uranium and plutonium generated within the fuel in the reactor
8. Waste treatment and disposal.

One ton of slightly enriched fuel will produce about 200 million kWh of electricity (enough to meet the annual needs of about 70,000 people). When this fuel is reprocessed, it will produce 0.4 to 0.8 m^3 of high-level liquid wastes, or about 0.04 m^3 of solidified waste (5). Fuel reprocessing and waste disposal are potentially dangerous steps in the cycle.

Some 99.9 percent of the radioactivity generated in power reactors is retained within the fuel elements until they are reprocessed.

4.3.2 Mining and Milling

Air pollution in uranium mining derives principally from radioactive dust particles, which are an occupational hazard to the miner. These dusts can contain naturally occurring radionuclides mixed with radon, a radioactive gas. Radon forms by the decay of Ra-226, which is in turn formed by the decay of U-238. According to the International Atomic Energy Agency (IAEA),

> before the development of current practices, which reduce the exposure of miners, overexposures were experienced, and a number of deaths due to lung cancer resulted. During the past 20 years more than 100 uranium miners have died of lung cancer in the United States, and it is estimated that 500 to 1500 miners who were exposed prior to the establishment of present-day occupational safety standards may die similarly of radiation-induced-disease. (5)

Milling the uranium ore can also result in radioactive dust, not just from operations, but from the large piles of "tailings," particularly in dry, windy places. The ores are ground into fine sand before extraction of the material. Perhaps 99 percent of this material is discarded in the form of solid or slimelike wastes. If these are deposited in a pile, seepage and runoff may transport acid, heavy metals, and other toxic chemicals into the surrounding area. In Ontario, uranium ore is found in conglomerates containing large amounts of iron sulfide, so acid mine drainage can be a serious problem. Also, radioisotopes could be transported into local water supplies. The solution to most of these problems lies in isolating the mill tailings sufficiently from interaction with wind and water. Drying these wastes and storing them in dry mines is recommended, but other kinds of treatment may be needed.

4.3.3 Enrichment

For reactors that are not designed to use natural uranium, the fuel must be enriched: the fraction of the fissile isotope U-235 must be increased from 0.7 percent to 2 or 3 percent. This is done by converting the fuel to a gaseous compound, UF_6; the lighter isotope U-235 can then be partially separated from U-238 in a large number of diffusion or centrifugal separation cells. Gaseous wastes are small, and the danger to the public from gaseous releases is negligible under the best conditions. Control of occupational hazards has been well established for the enrichment of fuels containing only uranium. Enrichment of material obtained from spent fuel, and containing plutonium, is more dangerous, since plutonium is not only highly toxic but remains radioactive for many thousands of years.

4.3.4 Fuel Reprocessing

Management of wastes from fuel reprocessing plants is a major problem for the breeder reactor. Fuel reprocessing wastes require special treatment for many different radioactive substances. Low-level radioactive wastes are treated simply by dilution and dispersion in water or air. Other wastes of higher radioactivity but shorter half-life are retained within the plant for radioactive decay. In subsequent stages they are chemically treated, concentrated, and contained. A completely acceptable method of containment has still not been found.

Krypton, tritium, and iodine isotopes are three radionuclides that are emitted from fuel reprocessing plants. Krypton 85 has a half-life of 10.8 years and is typically discharged through high stacks. The worldwide concentration of krypton in the atmosphere is being watched closely. Though it is a noble gas and not subject to chemical treatment, it could be trapped and retained by absorption, liquid filters, or by a cold trap. If necessary, krypton could be trapped and stored for years of decay and then released.

Tritium has a half-life of 12 years. Most of it is released in the form of tritiated water in liquid discharges, and does not appear to be a serious environmental threat (5).

The isotopes of iodine are a threat in that they concentrate in the human thyroid and can travel fairly quickly through the food chain (from grass to cow to milk). Children not only consume a lot of milk, but their thyroids are thought to be more affected by radiation than adults. Iodine 131 has a half-life of only 8 days, so it will substantially decay inside the fuel element before any reprocessing is carried out. Spent fuel elements are normally stored for at least three months before reprocessing. Under accident conditions, however, I-131 contamination could be serious. Other iodine isotopes have quite different half-lives and require special treatment.

4.3.5 Reactor Emissions

In normal operation, nuclear power plants emit gaseous, liquid, and solid wastes. Noble gases, tritium, and I-131 are among the radionuclides commonly emitted. Small quantities of I-131 are released due to defects in the fuel elements. The noble gases argon, krypton, and xenon are formed either by fission or by activation, that is, as the result of neutron bombardment. Tritium forms in water by neutron collision with deuterium in the water. Other activation products include carbon (in gas-cooled reactors), iron, cobalt, and manganese. These substances may be released by corrosion to circulate through the core of the reactor, where they become activated in regions of high neutron density.

Gaseous reactor wastes can be treated in a variety of ways, including filtration and absorption, and storage filters are used to collect radioactive dust particles. Charcoal beds can be used to absorb iodine isotopes. Cold traps are used for noble gases. Liquid wastes are filtered to remove suspended solids and are demineralized in ion exchangers. Evaporation is used to reduce the volume of the liquid waste and the radioactivity of the evaporated part. At sufficiently low levels of activity, liquid wastes may be discharged with plant cooling water. A general description of methods for treatment of radioactive gas and liquid waste from various reactors is provided in Ref. (5).

4.4 FOSSIL FUEL AND NUCLEAR POWER GENERATION

Since the environmental effects of power production are manifold, it is not easy to grasp the relative magnitudes of the important quantities. Table 4.2 provides comparative magnitudes of the daily flows of fuel, energy, and emissions for oil, coal, and nuclear power plants with controlled output of pollutants. These plants were considered as possible candidates by Northeast Utilities for the generation of 1150 MWe. The relative magnitudes of processing and transport facilities show a great advantage in favor of nuclear power. The nuclear plant site itself would require a relatively large area, but this disadvantage seems small relative to the shipping, refining, and storage requirements for supply of an oil-fired plant or for the transport and waste disposal needs of a coal-fired plant. Further, the oil and coal plants are characterized by vast volumes of gaseous emissions. The tonnages of SO_2 emitted are impressive even with the use of low-sulfur oil or with stack gas cleaning of coal combustion products. Waste heat is significantly larger for the nuclear plant (due to lower efficiency), but the seriousness of this point will depend heavily on local factors such as the availability of cooling water, climate, terrain, and ecology. If radioactive emissions and spent fuel transport and disposal are regarded as "acceptably safe," the total environmental impact of the nuclear plant will be regarded as substantially more agreeable than that of the corresponding fossil fuel plant.

Table 4.3 shows another comparison between estimated magnitudes of environmental effects from various modes of electrical power generation. The plants mentioned in Table 4.3 are assumed to operate with low levels of environmental controls. Table 4.4 shows effects and cost increases due to the application of emissions controls.

4.4.1 Gaseous Emissions

Air emissions estimated for coal are mostly power plant emissions. With coal of 2.6 percent sulfur content, it is estimated (7) that a 1000-MWe coal-fired plant would emit the following tonnages per year of pollutants (largely in the absence of controls):

Particulates:	203,200	tons
SO_2:	124,500	tons
CO:	1270	tons
HC:	381	tons
NO_x:	22,860	tons
Aldehydes:	6	tons.

Total pollutants shown in the above list are 352,217 tons. Most of the difference between this total and the figure in Table 4.3 is due to coal blown away in transport (25,700 tons). About 4700 tons of coal dust are lost in coal cleaning and drying processes. Table 4.3 indicates that, with controls, these emissions can be reduced 81 percent, at a corresponding increase in electrical power cost of 23 percent. It is estimated that 85 percent of the SO_2 emissions from the power plant can be eliminated by wet limestone scrubbing (see Chapter 9), and that particulate emissions can be reduced by 99 percent.

The air emissions shown in Table 4.3 for oil fuel are mainly due to processing (87,524 tons) and power plant emissions (69,321 tons), with small amounts due to emissions during transport. The emissions for imported oil are lower because processing is not included. The total power plant emissions were evaluated (7) for 1.5 percent sulfur oil and consist of:

Particulates:	1578	tons
SO_2:	46,438	tons
CO:	8	tons
HC:	395	tons
NO_x:	20,705	tons
Aldehydes:	197	tons.

With controls, emissions during processing can be reduced by 96.3 percent. Using 0.6 percent sulfur oil lowers the total tonnage of power plant emissions by 41.3 percent. Table 4.4 indicates an overall power cost increase of 31 percent corresponding to the application of these controls.

Table 4.2
Environmental Features of 1150-MWe Power Plants

	Oil-Fired Plant	Coal-Fired Plant	Nuclear Plant
Fuel Supply			
Production	Production and refining of sufficient crude oil to yield 40,000 bbl of fuel oil per day. This corresponds roughly to the fuel oil output of a large oil refinery.	Mining of 8000 tons of coal per day.	Mining and milling of 75,000 tons of uranium ore per year. Processing and fabrication of 150 tons of uranium metal per year. If by-product plutonium is recycled, 1000 1b/yr of plutonium would require processing and fabrication under equilibrium conditions.
Transport	One supertanker delivery of crude oil every 3 or 4 weeks, or, in the case of the larger tankers now serving U.S. ports, 1 delivery every 2 or 3 days.	Daily unit-train delivery (105 rail carloads).	Six truckload deliveries per year.
Storage	Storage of a 3-month fuel oil supply would require 20 large oil storage tanks (10^6 ft^3 each) occupying 20 acres.	50-acre coal pile, assuming 2-month reserve.	Nominal.
Power Plant			
Installation	70-acre plant site (assuming cooling towers used).	250-acre plant site (assuming cooling towers used).	500-acre plant site, mostly undeveloped.

Operation	Discharge of 120 × 10^9 Btu of waste heat per day; emission of 70 tons/day of sulfur or SO_2 (assuming low-sulfur fuel oil being burned), 30 tons/day of nitrogen oxides and other gaseous effluents, and 0.5 tons/day of particulates.	Discharge of 120 × 10^9 Btu of waste heat per day; emission of 150 tons/day of sulfur or SO_2 (assuming use of stack gas desulfurization system, not yet commercially available), 60 tons/day of nitrogen oxides and other gaseous effluents, and 9 tons/day of particulates (assuming use of highly efficient precipitators and scrubbers).	Discharge of 160 × 10^9 Btu of waste heat per day; emission of trace amounts (~ 10^{-5} g per day) of radioactive substances containing 2 Ci of comparatively long-lived radioactivity. Shipment of 60 casks of spent fuel per year (60 truckloads or 10 railroad flat carloads).
Waste Disposal	Minor problems.	Disposal of 1200 tons/day of fly ash and 1600 tons/day of sulfur, based on 3.5 percent sulfur coal and assuming 80 percent stack gas desulfurization efficiency.	"Perpetual" storage of solidified high-level radioactive waste concentrates from spent fuel reprocessing, which, in calcined form and with inert diluents, accumulate at a rate of 100 ft^3/yr. Also land burial of 200 ft^3/yr of miscellaneous low-level radioactive waste materials.

Source: Ref. (6), courtesy of Northeast Utilities.

Table 4.3
Comparative Environmental Impacts of 1000-MW Electric Energy Systems Operating at a 0.75 Load Factor with Low Levels of Environmental Controls or with Generally Prevailing Controls

	Air Emissions			Water Discharges				Solid Waste			Land Use	
System	Tons ($\times 10^3$)	Curies ($\times 10^3$)	Severity	Tons ($\times 10^3$)	Curies ($\times 10^3$)	Btu's ($\times 10^{13}$)	Severity	Tons ($\times 10^3$)	Curies ($\times 10^8$)	Severity	Acres ($\times 10^3$)	Severity
Coal												
Deep-mined	383		5	7.33		3.05	5	602		3	29.4	3
Surface-mined	383		5	40.5		3.05	5	3267		5	34.3	5
Oil												
Onshore	158.4		3	5.99		3.05	3	NA		1	20.7	2
Offshore	158.4		3	6.07		3.05	4	NA		1	17.8	1
Imports	70.6		2	2.52		3.05	4	NA		1	17.4	1
Natural Gas	24.1		1	0.81		3.05	2			0	20.8	2
Nuclear		489	1	21.3	2.68	5.29	3	2620	1.4	4	19.1	2

NA = not available. Severity rating key: 5 = serious, 4 = significant, 3 = moderate, 2 = small, 1 = negligible, 0 = none.
Source: Ref. (7).

Table 4.4
Cost of Controls and Changes in Environmental Impacts of 1000-MW Electric Energy Systems Operating at a 0.75 Load Factor with a High Level of Environmental Controls

	Air		Water		Land		Solid Waste	
System	Tonnage Change (percent)	Cost Increase (percent)	Tonnage Change (percent)	Cost Increase (percent)	Acreage Change (percent)	Cost Increase (percent)	Tonnage Change (percent)	Total Cost Increase (percent)
Coal								
Deep-mined	−81.3	23	−96.2	4	+1	0	+159	28
Surface-mined	−81.3	23	−92.4	4	−37	4	+29	31
Oil								
Onshore	−73.0	31	−38.8	5	0	0	0	36
Offshore	−73.0	31	−38.8	5	0	0	0	36
Imports	−39.4	28	−77.2	5	0	0	0	34
Natural Gas	0	0	0	5	0	0	0	5
Nuclear	−29.3	1	0	4	0	0	0	5

Source: Ref. (7).

As shown in Table 4.3, uncontrolled air emissions, waste water discharges, and solid wastes from natural gas plants are relatively small compared with coal and oil plants.

4.4.2 Water Discharges

The waste water discharges for coal-fired generation do not include the cooling water, which is around 10^9 tons/yr. The total water discharge in Table 4.3 is mainly comprised of:

Deep mine draining (mostly sulfuric acid):	2661	tons
Surface mining siltation:	35,820	tons
Coal washing:	3853	tons.

Table 4.4 indicates that most of these discharges can be satisfactorily treated, with a power generation cost penalty of about 4 percent.

Waste water discharges for oil-fired generation are mostly associated with processing (4631 tons) and transport of imported oil (1711 tons).

4.4.3 Solid Wastes

The largest quantity of solid waste shown in Table 4.3 is for coal-fired generation, with the figure for nuclear power close behind. These figures refer mainly to solid waste in mining (2,762,000 tons for surface mining of coal and 2,620,000 tons for uranium mining). The figure for surface mining of coal does not include the overburden, which is deposited on the site. Solid waste from deep mining of coal totals 97,141 tons. The solid waste from the power plant, 150,800 tons, corresponds to combustion of coal with 10 percent

Table 4.5
Land Requirements for Fossil Fuel and Nuclear Power (1000-MWe Plant Operating at 0.75 Load Factor)

	Coal (acres)	Oil (acres)	Nuclear (acres)
Extraction[a]	9,120 (deep) 14,010 (surface)	1,572 (onshore) 164 (offshore)	785
Transport	2,213[b]	1,656 (onshore) 168 (offshore) 10 (import)	0
Processing	161	76	9
Conversion	696	250	300
Transmission	17,188	17,188	17,188

[a]Refers to the average quantity of land which has been used over a 30-year plant life, or 15 times the annual increment.
[b]Calculated as the total railroad right-of-way in the United States (365,000 miles × 50 ft) multiplied by 0.001, the fraction of annual ton-mileage accounted for by coal transported 300 miles to a 1000-MW power plant.
Source: Data from Ref. (7).

ash, most of which (203,200 tons) is emitted as fly ash in uncontrolled plants. With capture of 99 percent of this fly ash, and discharge of 742,000 tons of limestone sludge from a wet limestone SO_2 stack gas scrubbing process, power plant solid wastes will rise to 994,000 tons. Table 4.4 shows the corresponding changes in total solid waste quantities for coal-fired generation; the associated control costs are included with those of the air and water emissions.

4.4.4 Land Requirements

The land requirements for coal, oil, and nuclear power are given in more detail in Table 4.5. The largest single entry is land set aside for right-of-way of transmission lines (but not necessarily removed from other use). The next largest entry is for surface mining of coal. Coal and oil transport are also substantial, but the areas occupied by the power plants are relatively small.

4.4.5 Occupational Hazards

Table 4.3 includes a summary of occupational hazards in activities related to power generation. The details of these hazards are shown in Table 4.6, which indicates the severe hazard in underground coal mining relative to surface mining and other occupations associated with power production. The relatively large death and

Table 4.6
Occupational Hazards for Fossil Fuel and Nuclear Power (per year per 1000-MWe plant operating at 0.75 load factor)

	Coal	Oil	Nuclear
Deaths			
Extraction	1.67 (deep) 0.308 (surface)	0.21	0.16
Processing	0.024	0.08	0.01
Transport	2.30	0.05	0.02
Conversion	0.012	0.01	0.009
Transmission			
Total	4.0 (deep) 2.64 (surface)	0.35	0.20
Workdays Lost			
Extraction	12,688 (deep) 499 (surface)	2,135	1,629
Processing	99.5	785	92
Transport	2,340	562	145
Conversion	152.9	127	120
Transmission			
Total	15,280 (deep) 3,091 (surface)	3,609	1,986

Source: Data from Ref. (7).

injury toll for coal transport is mostly accounted for by accidents at level crossings.

4.5 ENVIRONMENTAL HEATING

All of the energy used in human activities eventually becomes a heat input into the environment, almost all of which enters the atmosphere and can affect urban (and global) climate. Waste heat discharged via cooling water to rivers and lakes may also have important effects on local biology and water quality.

4.5.1 Global Climate

Table 4.7 shows the ratio between heat release due to human activity in energy conversion and the rate at which solar energy is absorbed at the earth's surface. For the world as a whole, the ratio is currently much less than 1 percent. Such a level would be approached only if total energy consumption were about 40 times what it is today. Averaging for the United States only, the 1 percent level could be reached by the end of the century.

The effects of artificial heat release on global climate have been considered by a number of workers. Simple numerical models of global climate suggest that a 1 percent change in the absorption of solar energy at the earth's surface may induce a significant change in the earth's climate, or a 1°C increase in mean surface temperature

Table 4.7
Heat Release to the Environment Due to Human Activity in Energy Conversion

	Total Heat Release Rate (10^{12} W)	Heat Release Density (W/m^2)	Percentage of Solar Energy Absorbed at the Earth's Surface[a]
World			
In 1975	8	0.053	0.03
Possible by A.D. 2000[c]	25	0.17	0.09
Possible Steady State[d]	300	2.0	1.0
United States			
In 1975	2.4	0.3[b]	0.2
Possible by A.D. 2000[c]	8.1	1.0	0.6

[a]Average solar energy absorption at the earth's surface is taken to be 170 W/m^2 for the United States, and 200 W/m^2 for the world (27×10^{15} W and 1.5×10^8 km^2 of land area).
[b]Current energy release rate 2.4×10^9 kW over 8×10^{12} m^2
[c]Allowing for 5 percent annual growth rate over 25 years.
[d]Allowing for a possible world population of 20×10^9 with average energy consumption 15 kW/person.

(8). A small warming effect due to artificial heat release could reduce polar ice and snow, which would mean more absorption of solar energy in polar regions and enhanced warming. However, any warming that took place might also considerably affect cloud cover and the interaction between the oceans and the atmosphere. These effects are not yet sufficiently understood to be predictable.

Dr. Warren M. Washington of the National Center for Atmospheric Research has used a numerical model of the earth's atmosphere to explore the effect on climate of an artificial heat input corresponding to 300×10^{12} W distributed roughly according to the present geographical population pattern of the world (9, 10). This magnitude of heat release would correspond to 15 kW/person for a world population of 20 billion. In certain densely populated areas this could amount to a significant fraction of the local solar flux. In total it is about 40 times the present value of global heat release caused by man.

The numerical model of the atmosphere used by Washington subdivides the lower part of the atmosphere into six 3-km layers. For each, the momentum, energy, and mass conservation equations are solved in numerical form. The model estimates several modes of energy transport: radiation (both short-wave solar radiation and long-wave or infrared radiation from the earth's surface), condensation, and diffusion (both vertically and horizontally). Starting from any assumed initial state of the atmosphere, the model computes its development in time toward some state that will be steady if the boundary conditions are unchanged. The asymptotic state will depend somewhat on the assumed initial state. Washington emphasizes that the model is still under development and makes many simplifying approximations. For example, the ocean surface temperatures were considered fixed, and many other mechanisms for changing climate, such as deflection of solar radiation by aerosols, were not included.

Though approximate, model results suggest that the large heat release mentioned earlier (300×10^{12} W) will not significantly affect global climate (10). In a separate calculation with 25 W/m^2 of artificial heat input uniformly distributed over land areas of the globe (a total heat input nearly 500 times higher than the current actual total), the January global climate predicted by the model was quite noticeably affected by the heat input, with an average temperature increase of 5°C over the surface of the globe (9).

Washington's basic conclusion is that

> with the expected levels of man's thermal energy production, there is a relatively small modification of the model earth–atmosphere heat balance. The differences in numerical experiments with and without thermal energy input produce changes of the same order as the natural fluctuation of the model. (10)

He also points out that natural climatic fluctuations which are not yet understood could easily mask the effects of global heat release due to human-caused energy conversion. The delicate radiation balance in the atmosphere might be much more affected by changes in cloud cover, aerosols, or CO_2 buildup than by the enhanced terrestrial heat release (11).

As long ago as 1863, John Tyndall pointed out that CO_2 in the atmosphere could well have a "greenhouse" effect, since it absorbs infrared radiation emitted at the earth's surface and reradiates part of it back to earth. The CO_2 content of the atmosphere is building up slowly but definitely (11). Since the International Geophysical Year (1958), when worldwide systematic measurements were begun, CO_2 has been increasing in concentration at about 0.2 percent per year. The present concentration is around 320 ppm. At least half of the CO_2 formed during fossil fuel combustion remains in the atmosphere; the remainder is absorbed by the ocean and the biosphere (11). Simple numerical models of global climate that ignore the effect of the temperature increase on evaporation, cloudiness, and interaction of the atmosphere with the ocean reservoir, suggest that the buildup in CO_2 by the end of the century could result in a rise in mean annual temperature of 0.3°C (12).

Between 1880 and 1940, the annual mean temperature did increase rather steadily at a rate of about 0.8°C per century. However, from 1940 until recently the mean annual temperature of the earth's surface has gone down at about the same rate as it formerly increased, so that available data provide no simple association between temperature increase and CO_2 concentration; the effect is apparently masked by other complex effects that determine the overall radiation energy budget at the earth's surface.

Changes in mean annual temperature as small as 0.5°C can be important. The warming encountered between 1880 and 1940 brought a significant northward movement of frost and ice boundaries, an increase in aridity in central parts of Asia and North America, and strong northern hemispheric zonal circulations (11). Gordon J. F. MacDonald reports that the average temperature decrease of about 0.2°C in the last 25 years has been associated with substantial effects; for example, sea-ice coverage in the North Atlantic in 1968 was the most extensive in over 60 years (13). In contrast, the rains in central continental regions, particularly in India, led to high wheat yields during part of this period.

R. M. Rotty has pointed out that the observed decrease in mean global temperature since 1940 has been accompanied by a decrease in incident solar radiation, which may be due to the presence of dust in the upper atmosphere (14). MacDonald indicates that, in addition to CO_2 concentrations, man's activities could otherwise perturb the atmospheric heat balance and the climate by reducing atmospheric transparency with aerosols from industry, automobiles, and home-heating units, and dust from agricultural activity; direct heating; changing the albedo (percentage of incoming solar radiation directly reflected outward) on the earth's surface through urbanization, agriculture, and deforestation; and causing oil films to develop on the oceans due to incomplete combustion or oil spills, which would significantly affect transfer of mass and energy between the ocean and atmosphere (13).

In the absence of reliable models of the dynamics of the global atmosphere that take all of these effects into account, there appears to be no reason to conclude that human-caused environmental heating or the buildup in atmospheric concentration of CO_2 will have serious effects on global climate.

4.5.2 Urban Climate

Table 4.8 shows the relation between local heat release and average solar absorption for selected areas in the United States and the world. In Manhattan, local energy consumption is more than six times higher than solar absorption. L. Lees estimates that the magnitude of artificial heat production in the 4000-square-mile area of the Los Angeles basin now equals 5 percent of the solar energy absorbed at the ground surface, and that this will increase to about 18 percent by the year 2000. For the area of the United States bounded by Boston, Washington, St. Louis, and Chicago, where 40 percent of the national energy utilization occurs, it is estimated that

Table 4.8
Energy Consumed and Average Sun's Energy Absorbed at Typical Cities in the United States and the World (1970)

	Area (km^2)	Average Solar Absorption (W/m^2)	Energy Consumption (W/m^2)	Percent of Solar Intensity
New York (Manhattan)	59	93	630	680
Los Angeles	3,500	108	21	19
Los Angeles County	10,000		7.5	5
Cincinnati		99	26	26
22 Metropolitan Areas, Washington–Boston		90	4.4	5
Fairbanks, Alaska		18	18.5	100
Moscow	880	42	127	290
West Berlin	234	57	21.3	37
United States Average		168	0.297	0.18
Global Average (land areas)		238	0.05	0.03

Source: Ref. (15).

artificial heat release is now 1 percent of the absorbed solar energy and that by the end of the century it will rise to 5 percent (11). R.T. Jaske has estimated that by A.D. 2000 artificial heat release in the 12,000-square-mile Boston–New York–Washington urban corridor will average as much as 50 percent of solar absorption over the entire area in the winter, and about 20 percent in the summer (16). Lees has predicted that in Los Angeles County the effect of artificial heat release may raise the December temperature by 13°F in Los Angeles and 5°F in New York if the heat rejections predicted for the end of the century are realized. A rise in summer temperature in these cities would naturally lead to greater demand for air conditioning and thus increased heat rejection, which would aggravate the problem.

Temperature increases in cities (relative to rural areas) are due not only to heat release but also to reduced convective cooling from winds as a result of increased effective surface roughness or buildings. Building surfaces also absorb radiant energy from the sun more easily than does vegetation.

Pollution may have a substantial influence on urban climate (17, 18). According to James T. Peterson, average urban–rural temperature differences are 1 to 2°C, and occasionally reach 10°C (8). He lists the major features of urban climate that are related to energy use as:

1. longer frost-free growing season
2. less snowfall because of snow melting as it falls through the the warmer urban atmosphere
3. lower relative humidity
4. less fog because of the lower humidity, a feature which may be offset by increases in particulates which serve as condensation nuclei
5. about 5 to 10 percent more precipitation downwind of cities, a phenomenon at least partially due to increased convection (vertical motion)
6. a slight component of the wind directed toward the city center as a result of the horizontal temperature contrast
7. a noticeable increase in the death rate during prolonged heat waves (in cities such as New York and St. Louis).[1]

Peterson concludes that both the climatic and health effects of waste heat on urban areas will continue to be tolerated. At some point, however, concentration of energy conversion may have to be limited in certain areas.

4.5.3 Rivers and Lakes

The total cooling capacity of river water on the North American

[1]Quoted with permission from E. S. Barrekette, ed., *Pollution: Engineering and Scientific Solutions* (New York: Plenum Publishing Corp., 1973).

continent is limited in relation to the quantities of waste heat that could be generated by power plants by the year 2000. Extrapolating from Figure 3.12, the total waste heat flow rate by the year 2000 could be 1.5×10^9 kW. If the total flow in continental rivers and streams is estimated as 2.5×10^6 ft^3/sec, and all this flow were used once for power plant cooling, the average temperature increase would be about 9°F (5°C). Very much less than this total flow will be physically accessible to future power plants, though it is true that the same river could be used by several power plants in line; downstream of each plant the water cools by evaporation and convection.

To keep power plant costs down, it is desirable to warm the cooling water by 15 to 20°F (8 to 11°C), but this may be harmful to local plant and fish life in the river. If a river or lake is small enough and is already warm, further heating will hasten the processes of biological decay. Algae and aquatic plants grow faster in warmer water, so that there is greater absorption of oxygen. The decomposition of organic waste (sewage) is accelerated by elevated temperature, and this reduces the oxygen supply. Heating also reduces the capacity of the water to retain oxygen and can disrupt the life cycles and migration patterns of fish. Many of these effects are exceedingly complex and depend on local conditions (19). Field data are necessary to decide whether they are important in a given situation.

After study of many possible effects, the United States National Technical Advisory Committee on Water Quality Standards (20) has recommended that heat addition to any river should not raise the water temperature more than 5°F for streams with warm water fisheries and 3°F for the surface layers of lakes and reservoirs, and that the temperatures of estuaries should not be raised more than 4°F during fall, winter, and spring, or by more than 1.5°F during summer. These restrictions are so severe that future power stations in many parts of North America must resort to the use of either cooling ponds or cooling towers. The extra expense of building cooling towers rather than using river water for once-through cooling may be only 5 percent of plant capital cost. However, for a 1000-MWe plant this could amount to $40 to $50 million. Cooling towers can also introduce environmental problems. They can be the cause of fog and drizzle, and may initiate clouds and enhance rainfall. Cooling tower mists could reduce visibility on highways, and could interact with stack emissions of SO_2 to produce acidic mists. However, by siting towers away from highways and in zones of favorable meteorology, serious effects can usually be avoided (21). Cooling towers that rely on air cooling rather than evaporation are an alternative, but relatively costly, solution.

Estimates of waste heat flows from all sources into the Great

Table 4.9
Thermal Discharges from All Artificial Sources into the Great Lakes

	1968		2000	
Lake	Btu/ft²/hr	W/m²	Btu/ft²/hr	W/m²
Ontario	0.093	0.29	1.26	3.9
Erie	0.126	0.40	1.20	3.77
Huron	0.0077	0.024	0.289	0.91
Michigan	0.046	0.145	0.52	1.63
Superior	0.0021	0.0066	0.022	0.069

Source: Ref. (22).
Note: Solar radiation is approximately 50 Btu/ft² = 157 W/m².

Lakes are provided in Table 4.9. These flows per unit area are all small compared with the average solar radiation flux. However, such a comparison does not provide a satisfactory indication that the Great Lakes would be able to accommodate greatly increased thermal discharges. Waste heat inputs are concentrated at specific points on the shore. Warm water discharged from power plants tends to stay in fairly well defined plumes, mixing only slowly with the lake water. These plumes drift with the wind and may have persistent influence over certain regions of the shoreline. The near-shore areas are particularly important for many forms of fish and plant life. For these reasons a calculation of the effect of artificial heat input on average lake temperature (presupposing complete thermal mixing) may be quite irrelevant. Local effects on the shores of Lake Erie and Lake Michigan have already been severe enough that cooling towers are considered appropriate for all new power plants.

4.6 SUMMARY

Power generation leaves its effects on the physical world in many ways. The mining of fuels has devastated certain landscapes; air and water pollution have been widespread; noise, dust, and ugliness have been all too common. However, it would appear that many of these effects are controllable, or even preventable, at reasonable cost. It is hardly inevitable that greatly increased power generation will make the land uninhabitable.

The central problem is human safety, and the chief areas of concern are air pollution due to fossil fuels and nuclear radiation. We can now compare the gross effects on the environment of fossil fuel and nuclear power, but we are still uncertain about the precise mechanisms and health dangers of chemicals such as SO_2 and plutonium. In the absence of precise data, attempts to define social costs and benefits of fossil and nuclear power must be regarded as tentative.

Although they are in relatively primitive stages of development, computer models of the atmosphere suggest that total heat release due to energy use will not affect global mean temperatures significantly. Global radiation is in delicate balance and there are possibilities of negative feedback, so present conclusions are tentative. These matters must be kept under close scrutiny, but so far it is not clear that they should limit energy use.

If atmospheric CO_2 content continues to increase, mean climate temperature could conceivably rise. This is not in accord with recent observations, and the discrepancy may be due to the increasing dust or aerosol content of the atmosphere or other effects. Until these are explained or serious consequences are observed, there appears to be no clear basis in climatology for restricting total use of fossil fuels, at least in the next few decades.

Urban climates differ from rural ones, but this is not due solely to high local rates of heat release. The existence of urban heat islands suggests the possible need to limit the concentration of energy in specific areas.

The effects of waste heat on relatively warm river and lake waters can be detrimental to fish and plant life, and have led to restrictions on the temperature rise of water used in once-through cooling of power plants.

References

1. Environmental Protection Agency. *The Economics of Clean Air*, Annual Report of the Administrator. Washington, D. C.: U. S. Government Printing Office, March 1972.

2. Harry Perry. "Environmental aspects of coal mining." Chapter 18 in *Power Generation and Environmental Change*, David A. Berkowitz and Arthur M. Squires, eds. Cambridge, Mass.: MIT Press, 1971.

3. E. A. Nephew. *Surface Mining and Land Reclamation in Germany*. Report ORNL-NSF-EP-16, Oak Ridge National Laboratory, May 1972.

4. J. T. Kitchings, H. H. Shugart, and J. D. Story. *Environmental Impact Associated with Electric Transmission Lines*. Report ORNL-TM-4498, Oak Ridge National Laboratory, March 1974.

5. International Atomic Energy Agency. *Nuclear Power and the Environment*. Vienna, 1973.

6. Arthur D. Little, Inc. *A Study of Base Load Alternatives for the Northeast Utilities System*. Cambridge, Mass., July 5, 1973.

7. Council on Environmental Quality. *Energy and the Environment: Electric Power*. Washington, D. C.: U. S. Government Printing Office, August 1973.

8. James T. Peterson. "Climatic aspects of waste heat." In *Pollution: Engineering and Scientific Solutions*, E. S. Barrekette, ed. New York: Plenum Press, 1973, pp. 10–17.

9. Warren M. Washington. "On the possible uses of global atmospheric models for the study of air pollution." In Wesley E. Brittin et al., *Air and*

Water Pollution. Boulder, Colo.: Colorado Associated University Press, 1972, pp. 599–613.

10. Warren M. Washington. "Numerical climatic change experiments: The effect of man's production of thermal energy." *Journal of Applied Meteorology*, vol. 11, no. 5 (August 1972), pp. 768–772.

11. *Inadvertent Climate Modification: Report of the Study of Man's Impact on Climate (SMIC)*. Cambridge, Mass.: MIT Press, 1971.

12. S. Manabe and R. T. Wetherald. "Thermal equilibrium of the atmosphere with a given distribution of relative humidity." *Journal of the Atmospheric Sciences*, vol. 24, no. 3 (May 1967), pp. 241–259.

13. Gordon J. F. MacDonald. "Climatic consequences of increased carbon dioxide in the atmosphere." Chapter 14 in *Power Generation and Environmental Change*, David A. Berkowitz and Arthur M. Squires, eds. Cambridge, Mass.: MIT Press, 1971.

14. R. M. Rotty. "Global production of CO_2 from fossil fuels and possible changes in the world's climate." Paper 73-Pwr-11 presented at the ASME–IEEE Joint Power Generation Conference, New Orleans, Louisiana, September 16–19, 1973.

15. L. Lees. Quoted in *Man's Impact on the Global Environment*, William H. Matthews, ed. Cambridge, Mass.: MIT Press, 1970.

16. R. T. Jaske. Statement submitted to Hearings before the Subcommittee on Science Research and Development of the Committee on Science and Astronautics, U.S. House of Representatives, May 1972.

17. H. Landsberg. *City Air—Better or Worse?* Cincinnati, Ohio: U.S. Public Health Service, 1961.

18. S. Tilson. "Air pollution." *International Science and Technology*, June 1965.

19. Loren D. Jansen. "Spectrum of concerns from power plant thermal discharges." In *Electric Power and Thermal Discharges*, Merril Eisenbud and George Gleason, eds. New York: Gordon & Breach, 1969, pp. 47–74.

20. National Technical Advisory Committee to the Secretary of the Interior. *Water Quality Criteria.* Washington, D.C.: Federal Water Pollution Control Administration, 1968.

21. John R. Hummel. "Cooling towers and weather modification." In *Pollution: Engineering and Scientific Solutions*, E. S. Barrekette, ed. New York: Plenum Press, 1973, pp. 19–25.

22. H. G. Acres. *Thermal Input to Lake Ontario to 2000.* Report to the Inland Waters Branch, Energy, Mines and Resources Canada, 1970.

Supplementary Reference

A. M. Weinberg and R. P. Hammond. "Global effects of the use of energy." Paper A/Conf 49/P/033, *Proceedings of the Fourth United Nations International Conference on the Peaceful Uses of Atomic Energy*, Geneva, September 6–16, 1971.

Problems

4.1 a. Estimate the rates of once-through cooling water flow (ft^3/sec) for 1000-MWe coal and nuclear plants, for temperature rises of

(i) 15°F and (ii) 10°F. Assume 39 percent overall efficiency for the coal plant and 31 percent for the nuclear plant. Stack gases and other losses account for 14 percent of the energy input of the coal plant; the remaining heat rejection is via cooling water. For the nuclear plant essentially all heat rejection is to the cooling water.

b. Given two 500-MWe turbine units in each plant, each unit with its own condenser, calculate the required diameter of a single cooling water inlet for each unit, with

i. 15°F temperature rise in cooling water and 15 ft/sec intake velocity;

ii. 10°F temperature rise in cooling water and 2.5 ft/sec intake velocity.

4.2 By 1985, it is estimated that surface mining for coal in the United States may have expanded to the rate of 780×10^6 tons annually (Ref. (9) of Chapter 2). Using Table 4.5, estimate the following:

a. The average land required for surface mining of the 1985 coal requirement.

b. The total land use as a fraction of land area in the United States (5×10^6 square miles) for 1985 and from then to the end of the century, if surface mining grows at 10 percent per year.

c. The total cost for reclaiming the land used in 1985, at $500/acre.

d. The cost of reclamation per ton of coal.

4.3 Commercially exploitable resources of oil shales or bituminous sands generate from 3 to 3.5 m^3 of solid waste for each cubic meter of oil produced (Ref. (1) of Chapter 1). Surface mining is normally restricted to deposits no more than 30 to 45 m deep.

Assume that shale and oil sand resources are developed at a rate sufficient to supply half the projected 1980 North American oil consumption (see Figures 2.11 and 2.12) from surface mining, and estimate the following:

a. The volume of oil that would be processed each year (km^3).

b. The volume of waste material per year.

c. The land area (km^2) required each year for surface mining.

d. The land area required to collect the same amount of solar energy each year, assuming a rate of 200 W/m^2.

4.4 Suppose that total average U.S. electric power requirements by the year 2000 are 1500 GWe. With an average load factor of 0.75, this would require an installed capacity of 2000 GWe, plus reserve for emergencies. If this total capacity is 50 percent nuclear, 40 percent coal, and 10 percent oil, estimate (using Table 4.5) the total land requirements for extraction, processing, transport, conversion, and transmission.

Assume that the coal is all surface mined and that the oil is onshore. Show the fractional use of total U.S. land area (5×10^6 square miles). For surface mining of coal use the average area of land required over the 30-year plant life (Table 4.5).

4.5 A city with 1 million inhabitants occupies an area of 1000 km^2. If the rate of energy consumption per person is equal to the 1975 average for North America, estimate the average artificial heat release rate (W/m^2) within the metropolitan area. Approximately what fraction is this of average solar energy absorbed?

If one were to imagine a day of nearly stagnant air above the city, with no cooling of the air mass, by how much would artificial heat input raise the temperature in a layer of air 1000 m thick lying on the city surface? The sea level density of air is 1.2 kg/m^3 and its specific heat is 0.24 cal/g°C (1 cal = 4.18 J = 4.18 W-sec). Suppose one day's solar energy were nearly all absorbed at the earth's surface and then transmitted to this same stagnant layer of air. By how much would the temperature rise?

4.6 Estimate the rate of increase in the concentration (ppm by volume) of CO_2 in the global atmosphere for the year 1970, using the total consumption rates indicated by Figure 1.1. Assume that half of the CO_2 produced (Table 4.A) remains in the atmosphere, and that the atmosphere has approximately 5×10^{21} g of air.

Table 4.A
Carbon Dioxide Production of Various Fuels

	Average Percent Carbon by Weight	Average Heating Value (10^6 J/kg)
Coal	80	36
Oil	80	43.3
Gas	75	57.3

For comparison, the measured increase in the year-round average level of CO_2 in the atmosphere is around 0.7 ppm (13). Superimposed on the gradual increase is an annual periodic variation of ± 3 ppm.

4.7 This exercise provides a review of chemical stoichiometry needed in the solutions of problems 4.8 and 4.9.

A gaseous fuel has the composition shown in Table 4.B.

a. Let V_i be the partial volume of constituent i (i.e., the volume it would occupy if it were separated from the mixture and held at the pressure P_m and temperature T_m). Show that if each constituent of the mixture is a perfect gas, the volume fraction equals the mole fraction: $V_i/V_m = N_i/N_m$.

Table 4.B
Composition of a Gaseous Fuel

Constituent	Percent by Volume	Percent by Mass
N_2	30.2	37.8
CO_2	10.7	21.1
CO	10.7	13.4
H_2	15.7	1.4
CH_4	4.4	3.2
H_2O	27.8	22.4
H_2S	0.5	0.8
Total	100	100

b. Verify that the second column in Table 4.B can be derived from the first using atomic weights for N, C, O, H, and S as 14, 1, 16, 1, and 3, respectively. If one assumes 100 moles of mixture, then the numbers in the first column represent the number of moles of each constituent.

c. Taking 100 moles of gas mixture, and representing air as 21 percent O_2 and 79 percent N_2 by volume, show that if the products of combustion include only CO_2, H_2O, SO_2, and N_2, the reaction may be represented by

$$30.2\ N_2 + 10.7\ CO_2 + 10.7\ CO + 15.7\ H_2 + 4.4\ CH_4$$
$$+\ 27.8\ H_2O + 0.5\ H_2S + A(O_2 + 79/21\ N_2)$$
$$\rightarrow W\ CO_2 + X\ H_2O + Y\ SO_2 + Z\ N_2$$

By balancing the number of C, H, S, and O atoms, respectively, show that $W = 25.8$, $X = 52.8$, $Y = 0.5$, $A = 22.75$, $Z = 114.9$, and that the composition of the products is as shown in Table 4.C.

Table 4.C
Products of a Gaseous Fuel

Constituent	N_i	Percent by Volume
CO_2	25.8	13.3
H_2O	52.8	27.1
SO_2	0.5	0.3
N_2	115.8	59.4
Total	194.9	100

d. If twice as much air had been used as in Table 4.C (with the excess O_2 appearing in the exhaust), show that the exhaust product composition would have been as shown in Table 4.D.

Table 4.D
Exhaust Products of a Gaseous Fuel

Constituent	N_i	Percent by Volume
CO_2	25.8	8.5
H_2O	52.8	17.4
SO_2	0.5	0.2
N_2	21.4	66.4
O_2	22.75	7.5
Total	303.2	100.0

4.8 Bituminous coal for a 100-MWe power plant has the following total composition by weight:

Carbon: 78.4 percent
Oxygen: 5.0 percent
Hydrogen: 5.0 percent
Nitrogen: 1.5 percent
Sulfur: 2.6 percent
Ash: 7.5 percent.

The heating value of the coal is 25×10^6 Btu/ton of dry coal and the plant thermal efficiency is 39 percent. In addition to these constituents, natural moisture in the coal is 7 percent. The plant operates with 0.75 load factor.

a. Consider the combustion of 107 lb of undried coal. Assume that the chemically correct proportion of air to coal is used and that the ash is completely removed from the stack gases. Neglecting the formation of NO_x, sulfur oxides other than SO_2, and CO, and assuming complete removal of ash, show that the stack gas composition is as shown in Table 4.E.

Table 4.E
Stack Gas Composition of Coal

Constituent	Moles	Mole Percent
CO_2	6.53	17.2
H_2O	2.89	7.6
N_2	28.44	75.0
SO_2	0.08	0.2

b. Determine the relative concentration by volume (ppm) of SO_2 in the stack gases. Estimate the volumetric flow (ft^3/sec) of stack gases and SO_2 for a 1000-MWe coal plant if the temperature at the base of the stack is 350°F and the pressure is nearly atmospheric.

c. If no SO_2 removal process is employed, by what ratio is it necessary to dilute the stack gases to achieve a concentration by volume of no more than 0.01 ppm (see Table 6.6, for example).

Note: The atomic weight of sulfur is 32.

4.9 Wet limestone scrubbing of stack gases (see Chapter 9) can remove 85 percent of the SO_2. Pulverized limestone, which can be considered as mainly $CaCO_3$, is mixed with H_2O and sprayed into the stack gases. The liquid collected after spraying contains a mixture of $CaSO_3$, $CaSO_4$, and unreacted limestone. After removing much of the H_2O from the mixture, a wet sludge must be discharged as waste. According to Ref. (7), 85 percent removal of SO_2 from a 1000-MWe coal plant operating with a load factor of 0.75 results in a discharge of 742,000 tons of limestone sludge per year.

a. Estimate the additional depth per year of this waste in a 100-acre storage area, assuming a waste density of 90 lb/ft^3.

b. Estimate the tonnage of limestone required annually for 85 percent sulfur removal if twice the stoichiometric quantity is required and if the fuel and combustion products are the same as in Problem 4.8.

Note: The molecular weight of $CaCO_3$ is 80.

5
Nuclear Radiation Hazards

5.1 INTRODUCTION

There is no doubt that pollution arising from energy conversion could have tragic effects on man, and one of the most serious forms of such pollution is nuclear radiation. The key questions are whether the adverse effects of this or any other kind of pollution are controllable, and at what costs.

Butler and Barry (1) have classified the general effects of pollution on man as direct and indirect:

Direct Effects

1. Deleterious to health and well-being
   a. Acute Effects. Immediate results of short-term exposure to high levels of damaging agent leading to temporary or permanent incapacity or death.
   b. Delayed Effects
      (1) Effects on the exposed person. Manifestation of damage may not be evident for many years or even decades; these effects may include cancer, disorders of the respiratory tract, damage to the central nervous system, etc.
      (2) Effects on progeny of the exposed persons. These include genetic effects, which may range from the most trivial to the most serious, including permanent disability or premature death.

2. Annoyances. These include noises, odors, loss of amenities, and degradation of the quality of living.

Indirect Effects

1. Damage to the means of food production, e.g., farmland, domestic animals, crops and productive waters, etc.
2. Damage to property.
3. Damage to wildlife.[1]

It is expected that in North America by the year 2000, about half of the energy required for electricity production will come from nuclear fuels. This means that the total nuclear power level could rise by a factor of 30 to 50 by the end of the century. Since radiation sources will be widely distributed, the possible population exposure to radioactive emissions is of serious concern.

Radioactive emissions from nuclear power are of two types. First, minute quantitites of radioactive isotopes are emitted in gaseous and liquid streams from reactors in normal operation. These isotopes characteristically emit low-level (weak) radiation. Second, highly radioactive fuel emits radiation on removal from the reactor and during storage, transport, and reprocessing. The chemical and physical treatment of spent fuel (to recover unused fissionable material for fuel or to treat waste material before ultimate disposal) causes emissions of radioactivity in gaseous and liquid waste streams.

Radioactivity is generated within reactors in different ways. Nuclear fission produces radioactive fission fragments that mostly stay in place within the fuel. Small amounts may leak through the cladding material that surrounds the fuel and into the reactor coolant, which transfers the thermal energy to the power cycle. Nuclear fission also produces neutrons that can cause structural or coolant particles to become radioactive. Even though much of this induced radioactivity decays within the reactor, small amounts of liquid and gaseous radioactive matter are normally emitted. Fission neutrons also produce other radioactive materials from the uranium fuel; these so-called transuranium elements (plutonium and neptunium) are either stored within the spent fuel itself or are recovered during fuel reprocessing.

Many radioisotopes continue to radiate for long periods after they are emitted. Plutonium, with a half-life of 24,000 years, is an example of an isotope whose radiation could persist for thousands of generations.

Another serious feature of some radionuclides is that they concentrate in the food chain. The pathways by which radioactive sub-

[1]Reprinted with permission from *Proceedings of the International Symposium on Radioecology Applied to the Protection of Man and His Environment* (Commission of the European Communities, 1972).

stances travel from nuclear plants to human beings are complex. Gaseous radioactive releases can be inhaled directly by man or can reach him indirectly through food plants and animals. Liquid emissions can reach man directly, through water intake, or indirectly, through the intake of plants, animals, and fish. Concentration also occurs within the body, such as I-131 in the thyroid, Pu-239 in the lung, liver, or bone. Strontium 90 concentrates in the bone, as do the isotopes of radium. Continuing studies are being made of the most important of these pathways to identify and evaluate the natural processes of diffusion, sedimentation, absorption, and biological concentration of various radioactive isotopes.

Perhaps the most serious feature of all is the absence of any known safe "threshold," or level of intensity below which radiation can be said to be harmless. The International Commission on Radiological Protection (ICRP) has recommended that all exposure to radiation be considered potentially damaging to man (2, 3). This recommendation, which seems to be universally accepted, has had a profound effect on the way we view the radiation hazard. It means that no matter how weak the radiation source, it is considered to cause significant human damage if the total man-hours of exposure are sufficient. It does not necessarily imply that no man-made nuclear radiation should be permitted, but it does suggest that radiation exposure over which man has control should be permitted only if the benefits associated with the exposure outweigh the total costs. The ICRP has recommended that radiation exposure of the population from industrial sources be permitted only if it can be justified in terms of risk–benefit ratios.

From studies of persons exposed to internal radiation, as at Hiroshima or in certain occupations, the health risks of what is called "high dose rate" radiation are quite well known. The problem is the lack of data for human populations exposed over long periods to relatively weak or low-level radiation. The statistics of the effects are difficult, if not impossible, to discern. Given this uncertainty about the effects of low-level radiation, the experts have agreed that it is prudent to assume that all radiation is damaging and that the total health damage (for example, the number of cancers) occurring in a population is directly proportional to the total radiation received, independent of the rate of radiation dose or the size of the population exposed. It is widely believed that this "no-threshold" assumption relating health damage to radiation dose is conservative (in that it estimates the maximum possible health damage). The assumed proportionality between health damage and total radiation dose can be expressed by a "risk coefficient"; such coefficients are available for various kinds of health and genetic effects of nuclear radiation.

The risk coefficients can be used to estimate the upper limits of health and genetic damage from nuclear power, provided that power plant emissions are known, the important pathways to man of each kind of radioactive particle are known, and the time integral of all future human exposure to radiation from a given set of radioactive emissions can be calculated.

The results of studies of these matters over the past 25 years show that risks from exposure to the radiation caused by power plants in normal operation are much smaller than risks associated with natural causes (cosmic and terrestrial radiation and emissions from radioactive sources inside the human body). Moreover, risks due to nuclear power are also less than those associated with natural disasters such as earthquakes. While these findings would satisfy many people that nuclear risks are acceptable, they are insufficient to prove that the risks are justified by the benefits of nuclear power.

Unfortunately, the benefits of nuclear or any other kind of power are so difficult to quantify that a widely agreed-upon calculation method has yet to be developed. We must therefore rely on the judgment of experts as to the levels of risk due to artificial nuclear exposure to which individuals and populations should be exposed. The recommendations of the ICRP on maximum allowable human exposure to radiation have been generally accepted, and operating experience has now shown that reactors can operate economically with radiation exposure as low as one-hundredth of what the recommendations imply. Though this provides some indication that nuclear power may be generated with an acceptable level of safety, it does not treat questions about major plant accidents and waste disposal. Nor does it remove the notion that all power generation can be expected to have a finite toll in human life.

5.2 HEALTH EFFECTS OF RADIATION

5.2.1 Definition of Dose

The radiation from radioactive substances damages biological tissue by ionization. The damage depends on the energy of the radiation received per unit mass of material. This quantity is generally measured in rads. (One rad equals 10^{-5} joules of energy absorbed per gram of material.)

The physical damage caused by radiation depends not only on the number of rads, but also on the type of radiation (alpha, beta, or gamma) and on the location of the radiation source (inside or outside the body). Since it is desirable to have a measure of radiation intensity that directly indicates biological damage from any type or source of radiation, another unit, the rem, has been defined by

$$N\,(\text{rems}) = N\,(\text{rads}) \times QF \times DF,$$

where N indicates number, QF is a quality factor, and DF is a distribution factor. The quality factor varies between 1 and 20, depending on the density of ionization along the path of radiation through the tissue. For gamma and beta rays *QF* is 1; for alpha particles it is 10. The distribution factor depends on the location of the radioactive source; for example, a value of 5 may be used for the *DF* for radioactive particles located within the human body.

Both rads and rems are commonly used as units of radiation *dose* to biological tissue. While the rem unit may seem preferable (in that it is related to equivalent tissue damage from all types and sources of radiation), the values of *QF* and *DF* are not easy to assign. For this reason, rad units are often used, as by the United Nations Scientific Committee on the Effects of Atomic Radiation (UNSCEAR) (4).

For reference, the intensity of radiation dose (from cosmic and terrestrial sources as well as from sources internal to the body) is usually quoted in the range 0.10 to 0.15 rems/person-yr, or 0.25 to 0.30 rads/person-yr.

The Curie unit of radiation intensity is quite distinct from rads and rems, and pertains only to the radioactive source. It is a measure of source strength or activity of the radionuclide, not of dose rate. One Curie (Ci) is defined as 3.7×10^{10} disintegrations per second, the rate of decay of radium nuclei in one gram of radium (one nanoCurie is 10^{-9} Ci; one picoCurie (pCi) is 10^{-12} Ci).

5.2.2 Somatic and Genetic Effects

Radiation damage to man can be classified as either somatic or genetic. Somatic effects include damage to particular parts of the body but not to genetic makeup. These may be immediate (acute) or delayed (chronic).

At high dose rates there are many effects of radiation on the body. No effects are observed immediately for doses up to 25 rems. Doses in the range 25 to 100 rems produce slight and reversible blood changes, but no other observable effects. A dose of 500 rems causes a 50 percent probability of death within days.

Delayed effects of high-dose-rate radiation may not be observed for years. One such effect is leukemia. A single dose of 200 rems is said to triple a person's chances of developing leukemia within 5 to 25 years. Lung cancer is another effect that can be long delayed.

At low radiation dose rates, the effects are necessarily delayed. The greatest concerns are cancer induction and genetic effects. Genetic effects include damage to reproductive cells that might result in harm to future generations. Radiation of the genes and chromosomes of the reproductive cells can affect the offspring in many ways, ranging from mildly debilitating to seriously disabling or lethal effects. The frequencies of radiation-induced changes (mutations) in the genes and chromosomes are fairly well known.

The problem that geneticists face is in deciding the extent to which these mutations impose a social burden. For example, how great an increase in radiation dose would be required to double the incidence of mongolism in a population? The essential problem is the absence of data on hereditary disease induced by low-dose-rate radiation in man. The available risk estimates have been obtained by studying data on other organisms exposed to relatively high dose rates. Applying these data to predict genetic effects in man of continued low-level radiation requires the use of important assumptions which cannot be tested.

5.3 MEASURING THE RISK OF HEALTH DAMAGE

Biological risk coefficients are derived from studies of the observed incidence of leukemia and malignancies among the survivors of the bombing of Hiroshima and Nagasaki, and among patients receiving radiation therapy for nonmalignant diseases. Risk estimates have been made by the ICRP (1, 2), and more recently by UNSCEAR (4) and by the Committee on Biological Effects of Ionizing Radiation (BEIR) of the United States National Academy of Sciences (5).

Table 5.1 summarizes the main findings of UNSCEAR on radiation-induced cancer in man. A large fraction of these cancers can be assumed to be fatal. A risk coefficient of 100 million man-rads means a probable incidence of 100 cancers in a population of 1 million if each person receives a dose of 1 rad; the same number of cancers would be expected in a population of 100 million if the dose per person were only 0.01 rad. Thus the risk of death by radiation-induced cancer is assumed to be of the order of 10^{-4} per rad of osed. This refers to the delayed (chronic) effects of low-level radiation, as estimated with the assumptions mentioned earlier. Applying this to the level of natural radiation dose (0.3 rad/person-yr) yields a

Table 5.1
A Summary of Risk Estimates in Man for Somatic Effects of Radiation Exposure

Type of Malignancy	Number of Cases (per rad per million people exposed)
Leukemia	15–40
Lung Cancer	10–40
Breast Cancer	6–25
Thyroid Cancer	40
Other Cancers	40
Total	100–200

Source: Data from Ref. (4).

risk of 30 to 60 cancers per million population per year. The normal incidence of cancer from all possible causes is about 1600 per million population.

Genetic risk estimates are, as mentioned earlier, quite uncertain. A calculation (6) based on the UNSCEAR report (4) and experiments with mice indicates that the ordinary rate of occurrence of serious genetic disorders is about 3 percent, or 30,000 per million live births, and that 1 rad of ionizing radiation may increase this by 1 percent, or 300 per million live births. Of these 300, perhaps 20 will show up in the first generation after the radiation dose has been received. The "doubling dose" is defined as the increase in dose necessary to double the incidence of naturally occurring genetic disorders. A 1 percent increase per rad would correspond to a doubling dose of 70 rads; that is, $1.01^{70} = 2$.

The BEIR Committee has provided the estimates of cancer risk indicated in Table 5.2 for total radiation dose. The risk of cancer mortality induced in a period 25 to 27 years after irradiation is estimated to be about 100 deaths per million man-rems. The BEIR Committee concludes that cancer is the only somatic risk that needs to be taken into account in setting up radiation standards for the general population. Moreover, the somatic risk level is by no means unimportant relative to genetic risks, assuming that dose and effect are indeed linearly related.

Taking into account the differences between rems and rads, it may be said that UNSCEAR and BEIR risk estimates are roughly in accord, though both are only approximate due to gaps in scientific knowledge (most scientists believe they overestimate rather than

Table 5.2
BEIR Committee Biological Risk Coefficients

	Risk Coefficient[a] (per 10^6 man-rems)
Somatic Effects	
Excess Cancer Mortality	50–165
Genetic Effects (First Generation)	
Dominant Diseases[b]	10–100
Congenital Abnormalities[c]	1–100
Genetic Effects (Equilibrium)	
Dominant Diseases[b]	50–500
Congenital Abnormalities[c]	10–1000

[a]Assuming a doubling dose between 20 and 200 rems.
[b]Obvious abnormalities or deformities.
[c]Includes an increase in hereditary diseases.
Source: Ref. (5).

underrate the actual risk). Their significance in comparison with other risks will be discussed below.

5.4 RADIATION STANDARDS

Radiation standards are recommended maximum doses of radiation. These standards are set by experts in light of what is known about risks, the need for nuclear technology, and the availability of methods to control radiation dose in specific applications. The ICRP, which is composed of independent scientific experts from many countries, has assumed responsibility for recommending minimum safety standards for radiological protection, and its recommendations have been almost universally accepted. According to Butler and Barry, "it has formulated standards with a finite but acceptable measure of risk to the people involved, which has made possible the realization of many potential benefits to man of nuclear science and ionizing radiations" (1). The standards deal with maximum levels of exposure. Arguments that the standards are insufficiently restrictive typically involve calculations of risk which assume that everyone in the population is exposed to the maximum recommended dose for individuals.

The ICRP has recommended on the basis of genetic considerations that the total dose to each generation of the population be limited to 5 rems per person (from all sources other than nature and medical diagnosis). Taking 30 years as the time span of a generation, this is equivalent to 0.170 rem per person per year. However, as the BEIR Committee points out,

> Until recently it has been taken for granted that genetic risks from exposure of population to ionizing radiation near background levels were of much greater import than were somatic risks. However this assumption can no longer be made if linear non-threshold relationships are accepted as a basis for estimating cancer risks.(5)

Using the risk coefficient of 200 cancer deaths per million population per rem, and a uniform population exposure of 0.170 rem/person-yr, the annual cancer-induced death toll could be as high as 30 per million population. In the United States this would mean about 6000 cancer deaths annually. This would be equivalent to an increase of 2 percent in the spontaneous cancer death rate, and an increase of 0.3 percent in the rate of death from all causes.

In the United States and Canada the maximum exposure of a person just outside the site boundary of a nuclear plant has been limited by regulation to 0.5 rem/yr. Since the individual dose should decay substantially within 10 miles of the boundary, the observance of this limit would in itself reduce average population exposure from nuclear power to much less than 0.17 rem/person-yr. In practice, however, it appears quite feasible to limit maximum individual dose at the site boundary to 0.005 rem/yr. The dose limits for workers in the nuclear industry have been set much higher, at 5 rems/yr in the United States and Canada.

The BEIR Committee concludes that the present ICRP recommendation of 0.17 rem/person-yr is unnecessarily high, but has not offered a different recommendation, recognizing the complex technological, economic, and sociological factors that need to be considered. It recommends in principle that

—No exposure to ionizing radiation should be permitted without the expectation of a commensurate benefit.
—The public must be protected from radiation but not to the extent that the degree of protection provided results in the substitution of a worse hazard for the radiation avoided. Additionally there should not be attempted the reduction of small risks even further at the costs of large sums of money that spent otherwise would clearly produce greater benefit.
—There should be an upper limit of man-made non-medical exposure for individuals in the general population such that the risk of serious injury from somatic effects in such individuals is very small relative to risks that are normally accepted. Exceptions to this limit in specific cases should be allowable only if it can be demonstrated that meeting it would cause individuals to be exposed to other risks greater than those from the radiation avoided.
—There should be an upper limit of man-made non-medical exposure for the general population. The average exposure permitted for the population should be considerably lower than the upper limit permitted for individuals.
—Guidance for the nuclear power industry should be established on the basis of cost-benefit analysis taking into account the total biological and environmental risks of the various options available and the cost-effectiveness of reducing these risks. The quantifying of the "as low as practicable" concept and consideration of the net effect on the welfare of society should be encouraged.(5)[2]

5.5 ENVIRONMENTAL DOSE COMMITMENT

Table 5.3 shows approximate average exposure of the United States population in 1970 to radiation from natural and medical sources as well as from fallout and nuclear reactors. In 1970, the average radiation dose per capita due to environmental contamination from weapons testing (fallout) was 4 mrems, small compared with natural radiation. The latter averages 102 mrems, but varies quite widely with latitude and altitude. Contributors to natural radiation include cosmic rays (44 mrems) and terrestrial sources. Of the terrestrial sources, as much as 18 mrems per year can come from within man himself, from K-40, C-14, and other naturally occurring isotopes.

In 1970, diagnostic X-rays accounted for most of man-made

[2]Reproduced with permission of the National Academy of Sciences.

Table 5.3
Sources of Genetically Significant Radiation: Estimated Average Amounts in the United States, 1970

Source	Whole Body Exposure (mrem)	Genetically Significant Exposure (mrem)
Natural Radiation		
Cosmic Radiation	44	
Radionuclides in the Body	18	
External Gamma Radiation	40	
Total	102	90
Man-Made Radiation		
Medical and Dental	73	30–60
Fallout	4	
Occupational Exposure	0.8	
Nuclear Power (1970)	0.003	
Nuclear Power (2000)	1	
Radiation Protection Guide for Man-Made Radiation (Medical Excluded) to the General Population	170	

Source: Ref. (5). Reproduced with permission of the National Academy of Sciences.

radiation. It is hoped that in coming years this effect may be reduced by a factor of 10 or so by restricting use of X-ray equipment in public health surveys and by optimal operation techniques. A single chest X-ray could account for a dose in the 20 to 500-mrem range.

Estimated population exposure to radiation released by nuclear power was small in 1970 and is expected to remain below 1 mrem up to the year 2000. This would indicate quite clearly that reactors in normal operation pose a risk to the general population that is small even when compared with the risk due to naturally occurring radiation. But the fact that the risk is relatively small does not justify it. Even 1 mrem average dose per year for the population of the United States could lead to 300 cancer deaths annually.

Moreover, the nuclear era is just starting; much needs to be learned about the way radioactivity will accumulate in the environment in decades and centuries to come. Even in the next few decades, nuclear power levels could grow rapidly, from 6000 MWe in the United States in 1970 to possibly 800,000 MWe by 2000, at which time the world level of nuclear power could be as high as 4 million MWe. With the prospect of a large population of reactors and continuous buildup in the environment of many long-lived radionuclides, it is highly important that the long-range consequences of nuclear power be evaluated clearly now, and that they affect present control and monitoring methods and calculations of cost and benefit. It is to bring these future consequences into present consideration that the concept of environmental dose commitment has been developed.

The definition of "dose commitment" used in the UNSCEAR report is the infinite time integral of the average dose rate to individuals resulting from the quantity of material released during a given finite time interval. That is, the dose commitment of the radioactive material emitted in a given year is the sum of the annual doses caused by that same material in all future years. For power generation, dose commitment can conveniently be expressed in units of man-rads (or man-rems) per MWe-yr.

Calculation of the dose commitment requires knowledge of radioactive emissions and of the pathways and kinetics of migration of radioactivity to man through the entire period of radioactive decay.

When the number of reactors in a country is small and the accumulation in the neighborhood of a given reactor of radioactivity emitted in previous years from that reactor is negligible, it may be satisfactory to limit the annual emissions to some prescribed level. But when reactors are sufficiently close that their emissions may overlap, and when the buildup of radioactivity in the environment due to previous emissions of long-lived nuclides is sufficiently large, this practice will no longer be satisfactory. It would seem wise to set controls with a clear idea of an ultimate steady-state condition when the total annual population dose due to all emissions, new and old, is steady at an acceptable level. This steady state will be reached when the addition to dose in a given year (due to the total radioactivity emitted by reactors during that year) is just equal to the annual decrease in dose due to the decay, migration, and deposition of all the radioactivity present in the environment before that year. In the steady state, this total annual decrease in dose would necessarily be equal to the annual dose commitment.

P. J. Barry has provided an example of this principle (6). Suppose that the average annual dose in the future population is to be limited to 5 mrems and that ultimately the per capita demand for electricity rises to about five times what it is today, or about 5 kWe (yearly average). This would imply that the future population dose must be limited to 5 mrems/yr per 5 kWe/person, or 1 man-rem/MWe-yr. If, from now on, the annual dose commitment is equal to 1 man-rem/MWe-yr, the annual dose will necessarily be less than this value during the transient period, and only equal to it when a steady state has been reached. This means that plants whose emissions are acceptable now will not have to be modified for substantial reductions in emission level later.

The most important gaseous emissions of radionuclides from

water-cooled nuclear reactors include isotopes of the noble gases, tritium in the form of tritiated water vapor, and radioactive iodine compounds. Tritium is formed by interaction of neutrons with boron (used in reactors as a neutron absorber to control reactivity) or by interaction with deuterium in the water coolant. Tritium has a half-life of about 12 years, and could build up in the atmosphere, given a very large population of reactors. So could Kr-85 (with a similar half-life), one of the most important of the radioactive noble gas emissions. Although tritium and Kr-85 could accumulate in the environment, they do not concentrate in biological systems. However, if it is found that the presently minute and uncontrolled emissions are too large, they can be reduced. For example, gaseous emissions of noble gases could be virtually eliminated by the use of cryogenic entrapment.

Typical liquid wastes from water-cooled reactors contain tritium (in the form of tritiated water, HTO, which cannot be separated chemically from H_2O), fission products Co-134 and Co-137, and the radioactive products of neutron activation of iron, cobalt, nickel, and zinc.

The liquid-metal-cooled fast breeder reactor is now under development. It is cooled by liquid sodium and is used to generate plutonium fuel. Its emissions may include radioactive isotopes of plutonium and sodium.

Emissions of reactors are treated and continuously monitored to check that concentrations do not exceed permissible limits. Treatment includes processes for gaseous emissions (delay and decay, filtration, and low temperature adsorption on charcoal) and for liquid wastes (delay and decay, filtration, evaporation, and demineralization).

"Delay and decay" refers to the storage of wastes long enough for natural decay to reduce radioactivity significantly. The effectiveness of this technique depends very much on the isotopes present. Filtration implies the collection of radioactive solid particles in gaseous or liquid flows. Iodine can be removed from air by using charcoal filter beds. Another use of charcoal adsorption is in providing sites to delay radioactive noble gases, giving time for significant decay of radioactivity. Various filters are used in liquid wastes to remove suspended solids, some of which may be introduced by the water treatment processes. Evaporation serves to concentrate the wastes, and may reduce the activity of the flow (once vaporized) by two or three orders of magnitude. However, this depends on the concentrations of those materials, such as iodine and tritium, which can be carried over into the vapor. Demineralizers can also reduce the activity of liquid streams if the concentration of suspended solids is not too high. In general, the processes of gaseous and liquid emissions treatment have to be designed for the particular emissions from the reactor they are intended to serve. Solid wastes resulting from these treatment processes are concentrated, solidified with a bonding mixture, stored, and buried at a suitable site.

Detailed estimation of population dose rate from normal reactor operation is exceedingly complex. In 1973 the United States Atomic Energy Commission (USAEC) published the results of one of the first comprehensive computer models of the migration of radioactive materials in a large area, the Upper Mississippi region (7). The study assumes that in the year 2000 the region will have nuclear-electric capacity of 356,000 MWe (with associated fuel processing plants), and a population of 29.1 million. The study also assumes a specific distribution of nuclear plants of various types.

Radionuclide releases from these plants were estimated, and their transport throughout the region was simulated by an elaborate computer model. Some 45 radionuclides were considered in the study, but it was estimated that 95 percent of the radiation received by the body would be due to the isotopes of tritium, iodine, and cesium. For the specific organs considered (liver, lungs, gastrointestinal tract, thyroid, bone, and skin), over 80 percent of the total radiation burden would be due to the same three isotopes. Tritium, being the only gaseous isotope for which extensive waste removal was not assumed, was a major contributor. Most of the tritium produced in the reactor fuel was assumed to remain in the fuel until its release during reprocessing.

The results of the study indicated that the average radiation dose per person in the Upper Mississippi region in the year 2000 would be only 0.2 mrem. A small fraction of the population would receive up to 1.2 mrems/yr; a person living 1 mile from a fuel reprocessing plant could receive 7 mrems/yr.

Such studies are in early stages of development; much further knowledge regarding radionuclide transport in specific regions may be needed. Every area with a significant population of nuclear plants will need close scrutiny.

Table 5.4 shows a number of calculated dose commitments presented in the UNSCEAR report for various reactors and fuel reprocessing plants. It must be stressed that these are preliminary results, since transport coefficients are not well known. One of the experimental difficulties in determining these coefficients is that the environmental concentrations of certain isotopes due to nuclear power operation are much smaller than the concentrations due to weapons testing fallout.

An example of a calculation of possible future effects of nuclear power on human health has been provided by the EPA (8). The emphasis is on the sum of all doses to all individuals affected by emission of particular nuclides until they have decayed to harmless levels. Some radionuclides have half-lives of thousands of years,

Table 5.4
Dose Commitments and Population Doses from Nuclear Power Plant Operations

	Dose Commitment (man-rads/MWe-yr)
Global Doses	
Fuel Reprocessing Plants	
Kr-85 (gonads)	0.11
H-3 (whole body)	0.15
Local Doses	
Five Reactors, United States (built 1959–1963)	0.88
Six Reactors, United States (built 1967–1969)	0.021
Fuel Reprocessing Plants (Kr-85, 2×10^7 people)	0.0054
Occupational Doses	
Seven United Kingdom Reactors	0.2–1.4
Five United States Reactors	1.2–4.0
Three Italian Reactors	0.6–1.2
Five French Reactors	0.4–0.6
Three Canadian Reactors	0.6–8.9

Source: Refs. (4) (6).

and one might try to calculate the radiation health effects for the entire period of radioactivity. The EPA is content with a more modest objective: to estimate cumulative health effects, over a 100-year period following 1970, due to plant emissions in the period 1970–2020.

Today, any such analysis can only be approximate; as the nuclear age develops, however, there will no doubt be an increasing need for such detailed knowledge.

The radionuclides considered by the EPA were:

Tritium (half-life 12.3 years)
Krypton-85 (half-life 10.7 years)
Iodine-129 (half-life 1.7×10^7 years)
Actinides (including Pu-238, -239, -240, and -241, Am-241, and Cm-242, with half-lives up to 2.4×10^4 years)

Other long-lived nuclides, including Sr-90 and Cs-137, were not included in the study though they could have comparable health effects.

The amounts of radionuclides generated were estimated from projections of nuclear power developed in the United States from 1970 to 2020.

The fractions of these radionuclides released to the general environment were assumed as indicated in Table 5.5. The fraction of the actinides released, for example, is not expected to be greater than 10^{-7} of the total quantity handled in any one year. However, for public health planning a less favorable assumption is used; the

Table 5.5
Assumptions Used in the EPA Dose Commitment Analysis

		Values Used in Study	
Parameter	Range of Possible Values	Expected Minimum Performance	Public Health Planning
Actinide Release Fraction	10^{-6}–10^{-9}	10^{-7}	10^{-6}
Actinide Resuspension (m^{-1})	10^{-5}–10^{-11}	10^{-8}	10^{-6}
I-129 Release Fraction	10^{-1}–10^{-4}	10^{-3}	10^{-1}
Kr-85 Release Fraction	10^{0}–10^{-3}	10^{0}	10^{0}
Tritium Release Fraction	10^{0}–10^{-2}	10^{0}	10^{0}

Source: Ref. (8).

release fraction is assumed to be 10 times as great. "Resuspension" in Table 5.5 is the ratio of material concentration per unit volume in air to the surface concentration per unit area on the ground surface. This depends on many factors, including wind velocity, particle size, surface conditions, etc., so the uncertainty of six orders of magnitude in this parameter in Table 5.5 is perhaps not surprising. Resuspensions in the range 10^{-8} to 10^{-3}/m have been reported for newly deposited plutonium (8).

Following release, the radionuclides were assumed first to lead to short-term exposures of individuals within 50 miles and ultimately to be distributed over the world (K-85 and tritium) or over large portions of the United States (I-129 and the actinides).

The health effects were estimated with the risk coefficients of Ref. (5) and estimated populations for the United States and the world. It was supposed that the only significant pathway to man was due to particles initially suspended in air, deposited on soil, resuspended, and inhaled.

Figure 5.1 shows the results of calculations of the worst conceivable health effects. Two-thirds of these effects are assumed fatal for tritium and Kr-85, all are fatal for the actinides, and one-quarter are assumed fatal for I-129. The effects are cumulative in a particular sense. The topmost curve, for example, shows (at the year 2020) 23,800 health effects (deaths). This includes the effects of emissions of actinides up to 2020: first, in the period 1970–2020 (21,000 deaths), and second, in the period 2020–2070 (2800 deaths). It thus attempts to state the maximum irreversible commitment to health damage by actinide emissions in the next 50 years.

The analysis is riddled with uncertainty, yet it serves to indicate the conceivable seriousness of the effects of nuclear power on future generations. It underlines the need for detailed knowledge of radionuclide transport and deposition. It also warns that, even though deaths due to normal incidence of cancer in the United States alone

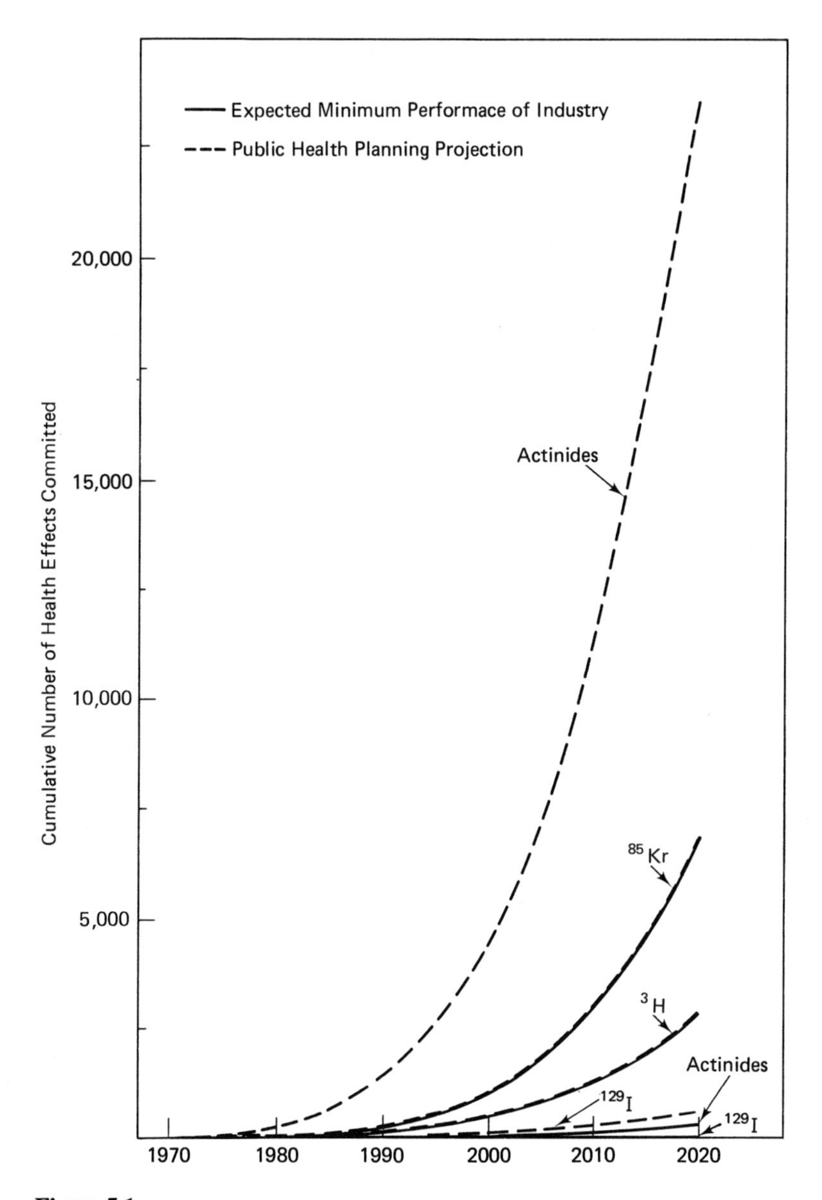

Figure 5.1
Estimated cumulative potential health effects of projected releases from the U.S. nuclear power industry (8).

could be 1000 times the largest number shown in Figure 5.1, extreme care should be taken to minimize release of radionuclides.

5.6 NUCLEAR PLANT ACCIDENTS

In general, we can assume that accidents have some probability of occurring; therefore either plants must be built far away from people or sufficient containment must be used so that if the accident occurs, the human population will be protected from radioactive contamination. The first of these alternatives is not generally feasible, so the designer relies on three barriers to separate radioactive matter from human exposure: the fuel cladding, the coolant boundary, and the reactor containment vessel. Are these adequate to reduce the risk of a major accident to an acceptable level? This question has been the subject of serious study and debate. The problem is aggravated by the possibility that reactors could be the targets of deliberate sabotage.

The hazards of nuclear power reactor accidents have been discussed in two different ways: analysis of maximum credible accidents, and analysis of probability of major accidents.

The first method supposes that the accident consists of a series of failures, each of which is individually credible, and that all take place together and produce the maximum possible radiation dose to the population. Safety requires that the dose from the accident be less than the maximum dose considered tolerable to individuals once in a lifetime. The probability of all of the events taking place together is not explicitly considered. The main difficulty with this method is the determination of whether possible events are "credible." If an accident occurs, is it credible that all of the various kinds of radioactive material will escape? Are seismic events credible?

The second method includes a more detailed judgment of credibility. The probability of the release of each radioactive constituent is estimated separately as a function of the amount released. The probability of dilution at sites of human exposure is then estimated. Finally, the probabilities of biological damage are estimated as a function of dose in order to arrive at an overall probability or risk of damage to the population. If the overall risk rate is viewed as acceptable, then the required containment effectiveness (fraction retained of radioactive material released during an accident) can be estimated. The great difficulty of this method is simply that not all the probabilities are known. Short of statistical experience, the probabilities of failure of various reactor arrangements and controls are only estimates.

The USAEC's most recent study of reactor safety, known as the Rasmussen Report (9) asserts considerable confidence in current abilities to analyze the probability of reactor safety. It advances the thesis that every possible sequence of events that might lead to failure can be analyzed. Thousands of possible accident sequences were analyzed in the study. Even though the probabilities of some faults are very uncertain, it could be shown in many of these cases that the probabilities were unimportant to overall risk. The arguments for confidence in the results of the study included: use of 20 years' experience in identifying and analyzing possible reactor accidents; comprehensive inclusion of all conceivable faults and events that might lead to failure (including natural disasters, but not sabotage); allowance for every source of radioactivity in light-water reactors, and recognition that large radioactivity release to the public could only occur with melting of the reactor fuel. The report concludes,

While there is no way of proving that all possible accident sequences which contribute to public risk have been covered in the study, the systematic approach used in identifying possible accident sequences makes it very unlikely that an accident which would contribute to the overall risk was overlooked. (9)

Calculated probabilities of reactor core accidents for a pressurized light-water reactor range from 7×10^{-7}/yr to 1×10^{-4}/yr (9). The first figure refers to the probability of a steam explosion (following core melt and dispersal of UO_2 in H_2O) that ruptures the pressure vessel and the containment. This type of accident releases the maximum quantity of radioactivity to the environment of all the accidents considered. The second, or upper limit probability (1×10^{-4}), refers to a core accident in which the containment vessel remains intact and only small fractions (4×10^{-4}) of the inventory of isotopes in the core escape. Radioactivity is mostly deposited within the reactor building, or removed by the reactor building filter system. Many other possible accident sequences between these two extremes were analyzed, and fractional releases of the core inventory of the important isotopes were estimated as well as the probability of each type of accident.

The population dose was estimated for each type of accident, and health effects were calculated using BEIR risk coefficients. The resulting probabilities of fatal accidents of various sizes are shown in Figures 5.2 and 5.3 in relation to other risks, which will be discussed later. Considering 100 reactors, the probable annual consequences were cited as:

Acute Fatalities:	4×10^{-2}
Acute Illnesses:	8×10^{-2}
Latent Cancers:	3×10^{-1}
Thyroid Injuries:	6
Genetic Damage:	3×10^{-1}
Property Damage:	$1.6 million.

Critical reviews of the Rasmussen Report have pointed to possible underestimates of fatalities from certain kinds of reactor accidents. A committee of the American Physical Society has estimated that the total fatality rate (including acute, or immediate, fatalities, and also fatalities occurring in later years but directly related to the accident) is a factor of 25 higher than that for acute fatalities alone (10). Displacing the "100 nuclear power plants" curves of Figures 5.2 and 5.3 to the right by this factor makes a marked difference in the comparison between nuclear accident risks and other risks. The total fatality rate, per reactor-yr, will be considered later in relation to other risks from nuclear power and from fossil fuel power.

Although the possible dangers are numerous, the safety record of nuclear reactors to date provides a degree of reassurance. World experience has now passed 8000 reactor-yr. Accidents have occurred, yet (excluding the Windscale accident in 1956, in an air-cooled natural uranium reactor that is not typical of more recent reactors) none of these has caused widespread environmental deterioration. It must be remembered, however, that the probability of a major accident is a matter of estimate than rather of statistical record.

5.7 NUCLEAR FUEL WASTE DISPOSAL

The safe disposal of nuclear wastes is an extremely important problem created by large-scale use of nuclear energy. Radioactive wastes with very long half-lives require processing to chemically stable form and storage in some isolated environment, protected from natural phenomena for thousands of years. F. K. Culler et al. (11) believe this is "probably the most important and limiting problem," but that conclusion is not generally agreed upon.

The spent fuel removed from a reactor is highly radioactive and may be processed to recover unused uranium, or plutonium produced by nonfission capture of neutrons by U-238. There are serious hazards in such processing because of the gaseous, liquid, and solid particle emissions that result from decladding, shearing, and dissolving the fuel in acid, followed by subsequent chemical processing. Minute quantities of gases and particles are emitted during purification and concentration of the recovered plutonium and uranium products. Radioactive emissions from fuel processing are controlled in general by dilution and dispersion of low-level gaseous and liquid wastes; delay and decay of radionuclides with short half-lives; concentration and containment of intermediate and high-level radioactive wastes with long half-lives (such as 24,000 years for plutonium). The devising of secure containment for these solid wastes over thousands of years is a serious problem.

Two general ideas on this problem have been much discussed. The first is that the wastes should be buried deep in the earth, in a very stable geological stratum, and sealed safely forever. The problem here is providing assurance that over thousands of years the wastes will not be disturbed, with subsequent dangerous leakage of radioactivity. Various possibilities have been considered, such as burial in abandoned salt mines or in rock. The USAEC seriously studied, but has since rejected, a salt mine site in Lyons, Kansas. The cost of storage and monitoring would have been reasonable—only around about 0.5 percent of the cost of the electricity generated from the fuel—but the site was found to be insecure.

Alternatively, it has been suggested by W. B. Lewis (12) that nations should make a continuing commitment to the monitoring and management of nuclear wastes, which could be stored in gigantic vaults on the surface. At various times the wastes could be treated by chemical separation to remove the inactive constituents and reduce their volume. An advantage of this method is that if future circumstances change or if new ideas for permanent deposition of the wastes were to develop, there would be freedom to change the procedure. The question here is whether we have the right to commit future generations to a continuing hazardous responsibility that could last for thousands of years even after the end of the age of nuclear power.

Other ideas being studied include the possibility of deposition in a deep cavity created by a nuclear explosion (13). The wastes would eventually melt and solidify with the surrounding rock. It is not clear that this method would guarantee satisfactory containment—faulting, earthquakes, or movement of groundwater might transport the radioactive material.

A. S. Kubo and D. J. Rose (13) point out that the problem of storing the high-level wastes can be greatly reduced in time scale if special measures are taken to recover plutonium and other actinides from spent fuel. At present, it appears only economical to recover about 99.5 percent of the uranium and plutonium from fuel for breeder reactors. The technological limit is perhaps 99.9999 percent for uranium, neptunium, and plutonium, and 99 percent for americium and curium. This would reduce the radioactivity of the waste by a factor of 100 and might imply an additional electricity production cost of only 2 or 3 percent. By recycling the extracted actinides through reactors, guaranteed safe storage would be required for only about 700 years, rather than for 1 million years (13). For the present, the storage of nuclear wastes in large surface vaults appears to be the most practical procedure, though time and effort will be needed to guarantee safety of all operations.

Nuclear waste disposal is undoubtedly a major problem, but seems unlikely to be one that will limit the amount of nuclear power developed. Since costs of disposal schemes now considered are approximately 1 percent of the value of the electrical energy generated, it is reasonable to contemplate more expensive methods, if needed.

The transportation of nuclear wastes has its own hazards. According to Alvin Weinberg (14), there may be 1 million MWe of nuclear power in the United States by the year 2000, of which two-thirds may be fast breeder reactors. These would require 7000 to 12,000 annual shipments of spent fuel from reactors to chemical plants. Each shipment, delayed for 30 days' cooling, would contain 1.5 tons of fuel, would generate 300 kW of heat, and would contain 75×10^6 Ci of radioactivity. With a railroad accident rate of one serious accident per million miles, and an average shipment of 100 miles, the rate of accidents during shipment of high-level wastes could be as high as one per year. With extremely strong fuel containers, perhaps only 1 in 1000 of these accidents would be serious enough to puncture the fuel containers. Nevertheless, improvement in the safety of transportation may be necessary for adequate protection of the population. Weinberg favors the creation of very large areas for both power production and fuel processing to minimize this transportation risk.

Weinberg also reports that in the United States, over a 19-year period, hundreds of thousands of radioactive shipments were made in special containers, but only 119 "incidents" occurred. No radioactivity was released from the container in 84 of these incidents; of the remaining 35, only one resulted in airborne dispersion and costly cleanup. In no case was there any serious resultant exposure to radiation. To date, the record for safe transportation of radioactive materials has apparently been satisfactory. Concern for the future centers on the rapidly increasing numbers of shipments and on more dangerous contents such as plutonium wastes.

5.8 NUCLEAR POWER RISKS COMPARED TO OTHER RISKS

Table 5.6 lists the major causes of accidental fatalities in the United States and their respective tolls. At the top of the list is automobile accidents, at 3×10^{-4}, or three fatalities each year for each 10,000 people in the country. This is greater than the sum of all other risks in the table. At the bottom of the table is the probability of death from pressurized light-water nuclear reactor accidents estimated in the Rasmussen Report (9). The estimate was based on the risk to the 15 million people living within 20 miles of nuclear plants. Had it been taken over the entire population of the United States, the risk level would have been 2×10^{-10}.

The probability of major accidents decreases with the number of fatalities, as indicated in Figure 5.2, which shows the comparable

Table 5.6
Individual Risk of Acute Fatality by Various Causes (Based on U.S. Population Average, 1969)

	Total Number for 1969	Individual Risk of Acute Fatality (probability/yr)[a]
Motor Vehicle Accidents	55,791	3×10^{-4}
Falls	17,827	9×10^{-5}
Fires and Hot Substances	7,451	4×10^{-5}
Drowning	6,181	3×10^{-5}
Poison	4,516	2×10^{-5}
Firearms	2,309	1×10^{-5}
Machinery (1968)	2,054	1×10^{-5}
Water Transport	1,743	9×10^{-6}
Air Travel	1,778	9×10^{-6}
Falling Objects	1,271	6×10^{-6}
Electrocution	1,148	6×10^{-6}
Railway Accidents	884	4×10^{-6}
Lightning	160	5×10^{-7}
Tornadoes	91[b]	4×10^{-7}
Hurricanes	93[c]	4×10^{-7}
All Others	8,695	4×10^{-5}
All Accidents		6×10^{-4}
Nuclear Accidents (100 reactors)	0	3×10^{-9}

[a]Based on total U.S. population, except as noted.
[b]1953–1971 average.
[c]1901–1972 average.
Source: Ref. (9).

frequency distribution for failures from accidents in light-water plants, again from the Rasmussen Report. Fatalities due to car crashes are not shown on Figure 5.2 because data for the frequency distributions are not available.

Figure 5.2 indicates that death due to light-water nuclear reactor accidents is 100 to 1000 times less likely than death due to man-caused events or disasters. Plant accidents are not, of course, the only potential cause of death associated with nuclear power. Hazards due to uranium mining and fuel processing are also significant for people in those occupations (15). But the results from what appears to be the most serious analytical study to date tell us that nuclear reactors (designed, maintained, and operated with high standards of workmanship) present a threat to human life that is small compared with many of the risks we ordinarily face.

Figure 5.3 compares the frequency distribution of fatalities due to natural disasters with the projected frequency of fatalities due to

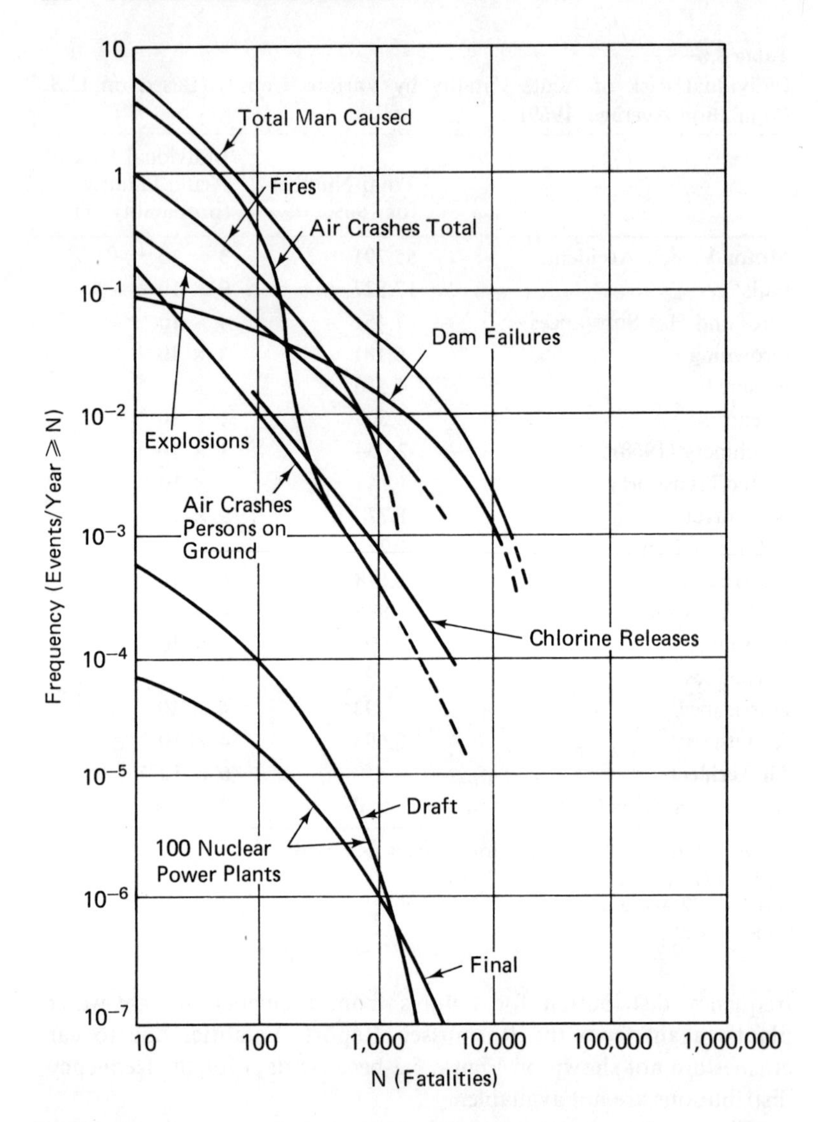

Figure 5.2
Frequency of man-caused events with fatalities greater than N (9). Only acute (immediate) nuclear fatalities are shown, and data for auto accidents (which cause about 50,000 fatalities per year) are not included.

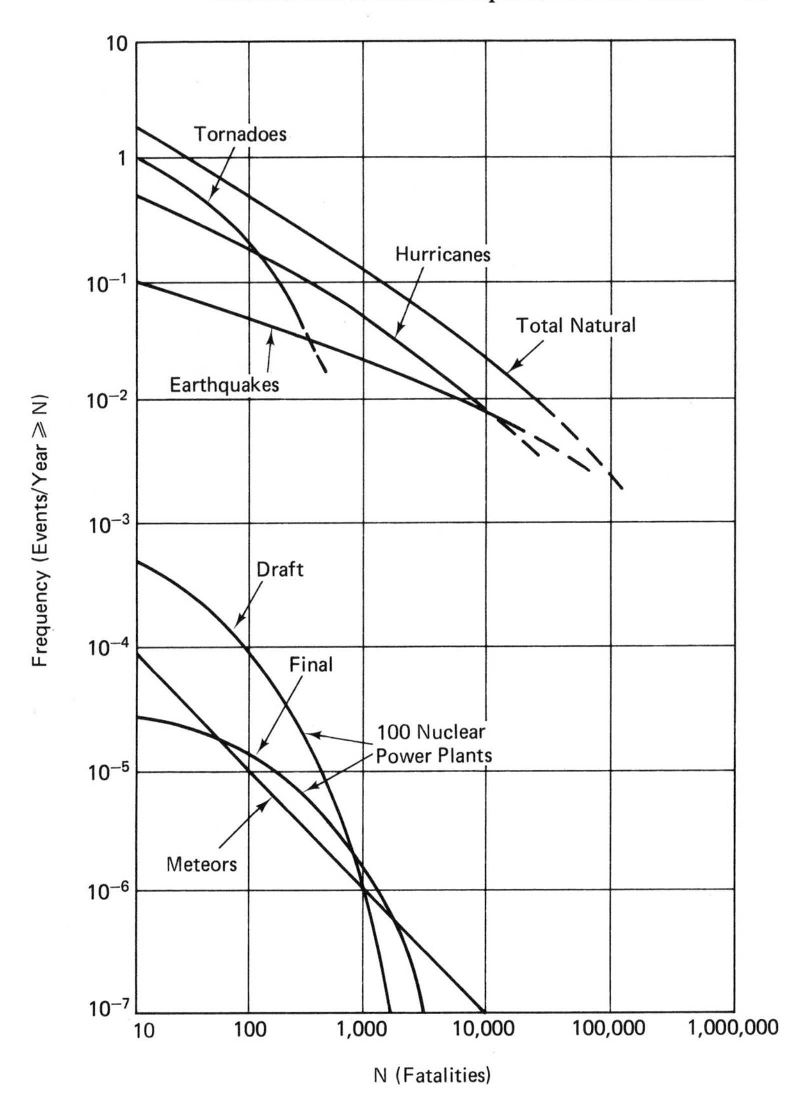

Figure 5.3
Frequency of natural events with fatalities greater than N (9). Only acute (immediate) nuclear fatalities are shown.

nuclear accidents, determined by the Rasmussen Report. It appears that risks due to pressurized light-water reactor accidents are relatively small.

The frequency curves of Figures 5.2 and 5.3 plunge rapidly with the number of fatalities. They show three orders of magnitude less probability of accident (or 10 times less risk of death) for accidents with 1000 fatalities than for those with 10 fatalities. This indicates why probability estimations cannot be avoided in assessing reactor safety. An early estimate of reactor accident consequences (the famous "WASH 740" report published by the USAEC in 1957) calculated the possibility of 3400 deaths and 43,000 acute illnesses from the failure of one 500-MWe reactor. The Rasmussen Report concludes that this event is not impossible, but that its probability is insignificant.

Is it in the economic interest of the owner of a nuclear plant to maintain very high standards of safety? Plants are normally designed for about 30 years of life, but on economic grounds alone the maximum acceptable probability of an accident that would virtually destroy the plant (and cause perhaps 10 fatalities) could be as high as 10^{-3}/plant-yr, or 10^{-1}/yr for a population of 100 plants. This would still mean that plant accidents were no more serious a risk than earthquakes or dam failures, but it is two orders of magnitude higher than the probability levels predicted by the Rasmussen Report. It suggests that close governmental regulation and control is necessary (in addition to commercial self-interest) to maintain the high degree of safety that the Rasmussen Report says is possible with present technology.

5.9 THE HEALTH COSTS OF NUCLEAR POWER

It may be questioned whether the total human cost of nuclear power can be properly calculated. Many effects must be considered, and some of these are not fully understood or well documented. For example, the incidence of cancer in the population due to a given low-level radiation source is only known to the extent that an upper limit can be calculated. Calculations of radiation dose commitments from particular reactors are still preliminary; assembling all the necessary data may require decades. Statistical knowledge of health damage to workers exposed to radiation, such as uranium miners, are still being gathered and correlated. At the same time, radiation exposure standards and actual exposure levels are subject to change with developing technology.

Despite these limitations, it is possible to make a rough estimate of the health damage and mortality that can be associated with production of a given quantity of electrical energy by nuclear plants. Such an estimate may serve to indicate the seriousness of particular hazards, and may also help in comparing relative risks to occupational workers and to the general population. It may suggest where money would best be spent to reduce risk, and could indicate the magnitude of the sum which a consumer of electricity would have to pay if he were directly to bear the full financial cost of the health damage which could be ascribed to the electricity he consumes. This may prove, as more data are acquired, to be a significant factor in choosing between nuclear and some other kind of power.

Table 5.7 presents a list of approximate human costs of nuclear power associated with operation of a plant producing 1000 MWe-yr of electricity annually. Such a plant would serve the needs of 500,000 to 1 million people. The data and assumptions on which the health cost estimates are based will now be briefly discussed.

The effects of nuclear radiation on the health of the workers in the nuclear field are becoming a matter of statistical record, but population doses and risks are still only estimates. Table 5.7 incorporates the preliminary results of UNSCEAR (4) on environmental and occupational dose commitment due to nuclear reactors and fuel processing plants.

For uranium mining and milling the fatal accidents and injuries were obtained as follows. One 1000-MWe-yr plant requires a fuel supply of about 140 tonnes of U_3O_8, which in turn requires 62 man-yr, or 109,120 man-hr, of work in mining and milling (15). Combining this with the quoted fatal and nonfatal accident rates produces the numbers of fatalities and lost days per 1000 MWe-yr shown in Table 5.7. The other significant factor is lung cancer, the incidence of which may be mostly fatal, though it could be delayed until fairly late in life. Working level month (WLM) is a unit of radiation exposure due to the presence of radium found in uranium ores. Radium decays to form radon gas, which in turn decays to what are called radon daughter products. One WLM is equivalent to a radiation dose from radon and decay products sufficient to produce emission of 1.3×10^5 MeV of alpha particle energy per liter of air. Uranium miners in the United States have been limited to a maximum annual dose of 4 WLM since 1971.

The statistics on lung cancer incidence among uranium miners, as presented by F. E. Lundin et al. (16), indicate that in a group of 10,000 there are likely to be 353 cases of lung cancer up to age 80 for exposure at 0 WLM/yr, 512 at 4 WLM/yr, and 684 at 12 WLM/yr. The estimate of lung cancer fatalities in Table 5.7 is based on the assumption that all the excess cancers are fatal, and that of the 62 men required to produce the annual fuel supply of a 1000-MWe reactor, 29 are working underground and are exposed to 4 WLM/yr.

Table 5.7 indicates that a significant factor in injury and death associated with nuclear power is fuel manufacture and reactor construction. Radiation exposure is not considered a significant

Table 5.7
Health Costs of Nuclear Power

	Estimated Rate	Nonfatal Injuries (man-days/ 1000 MWe-yr)	Fatalities (per 1000 MWe-yr)
Uranium Mining and Milling			
Fatal Accidents	0.892 fatality/ 10^6 man-hr (15)		0.1
Nonfatal Injuries	1065 days lost/ 10^6 man-hr (15)	116	
Lung Cancer	159 excess lung cancers to age 80 for 10,000 miners working 4 WLM/ yr[a] for 30 years (16)		0.015
Fuel Manufacture and Reactor Construction			
Fatal Accidents	0.2–0.4 fatality/ 10^6 man-hr		0.1–0.2
Nonfatal Injuries	500 days lost/ 10^6 man-hr	267	
Reactor Operation			
Fatal Accident	0.13 fatality/ 10^6 man-hr		0.05
Nonfatal Injuries	160 days lost/ 10^6 man-hr	64	
Occupational Dose Commitment	0.2–1.0 man-rads/ MWe-yr (4)		0.02–0.2
Population Dose Commitment	0.02 man-rads/ MWe-yr (4)		0.003
Major Plant Accidents	population risk (9)		0.01
Fuel Reprocessing			
Fatal Accidents	0.13 fatality/ 10^6 man-hr		0.018
Nonfatal Injuries	160 days lost/ 10^6 man-hr	23	
Occupational Dose Commitment	0.03 man-rads/ MWe-yr (9)		0.005
Population Dose Commitment	0.26 man-rads/ MWe-yr (9)		0.04
Total		470	0.35–0.64

[a]WLM = working level month (see text).

factor in these figures, which show the occupational hazards typical of the construction industry. If 1000 men are employed for 8 years to build a reactor that will serve for 30 years, and the fatality rate in this type of construction is $0.2/10^6$ man-hr, the fatality rate is 0.1/1000 MWe-yr. These fatal and nonfatal injuries, which are peculiar to manufacturing and construction rather than to nuclear power, are only rough estimates, but serve to provide some basis for comparison between nuclear and nonnuclear risks.

Fatal nonnuclear accidents in reactor operation were estimated from the rate supplied by R. Wilson for the electrical utilities industry in North America (17), with the assumption that, on the average, 200 men are required for operation of each 1000 MWe of plant. Estimates of fatalities due to radiation dose commitment were obtained from estimates in the UNSCEAR report of the dose commitments to plant personnel and to nearby populations from relatively recent light-water reactors. The assumption mentioned earlier of 100 to 200 fatal cancers per 10^6 man-rads was used in estimating fatalities.

The figures for major plant accidents were referred to earlier; 0.0004 acute fatalities/reactor-yr is cited in the Rasmussen Draft Report for light-water reactors. The higher figure results from multiplying this by 25. The result, 0.01, is small compared with the total human health cost of nuclear power, which, as the Table indicates, is of the order of 0.50 fatalities/reactor-yr.

Fatal and nonfatal injuries in fuel reprocessing were estimated from experience in the United States, which suggests that as many as 80 people could be employed in reprocessing the fuel for a 1000-MWe reactor (15). This estimate pertains to an early point in the development of nuclear power (1969) and might be a high estimate of the manpower required. The extent to which fuel reprocessing can be carried out will depend on economics and on the introduction of the breeder reactor (see Chapter 8), but it is acknowledged as a relatively risky process. Dose commitments to occupational groups and the general population were taken from the UNSCEAR report. A fatality rate of $100/10^6$ man-rads was assumed.

Table 5.8 shows an independent estimate of health damage due to nuclear power (18). This estimate differs considerably from that of Table 5.7, but if we recognize all the uncertainties involved, the totals are in reasonably close agreement. Fatalities for nuclear plant accidents are not included, but again, 0.01 per reactor-yr is small compared with the total. The relative estimates of the fatalities related to radiation are noteworthy. Tables 4.6, 5.7, and 5.8 show that independent estimates of the total fatalities and injuries associated with nuclear power may easily differ by factors of two or three.

These estimates lead to total health injuries due to nuclear power

Table 5.8
Summary of Health Effects of Civilian Nuclear Power

	Fatalities (per 1000 MWe-yr)			
	Accidents (not radiation related)	Radiation Related (cancers and genetic)	Total	Injuries (per 1000 MWe-yr)
Uranium Mining and Milling	0.173	0.001	0.174	330.5
Fuel Processing and Reprocessing	0.048	0.040	0.088	5.6
Design and Manufacture of Reactors, Instruments, etc.	0.040		0.040	24.4
Reactor Operation and Maintenance	0.037	0.107	0.144	158
Waste Disposal		0.0003	0.0003	
Transport of Nuclear Fuel	0.036	0.010	0.046	
Total	0.334	0.158	0.492	518

Source: Ref. (18).

of approximately 500 injury days and 0.5 fatality per 1000 MWe-yr. One can relate this directly to the monetary cost of the electricity produced only by making some sort of judgment concerning the value of human life and health. If one takes the view that a human life is worth $1 million, then the overall equivalent cost of this health damage is only 0.1 mill/kWh, which is less than 1 percent of the cost of generating electricity. But such a number can hardly be taken to indicate that the benefits of nuclear power outweigh the human costs and that nuclear power is satisfactorily safe. For one thing, it begs the question of what human life is really worth. For another, it provides no help on what it would cost to reduce health damage below the levels estimated. A much more significant consideration, as Sagan points out (15), is where money should be spent to improve health protection. He quotes estimates of $10,000 to $100,000/man-rem to reduce radiation due to nuclear reactors, whereas measures to reduce dose from medical X-ray equipment may cost only $7/man-rem. He also notes that the total population dose due to medical radiation could be two orders of magnitude higher than the population dose due to nuclear reactors. The question of where limited national resources should be spent to enhance human health is multifaceted, and these figures must also be compared to relative benefits from improvements in other areas of health care.

Some attempts have been made to compare the total risks to society of power from nuclear fuel with power from coal. For an adequate comparison, the total health effects of fuel extraction, processing, transportation, energy conversion, and waste disposal must be included. Using existing information, which is admittedly incomplete, L. B. Lave and L. C. Freeburg (19) have attempted to do this. Insufficient information was available on nuclear fuel processing, transport, and waste disposal to estimate the corresponding hazards. For each 1000 MWe-yr, their figures indicate accidental deaths in mining and milling to be 0.2 for uranium and 2.5 for coal. Disability days number 1100 and 11,000, respectively. Thus the mining of coal to produce electric power could conceivably represent 10 times the human hazard of uranium extraction.

How do population death risks due to gaseous emissions from coal and nuclear plants of equal size compare? This question cannot now be answered with confidence. In should be remarked, however, that statistical studies of health effects in large populations indicate that even with moderate levels of air pollution, health effects may be serious. In general, these statistically derived associations between human health damage and low levels of air pollution should not be taken as proofs of causality and are not supported by clinical evidence. Thus the effects of low levels of air pollution on human health are not well known; moreover, in contrast to nuclear radiation risk, the linear no-threshold assumption seems inappropriate here. An acceptable method of inferring risks from data remains to be established.

C. L. Comar and L. A. Sagan (20), summarizing the results of statistical studies of public health effects of the operation of coal and oil plants, found for the former a wide range of estimates of premature deaths, from 1 to 100 per 1000 MWe-yr. An example of an estimate at the upper end of the range is given by Lave and Freeburg (19), who assume a linear no-threshold relationship between dose and effect and, for both fossil fuel and nuclear plants, an exposure of the whole population to the EPA primary standard atmospheric concentrations for SO_2 and particulates (see Table 6.5). They used coefficients derived from a statistical study by Lave and E. Seskin (21) of the effects of air pollution on health in 117 cities. These coefficients are 3.9 and 8.5 additional deaths per million population per year per $\mu g/m^3$ of SO_2 and particulates, respectively, and the calculated death risks are 312×10^{-6}/yr for SO_2 and 22×10^{-6}/yr for particulates. This suggests, for SO_2 exposure under the given assumptions, which are clearly overestimates, a fourth-order death risk, comparable to that of car accidents.

C. Starr and M. A. Greenfield estimate the total average death risk to the population from fossil fuel power stations with present distributions to be 4×10^{-6}/yr in the United States (22). It seems that the total death risk for an equal capacity of nuclear power is an order of magnitude smaller. Taking the estimate derived earlier of

0.5 fatality/yr for all activities related to nuclear power from a 1000-MWe nuclear plant (which would serve about 1 million people), the risk would be about 0.5×10^{-6}.

5.10 SUMMARY

Hazard due to nuclear power generation is unavoidable, and precise prediction of the consequences of radioactive emissions is not possible.

There are uncertainties in predicting the degree of human exposure to emitted isotopes, which may migrate widely while they decay and may concentrate in living organisms. It is particularly difficult to deal with very long-lived isotopes such as plutonium. Despite these difficulties, methods have been developed for estimating the total radiation doses associated with particular emission quantities. Preliminary estimates of radiation dose commitment are available. In future years, when more detailed and reliable physical information is available, calculations of environmental dose commitment should be useful in deciding what emissions levels are acceptable in a world that may have many thousands of reactors.

The human health danger due to population exposure to a given dose of low-level radiation is unknown, except that an upper-limit estimate is available. The assumption that total risk (e.g., cancers per year per million people) is directly proportional to total dose is widely used, the proportionality coefficient being set by observed results of high-level radiation. This proportionality (or "no threshold" assumption) is believed by many researchers to overstate the actual danger of low-level radiation, but there is now no way of knowing the extent of the overstatement.

Risks due to radiation from nuclear power plant emissions or accidents appear small compared with risks from natural radiation sources, natural disasters, and disasters associated with other human activities. They may also be less than risks due to coal-fired plant emissions. However, they are still finite, and may amount to as much as 0.5 fatality and 500 injury days per 1000 MWe-yr of electrical energy (only a portion of these are radiation-related). A fully satisfactory analysis of both benefits and costs of nuclear power remains to be done.

References

1. G. C. Butler and P. C. Barry. "Surveillance of environmental contamination: Concepts developed for ionizing radiations." *Proceedings. International Symposium on Radioecology Applied to the Protection of Man and His Environment*, Rome, September 7–10, 1971. Commission of the European Communities, 1972.

2. International Committee on Radiological Protection. *The Evaluation of Risk from Radiation.* New York: Pergamon Press, 1966.

3. International Committee on Radiological Protection. *Recommendations of the International Committee on Radiological Protection.* New York: Pergamon Press, 1966.

4. United Nations. *Ionizing Radiations.* Vol. 1: *Levels.* Vol. 2: *Effects.* Report of the United Nations Scientific Committee on the Effects of Atomic Radiation. New York, 1972.

5. National Academy of Sciences, Committee on Biological Effects of Ionizing Radiation. *The Effects on Populations of Exposure to Low Levels of Ionizing Radiation.* Washington, D.C., November 1972.

6. P. J. Barry. Personal communication. July 31, 1975.

7. U.S. Atomic Energy Commission, Division of Reactor Development and Technology. *The Potential Radiological Implications of Nuclear Facilities in the Upper Mississippi River Basin in the Year 2000.* Report WASH-1209, UC-12, January 1973.

8. U.S. Environmental Protection Agency. *Environmental Radiation Dose Commitment: An Application to the Nuclear Power Industry.* Report EPA-520/4–73–002, February 1974.

9. Norman C. Rasmussen, Project Director. *Reactor Safety Study: An Assessment of Accident Risks in U.S. Commercial Nuclear Power Plants.* U.S. Atomic Energy Commission Report WASH-1400 (draft version), August 1974; WASH-1400, NUREG-75/014 (final version), October 1975.

10. *Report to the American Physical Society by the Study Group on Light-Water Reactor Safety*, April 28, 1975.

11. F. K. Culler, J. P. Bloneke, and W. G. Belter. "Current developments in long-term radioactive waste management." Paper A/Conf 49/P/839, *Proceedings of the Fourth United Nations International Conference on the Peaceful Uses of Atomic Energy*, Geneva, September 6–16, 1971.

12. W. B. Lewis. *Radioactive Waste Management in the Long Term.* Report DM-123, Atomic Energy of Canada Ltd., Chalk River, July 13, 1971.

13. A. S. Kubo and D. J. Rose. "Disposal of nuclear wastes." *Science*, vol. 182 (December 21, 1973), p. 1205.

14. Alvin Weinberg. "Social institutions and nuclear energy." *Science*, vol. 177 (July 7, 1972), pp. 27–34.

15. L. A. Sagan. "Human costs of nuclear power." *Science*, vol. 177 (August 11, 1972), pp. 487–493.

16. F. E. Lundin, J. K. Wagoner, and V. E. Archer. *Radon Daughter Exposure and Respiratory Cancer: Quantitative and Temporal Aspects.* National Institute for Occupational Safety and Health and National Institute for Environmental Health Sciences, joint monograph no. 1. Springfield, Va.: U.S. Department of Commerce, National Technical Information Service, June 1971.

17. R. Wilson. "Man-rem economics and risk in the nuclear power industry." *Nuclear News*, February 1972, pp. 28–30.

18. David J. Rose. "Nuclear electric power." *Science*, vol. 189 (April 9, 1974), pp. 351–359.

19. L. B. Lave and L. C. Freeburg. "Health effects of electricity generation

from coal, oil, and nuclear fuel." *Nuclear Safety*, vol. 14, no. 5 (September–October 1973), pp. 409–428.

20. C. L. Comar and L. A. Sagan. "Health effects of energy production and conversion." *Annual Review of Energy*, vol. 1 (1976), pp. 581–600.

21. L. B. Lave and E. Seskin. "An analysis of the association between U.S. mortality and air pollution." *Journal of the American Statistical Association*, vol. 68 (June 1973), pp. 284–290. See also L. B. Lave and E. P. Seskin. "Air pollution and human health." *Science*, vol. 169 (August 21, 1970), pp. 723–733.

22. C. Starr and M. A. Greenfield. *Public Health Risks of Thermal Power Plants*. Report UCLA-ENG-7, School of Engineering and Applied Science, University of California at Los Angeles, May 1972.

Supplementary References

P. J. Barry. "The siting and safety of civilian nuclear power plants." *C.R.C. Critical Reviews on Environmental Control*, vol. 1 (June 1970), pp. 193–220.

F. C. Boyd, "Containment and siting requirements in Canada." In *Containment and Siting of Nuclear Power Plants*. Vienna: International Atomic Energy Agency, 1967.

L. K. Bustad, M. Goldman, L. S. Rosenblatt, C. W. Mays, N. W. Hetherington, W. J. Bair, R. O. McClellan, C. R. Richmond, and R. E. Rolland. "Evaluation of long-term effects of exposure to internally deposited radionuclides." Paper A/Conf 49/P/081, *Proceedings of the Fourth United Nations International Conference on the Peaceful Uses of Atomic Energy*, Geneva, September 6–16, 1971.

G. Hake, P. J. Barry, and F. C. Boyd. "Canada judges power reactor safety on component quality and reliable systems performance." Paper A/Conf 49/P/150, *Proceedings of the Fourth United Nations International Conference on the Peaceful Uses of Atomic Energy*, Geneva, September 6–16, 1971.

S. H. Hanauer and P. A. Morris. "Technical safety issues of large nuclear power plants." Paper A/Conf 49/P/040, *Proceedings of the Fourth United Nations International Conference on the Peaceful Uses of Atomic Energy*, Geneva, September 6–16, 1971.

G. T. Seaborg. "Energy and the environment." *International Journal of Environmental Studies*, vol. 3 (1972), p. 305.

J. B. Storer and V. P. Bond. "Evaluation of long term effects of low level whole body external radiation exposures." Paper A/Conf 49/P/082, *Proceedings of the Fourth United Nations International Conference on the Peaceful Uses of Atomic Energy*, Geneva, September 6–16, 1971.

Study Panel on Nuclear Power Plants, Maryland Academy of Sciences. "Radiation dose limits." Chapter 8 in *Power Generation and Environmental Change*, David A. Berkowitz and Arthur M. Squires, eds. Cambridge, Mass.: MIT Press, 1971.

A. M. Weinberg and R. P. Hammond. "Global effects of the use of energy." Paper A/Conf 49/P/033, *Proceedings of the Fourth United Nations International Conference on the Peaceful Uses of Atomic Energy*, Geneva, September 6–16, 1971.

Problems

5.1 Estimate the required rate of dilution (km^3/hr) required for SO_2 and radioactivity in fly ash emitted from coal-burning power plants of 1000-MWe capacity under the following conditions.

From Chapter 4 it may be deduced that a 1000-MWe plant (operating with 0.75 load factor and coal of 2.6 percent sulfur and 10 percent ash) will discharge, each year, 18,700 tons SO_2 (with 85 percent sulfur removal from stack gases), and 2000 tons fly ash (with 99 percent removal).

The allowable annual mean atmospheric concentration of SO_2 is 80 $\mu g/m^3$ (see Table 6.5).

The radioactivity content of the fly ash may be assumed to be 3 pCi/g for each of Ra-226, Ra-228, and Th-228. The allowable atmospheric concentration of these isotopes is 1 pCi/m^3.

5.2 The AEC cites the release rates for radioactivity from pressurized light-water reactors shown in Table 5.A.

Table 5.A
Radioactivity from Pressurized Light-Water Reactors

		Radionuclide Release Rates (Ci/MWe-yr)	
Isotope	Half-Life	Boiling-Water Reactor	Pressurized-Water Reactor
H-3 (liquid)	12.3 yr	0.13	0.49
Kr-85 (gas)	12.3 yr	4.4	3.4
Kr-87 (gas)	76 months	5.4×10^{-4}	1.4×10^{-30}
I-129 (gas)	1.6×10^{7} yr	3.4×10^{-9}	6.33×10^{-10}
(liquid)		1.9×10^{-9}	2.2×10^{-10}
I-131 (gas)	8.05 days	1.9×10^{-2}	1.8×10^{-4}
(liquid)		1.6×10^{-3}	1.5×10^{-3}
Xe-133	5.27 days	28.0	3.5

Source: Ref. (7).

Determine the air flow rates required for dilution of gaseous emissions to maximum ambient concentrations of

Kr-85 (3×10^5 pCi/m^3 maximum),
Kr-87 (2×10^4 pCi/m^3 maximum),
Xe-133 (3×10^5 pCi/m^3 maximum),
I-131 (0.14 pCi/m^3 maximum).

Compare the maximum rate with that required for SO_2 (Problem 5.1).

Given the air concentration-to-dose conversion factor for continuous exposure to krypton of 1.5×10^{-8} (rem/yr)/(pCi/m^3 air), estimate the average yearly dose corresponding to the maximum concentration of Kr-85 cited in Table 5.A.

5.3 Emissions of Kr-85 and H-3 are dispersed throughout the world (8). Determine the average annual individual and total population dose due to krypton released in the year 2000 if all of this krypton stays in the atmosphere and is uniformly mixed. Assume the following:
Total nuclear capacity for the year 2000 is 5 million MWe
Average load factor is 0.7
World population for the year 2000 is 5 billion
Mass of world's atmosphere is 5.1×10^{18} kg
Sea-level air density is 1.24 kg/m^3
Dose conversion factor (8) is 1.5×10^{-8} (rem/yr)/(pCi/m^3 air) for the whole body, 50×10^{-8} (rem/yr)/(pCi/m^3 air) for the skin
Kr-85 release rate (mostly from fuel processing plants) is 500 Ci per MWe-yr
Fatality risk is 200×10^{-6}/man-rem (whole body).

Note: This calculation neglects dose due to isotopes released before the year 2000 or isotopes other than krypton, or the dose caused in all succeeding years by krypton released in the year 2000. The U.N. Scientific Committee on the Effects of Atomic Radiation estimates the total dose commitment due to nuclear power generated in the year 2000 to be 0.2 mrad (4).

5.4 The IAEA has projected a possible consequence of a major (and highly improbable) nuclear plant accident as release of several hundred million Curies from a 1000-MWe reactor (corresponding to a total body dose of 2.5×10^6 man-rems, and a total thyroid dose of 30×10^6 man-rems), resulting in possibly 500 deaths in a typical metropolitan area (Ref. (5) in Chapter 4).

For the above doses, estimate the total fatality risk, using the risk coefficients of Tables 5.1 and 5.2, assuming that only 25 percent of the thyroid cancers are fatal.

5.5 Using the rates of flow of coal, oil, natural gas, and nuclear energy into electrical power stations, as shown by Figure 2.11 (United States) and Figure 2.12 (Canada), calculate the occupational fatalities and electrical energy output for each type of power plant for the year 1980. The following energy consumption rates may be used:

Coal:	9700	Btu/kWh
Oil:	10,000	Btu/kWh
Natural gas:	10,300	Btu/kWh
Nuclear:	12,000	Btu/kWh.

For this calculation each plant may be assumed to have a load factor of 0.75. Suppose that 5 percent of the oil is offshore oil and that 65 percent of the coal is surface mined.

Note: 1 bbl crude oil $= 5.8 \times 10^6$ Btu.

5.6 UNSCEAR estimates the world dose commitments due to nuclear electrical power as 9×10^{-4} mrads and 0.2 mrad for the years 1970 and 2000, respectively (4). Installed power capacity is estimated as 20 GWe for 1970 and 4300 GWe for 2000. Using these estimates, determine the following:

a. The average percent annual growth in power capacity.

b. The average percent annual growth in dose commitment.

c. the integrated dose commitment (in man-rads) for the period 1970–1980, assuming world population growth at 2 percent per year from 3.2 billion in 1970.

d. The corresponding fatality risk.

e. The comparable risks over a 30-year period of cancer incidents not related to radiation, assuming this rate is as high as 1600×10^{-6} per year per million population.

5.7 Suppose the annual rate of emission of radioactivity into the atmosphere increases exponentially as

$$\dot{R} = \dot{R}_0 e^{rt},$$

where $\dot{R}_0$ is the emission rate (Ci/sec) at $t = 0$. Suppose also that the total radioactivity Q due to the emitted nuclides decays exponentially with time and that dose rate is proportional to this total quantity of radioactivity.

Show that at time t the number of Curies Q of radioactivity due to emissions in the time interval τ to $\tau + d\tau$ is

$$dQ = \dot{R}_0 \, e^{r\tau} \, e^{-\ln 2(t-\tau)/t_h} \, d\tau,$$

where t_h is the effective half-life of the radioactivity. Thus

$$Q(t) = \int_0^t \frac{dQ}{d\tau} \, d\tau = \frac{\dot{R}_0(e^{rt} - e^{-(\ln 2)t/t_h})}{r + (\ln 2)/t_h}.$$

As an example, assume that the total emissions from nuclear reactors in a given country increase 15 percent per year, and that the average half-life of the radioactivity in the environment is 10 years. If the annual dose rate at the end of the first year due to emissions released in that year is 0.003 mrem, and if the annual dose rate is considered to be proportional to Q, to what level could the annual dose rate increase over a 30-year period?

If the half-life were extended from 10 years to infinity, what would the annual dose rate be at the end of 30 years?

5.8 Consider the plutonium hazard in a steady-state world of population 20 billion and electricity consumption 5 kWe/person. Assume all this electricity to be generated by nuclear reactors producing about 33,000 MW-days (thermal) per ton of fuel and containing a plutonium inventory of 500 Ci/ton (8). The reactors are loaded annually, and the generation efficiency is 30 percent.

Table 5.5 suggests that the fraction of produced plutonium that is released to the atmosphere (mainly as a result of fuel processing) could be as high as 10^{-6}.

Estimate the following quantities:

a. Total reactor inventory (Ci) of Pu-239.

b. Average airborne concentration (pCi/m^3) of Pu-239 released as an aerosol.

c. Average annual dose per person.

Use the following assumptions:

—The aerosol remains in the atmosphere one month before being removed by rain.

—The aerosol is uniformly distributed in the earth's atmosphere (4×10^{18} m^3).

—Dose from Pu-239 absorbed in soil and water is negligible compared with dose from the airborne isotope.

—The air-concentration-to-dose conversion factor is 12 (rem/yr)/(pCi/m^3 air).

5.9 A 2000-MWe plant has 300 employees, 50 of whom do mechanical maintenance work and are permitted to receive an annual dose of 4 rems. Their total exposure is 40 percent of the total station dose.

a. Estimate the upper limit of the station death risk owing to occupational exposure.

b. Compare this with the death risk in the utilities industry (0.13 fatal accident/10^6 man-hr).

c. How much should be spent to reduce the total station exposure by 1 man-rem:

i. considering the economic benefit to the company of needing to employ fewer mechanical maintenance workers?

ii. considering the possible economic benefit to the family of sparing the life of a wage-earner?

Assume reasonable values for salaries and lifetime earnings.

5.10 Estimate the accumulated volume of solid wastes which could be generated by nuclear fuel processing in the United States between 1975 and 2000. One ton of fuel will yield around 25,000 MW-days (thermal) of energy and 0.04 m^3 of solid wastes. Nuclear capacity might be assumed to be 40,000 MWe in 1975, to grow at an average rate of 15 percent annually, and to operate at a load factor of 0.7. The thermal power of these wastes is about 10 kW thermal/m^3.

6
Air Pollution Due to Fossil Fuels

6.1 INTRODUCTION

Since air pollution is already nearly intolerable in many areas, there is serious concern for the future. Will it be feasible (at reasonable cost) to maintain adequate air quality over the next few decades? Are we approaching an absolute level of fossil fuel consumption, beyond which the environment must inevitably deteriorate? Can the worst effects of fossil fuels be avoided by switching to nuclear power?

Although nuclear power is developing rapidly, it may not supply more than half the energy for electricity production by the end of the century. Also, with current technology nuclear energy is not directly useful for most transportation and industrial needs. Thus if total energy demand were to continue to rise rapidly, there would be a continuing increase in demand for fossil fuels. By the year 2000 the total consumption rate of fossil fuels could be three times what it is today.

Severe air pollution due to fossil fuel use is hardly new. As early as 1273, the use of coal was prohibited in the city of London as being "prejudicial to health." Legislation continued over the centuries, although it was apparently very ineffective at times. In the reign of Elizabeth I the use of coal was prohibited in London while Parliament was sitting. Urban atmospheres were terribly polluted during the Industrial Revolution; of Manchester it was said that "a sort of black smoke covers the city; the sun seen through it is a disc without rays" (1). Several other British cities were just as bad. Manufacturers made little effort to reduce smoke nuisances; for over a century smoke, fogs, and the blackening of buildings were accepted as normal features of industrial towns.

In the late nineteenth century the worst emissions of the chemical industry had been controlled to a considerable extent by enforcement of the 1863 Alkali Acts. However, coal smoke and SO_2 continued for many decades to exact what must have been, in total, a terrible toll of human health and life. It was probably only the advent of widely reported disasters, such as the acute air pollution episodes noted in Table 6.1, that brought widespread awareness of the problem and remedial action.

The disastrous episodes shown in Table 6.1 were due to the buildup, in nearly stagnant air, of smoke particles, SO_2, and fog to such levels that people with lung problems (particularly the aged) were unable to survive. The cost of air pollution to human life has not been confined to these specific events. According to C. R. Lowe,

Table 6.1
Acute Air Pollution Episodes

Place	Date	Extent and Symptoms	Conditions and Probable Cause
Meuse Valley, Belgium (coke ovens, blast furnaces, steel, glass, zinc, and sulfuric acid plants)	1–6 December 1930	60 deaths, "thousands ill," coughing, breathlessness, chest pain, eye and nose irritation experienced	Inversion, stagnation in 15-mile river valley for 1 week; smoke and irritant gases; sulfur oxide, sulfuric acid mist, and fluorides suspected; estimated SO_2 of 25–100 mg/m^3 (10–40 ppm)
Donora, Pennsylvania (zinc smelter, wire coating mill, steel mill, sulfuric acid plant)	27–31 October 1948	6000 of 14,000 population ill, 1400 sought medical care, 17 died; coughing, sore throat, chest constriction, burning and tearing eyes, vomiting, nausea, excessive nasal discharge	Temperature inversion and fog along horseshoe-shaped valley of Monongahela River; sulfur oxides, smoke, and zinc compound particulates present; sulfuric acid mists likely: estimated SO_2 of 1.5–5.5 mg/m^3 (0.5–2 ppm)

London, England	5–9 December 1952	3500–4000 deaths in week of 5–12 December in excess of expected norm of like weeks; causes of death: chronic bronchitis, bronchopneumonia, and heart disease; increased hospital admissions for respiratory and heart disease	"Pea soup" fog and temperature inversion covered most of the U.K.; smoke and SO_2 accumulations in stagnated air; reported smoke highs of 4.5 mg/m^3 and sulfur oxide highs of 3.75 mg/m^3 (1.4 ppm)
London, England	January 1956	1000 excess deaths charged to a pollution episode	Extended fog conditions similar to 1952 episode; resulted in Parliament's passing Clean Air Act
London, England	5–7 December 1962	700 excess deaths and increased illness charged to a pollution episode; emergency medical care plan functioned	Severe fog and inversion; SO_2 levels higher than in 1952, but particulates were lower; alert system operated

Source: Ref. (2). Reproduced by permission of McGraw-Hill Book Company.

in 1970 "the annual contribution of atmospheric pollution to the cost of bronchitis in England and Wales may be about £5000 in deaths, £6,500,000 in working days, £3,500,000 in sickness benefits and also £3,500,000 in hospital, general medical and pharmaceutical services" (3).

Following passage of the 1956 Clean Air Act in the United Kingdom, the improvements have been dramatic (4). Since passage of the Act, smoke concentrations at ground level have dropped by a factor of three, and SO_2 concentrations have dropped by nearly 40 percent. December sunshine in central London has increased by almost 70 percent. Thus legislation controlling the use of coal coupled with a strong trend toward the domestic use of electrical heating can make a major improvement in air quality. However, the situation may still be viewed as marginal. In the United Kingdom severe problems may result from any large expansion of fossil fuel use in the future. New technology will almost certainly be needed to limit pollutant emissions to tolerable levels.

In the United States regulations on sulfur emissions from oil- and coal-fired power plants and limits on pollutant emissions from automotive engines offer hope that though control and prevention costs will be high, adequate protection will result. The EPA has estimated that in 1968 the total cost of air pollution in the United States was $16.1 billion ($6.1 billion for damage to human health and the rest mostly for material and residential property damage). Costs were determined by means of broad judgments where detailed evidence was lacking. Nevertheless, they provide some notion of what it may be reasonable to spend on air pollution control.

6.2 POLLUTANTS FROM FOSSIL FUELS

Relative rates of pollutant emission from various sources in the United States are shown in Table 6.2 for 1966 (prior to the imposition of stringent controls). The principal pollutants are:

1. Sulfur oxides, principally SO_2. Most of this pollutant is emitted by central station power plants and by burners for space and industrial heating.

2. Carbon monoxide (CO). By far the largest source is the car engine.

Table 6.2
Relative Pollutant Emissions from Combustion for Energy Conversion

	1966 Emissions (weight percent)				
Pollutant	Central Station Power Plants	Industrial Processing	Space Heating and Industrial Steam Generation	Gas Turbines	Reciprocating IC Engines
Products of Incomplete Combustion					
Combustible particulates	18	6	45	1	30
CO	<1	<1	1	1	98
Gaseous hydrocarbons	<1	<1	1	1	98
PNA	<1	4	90	nil	6
Nitrogen Oxides	21	3	17	1	58
Sulphur Oxides	63	5	31	nil	1
Noncombustible Particulates (ash)	72	3	25	<1	<1

Source: Refs. (5) (6).

3. Nitrogen oxides (NO_x). Both mobile and stationary sources are important emitters.

4. Unburned hydrocarbons (HC). The automobile is the main source of gaseous hydrocarbon emissions.

5. Particulates. While stationary power and heating plants are responsible for most of the particulate emissions, the internal combustion engine is responsible for a significant fraction. In many cities, coal-burning power stations are the greatest single source of particulates.

In various cities at particular times, concentrations of SO_2, CO, and particulates have exceeded levels believed to affect human health or function. In specific localities such as Los Angeles, concentrations of NO_x and unburned hydrocarbons have led to the formation of photochemical smog (under special atmospheric conditions) with significant effects not only on visibility but also on human comfort (eye irritation) and on vegetation. Close control of all of these pollutants will be necessary in the future, especially if fossil fuel use increases as expected in the next two or three decades.

Table 6.3 gives a more detailed view of pollutant emissions in a particular city, Toronto, Ontario, showing the relative contributions from small and large sources. The two largest electric power plants—the Richard L. Hearn station, operating on natural gas, and the Lakeview station, operating on coal—were together responsible for 80 percent of the emissions by weight of SO_2 into the Toronto atmosphere. They were, however, responsible for only about 13 percent of the total particulate emissions, about 0.2 percent of the CO emissions, and about 0.49 percent of the unburned hydrocarbon emissions. They were also responsible for about 53 percent of the total city-wide NO_x emissions. Residential and commercial heating units (numerous and difficult to control) are thought to be relatively small contributors to total emissions of SO_2 (13 percent), CO (1 percent), NO_x (14 percent), and particulates (24 percent). In contrast to those of central power plants, these emissions are close to ground level, even though widely dispersed. Other cities may show quite different contributions from various sources, but Table 6.3 shows that large fossil fuel plants can be heavy contributors to urban air pollution.

If all of these emissions were uniformly dispersed over the surface area of the earth, the pollution problem would be insignificant (8, 9). There is no evidence that the global average concentration of SO_2 is increasing; it is only about 0.0003 ppm, two orders of magnitude less than levels that might be expected to affect human health. World average ("background") level of NO_x is only about 0.001 ppm, which again is two orders of magnitude less than would affect human health. Natural sources are thought to emit perhaps 30 times as much NO_x as is released by combustion.

In contrast, global average concentration of CO_2 is known to be

Table 6.3
Emissions Summary, Metropolitan Toronto, 1 April 1971 to 1 April 1972

Source	Level of Emission (10^6 lb/yr and percentage of total)				
	SO_2	Particulates	NO_x	CO	HC
R.L. Hearn G.S.	34.64(7.33)	0.62(1.23)	26.44(13.14)	0.29(0.03)	0.12(0.06)
Lakeview G.S.	348.28(73.74)	5.94(11.82)	80.19(39.87)	2.00(0.23)	0.81(0.43)
Municipal Incinerators	0.55(0.11)	12.13(24.15)	1.35(0.67)	3.28(0.38)	0.16(0.08)
Industrial Sources	23.25(4.92)	12.08(24.05)	17.64(8.77)	4.89(0.57)	43.64(23.50)
Autos	3.49(0.73)	4.65(9.25)	43.65(21.70)	827.36(97.53)	126.42(68.09)
Railroads	0.43(0.09)	1.08(2.15)	2.19(1.08)	0.59(0.06)	1.33(0.71)
Shipping	0.90(0.19)	0.41(0.81)	0.66(0.32)	0.24(0.02)	0.32(0.17)
Aircraft	0.24(0.05)	1.18(2.34)	1.14(0.56)	2.21(0.26)	10.61(5.71)
Heating (residential)	19.82(4.19)	2.49(4.95)	5.91(2.93)	0.52(0.06)	0.76(0.40)
Heating (apartments)	10.73(2.27)	3.99(7.94)	5.68(2.82)	3.88(0.45)	0.84(0.45)
Heating (schools and universities)	8.29(1.75)	0.83(1.65)	2.69(1.33)	0.29(0.03)	0.11(0.05)
Heating (public and commercial buildings)	14.40(3.04)	2.83(5.63)	10.77(5.35)	0.74(0.08)	0.34(0.18)
Small Industries	7.19(1.52)	0.83(1.65)	2.54(1.26)	0.23(0.02)	0.10(0.05)
Incineration (apartments, schools, small industrial, public and commercial buildings)	0.07(0.01)	1.16(2.30)	0.26(0.12)	1.76(0.20)	0.11(0.05)
Total	472.29	50.22	201.08	848.26	185.66

Source: Ref. (7).

increasing slowly, apparently as a result of combustion. Although this does not threaten health, it could conceivably have a significant effect on climate, as mentioned in Chapter 4. Global atmospheric concentrations of lead increased markedly following the widespread introduction of lead in automotive fuel, but these concentrations are not thought to be near danger levels (though concentrations in city air may be potentially dangerous, and the problem could become serious if the present trend away from leaded automotive fuels is not continued).

The main problems with the pollutants named in Tables 6.2 and 6.3 are local or regional, rather than global; it is the pattern of diffusion of pollutants which is of primary concern. Much progress has been made in recent years toward understanding the processes of dispersion; it is now quite clear that natural processes of dispersion are generally inadequate to preserve urban air quality. Severe controls on emission rates must be imposed. There is certainly room for debate on whether emission rate controls should be adjusted to local conditions, but there can be no doubt that they are already necessary in many cities.

6.3 SULFUR DIOXIDE AND PARTICULATES

For many years sulfur oxide compounds have been of concern to combustion engineers, since conversion of a relatively small amount of SO_2 to SO_3 can lead to the formation of enough sulfuric acid to cause serious corrosion in boilers. In the atmosphere, SO_2 is fairly readily converted into sulfuric acid and sulfates; it associates readily with particulate matter in the air, including rain drops. For this reason it is believed that SO_2 lasts only two or three weeks in the atmosphere prior to deposition. For many soils the addition of sulfates is beneficial; for others it may be quite detrimental. It has been reported that in Sweden the productivity of forests may be reduced 10 to 15 percent over the next 30 years due to buildup of acidity in the soil by deposition of sulfur compounds; the acidity of certain rivers and lakes threatens to be hazardous to fish and animal life (10). Although these effects are still under study, there is concern that long-range transport of sulfur compounds may in time devastate large land areas. Thus the control of local concentration of SO_2 in the air near points of emission may not be enough. Methods of control of SO_2 emissions and dispersion will be discussed below.

6.3.1 Health Effects

Table 6.4 shows the effects associated by the U.S. Public Health Service with SO_2 and particulate concentrations in the air. Adverse health effects are associated with SO_2 concentrations of 115 $\mu g/m^3$ (annual average), or 300 $\mu g/m^3$ (24-hr average) for three to four days. A study of the effects of periods of high pollution in greater

Table 6.4
Some Observed Health Effects of SO_2 and Particulates

Concentration of SO_2		Concentration of Particulates		
$\mu g/m^3$	Measured as	$\mu g/m^3$	Measured as	Possible Effect
1500	24-hr average			Increased mortality
715	24-hr mean	750	24-hr mean	Increased daily death rate; a sharp rise in illness rates among bronchitics over age 54
300–500	24-hr mean	Low	24-hr mean	Increased hospital admissions of elderly respiratory disease cases; increased absenteeism among older workers
600	24-hr mean	300	24-hr mean	Symptoms of chronic lung disease cases accentuated
105–265	Annual mean	185	Annual mean	Increased frequency of respiratory symptoms and lung disease
120	Annual mean	100	Annual mean	Increased frequency and severity of respiratory diseases among schoolchildren
115	Annual mean	160	Annual mean	Increased mortality from bronchitis and lung cancer

Source: Refs. (2) (11).

London, in particular on the worsening of bronchitic patients attending chest clinics concludes that "the minimum pollution seen to be associated with worsening to any significant degree is that indicated by about 500 $\mu g/m^3$ SO_2 together with about 250 $\mu g/m^3$ smoke" (12). These numbers refer to 24-hr averages at a group of about seven sites in inner London. Studies on people in normal health, however, have shown "failure, despite many hundreds of experiments, to demonstrate any consistent effect of SO_2 in concentrations less than 1 ppm [= 2860 $\mu g/m^3$], alone or in combination with particles, on airway resistance or other measures of lung function."

Major plant damage has been reported for long-term SO_2 concentrations of about 500 $\mu g/m^3$; lichens will be killed at concentrations of about 50 $\mu g/m^3$, and rye grass growth can be reduced by half by the presence of about 200 $\mu g/m^3$ SO_2 (13).

Chronic bronchitis mortality rates are associated with atmospheric concentration of SO_2, although cigarette smoking is also recognized as a major contributor to bronchitis. In Great Britain chronic bronchitis "accounts for 10 percent of all deaths and over 10 percent of all absences for illnesses" (2).

The association between air pollution and health is a very complex matter and subject to considerable debate. Emil Chanlett concludes:

Faced with demands for precise replicate quantitative data on positive concentrations and resultant mortality, morbidity, and pathology, our best epidemiologists cannot satisfy those who will not act until tight evidence is provided. The nature of the case—for convicting dirty air as damaging to man's health is—

1. The occurrence of acute air pollution episodes
2. The relations between polluted air and several respiratory diseases
3. The evidence of damage from recognized toxicants
4. The interpretation of toxicological observations from various sources on the quality of the community air supply. (2)[1]

In relation to the danger levels from SO_2 indicated in Table 6.4, actual concentrations in congested areas have been worrisomely high. Over a six-year period it was shown that mean annual concentrations of SO_2 ranged from 0.01 ppm in San Francisco to 0.18 ppm in Chicago (14). New York had an annual average of 0.17 ppm and a peak 24-hr average concentration of 0.38 ppm. The problem was worst in the densely populated area defined by Boston, Washington, St. Louis, and Chicago, where most of the sulfur-bearing fossil fuels were consumed.

In the United Kingdom, annual concentrations of SO_2 in major cities and towns in 1970–1971 averaged over 100 $\mu g/m^3$; the level in London was 132 $\mu g/m^3$. These levels are at least 40 percent less than they had been a few years before introduction of the 1956 Clean Air Act, but could rise again with expanded fossil fuel use.

Annual average particulate concentrations in urban areas in the United States range from 60 to 200 $\mu g/m^3$ in urban areas. In nonurban areas the range is typically 10 to 60 $\mu g/m^3$. In the United Kingdom smoke concentrations have dropped in recent years from nearly 200 to 60 $\mu g/m^3$.

Table 6.5 shows the EPA's national air quality criteria and standards. These indicate that for adequate air quality, concentrations should not exceed 80 $\mu g/m^3$ of SO_2 and 75 $\mu g/m^3$ of particulate matter (annual average). In Canada, as Table 6.6 shows, concentration objectives are somewhat lower.

Currently, the danger to health from SO_2 is hypothesized to be due to the acid sulfates that form fairly rapidly from it in the atmos-

[1]Reprinted with permission from Emil T. Chanlett, *Environmental Protection* (New York: McGraw-Hill, 1973).

Table 6.5
National Air Quality Criteria and National Air Quality Standards in the United States, 1971

Pollutant	Threshold Concentrations for Adverse Health Effects	Proposed EPA Standard
Particulate Matter	80 μg/m³, annual mean	75 μg/m³, annual geometric mean 260 μg/m³, max. 24-hr value may occur once each year
SO_2	115 μg/m³, annual mean	80 μg/m³, annual arithmetic mean
	300 μg/m³, 24-hr average for 3–4 days	365 μg/m³, max. 24-hr value may occur once each year
CO	12–17 mg/m³ for 8 hr produces concentration of 2–2.5% carboxyhemoglobin.	10 mg/m³, max. 8-hr value may occur once each year
	35 mg/m³ for 8 hr produces concentration of 5% carboxyhemoglobin	40 mg/m³, max. 1-hr value may occur once each year
Photochemical Oxidants	130 μg/m³ hourly average impaired performance of student athletes; 200 μg/m³ instantaneous level increased eye irritation; 490 μg/m³ peaks with 300 μg/m³ hourly average increased asthma attacks	160 μg/m³, max. 1-hr value may occur once each year
Hydrocarbons	With nonmethane hydrocarbon, 200 μg/m³ in 3 hr, 6–9 A.M., produced (2–4 hr later) photochemical oxidant of up to 200 μg/m³ that lasted 1 hr; by extrapolation downward, HC concentration of 100 μg/m³ can produce lowest injurious level of photochemical oxidant	160 μg/m³, max. level that may occur 6–9 A.M. once each year

NO_x	118–156 $\mu g/m^3$, 24-hr mean over 6 months, produced increase in acute bronchitis in infants and schoolchildren; this is associated with a 24-hr max. of 284 $\mu g/m^3$; 117–205 $\mu g/m^3$, 24-hr mean over 6 months, and mean suspended nitrate level of 3.8 $\mu g/m^3$ or more produced increased respiratory disease in family groups	100 $\mu g/m^3$, annual arithmetic mean

Source: Refs. (2) (15).

phere, depending on the presence of moisture and particulates (16). Statistical studies relating excess mortality and chronic respiratory disease with acid sulfates suggest threshold concentration levels in the range 10 to 30 $\mu g/m^3$, below which significant health damage is unlikely. It is thought that concentrations of acid sulfates will not exceed this threshold if SO_2 concentration is kept to less than 80 $\mu g/m^3$. A 20 percent increase in sulfate concentration above the threshold level is associated with about 1 percent excess mortality (16).[2] The conclusion that this is the actual risk due to relaxing controls on SO_2 concentration is a tentative deduction from highly imperfect knowledge on this complex subject. A statistical association between health damage and acid sulfate concentration does not prove a causal link; other factors could be more important. This will only be revealed as better and more extensive data are acquired. At the least, however, one can say that typical levels of SO_2 (or sulfate) concentrations are close to levels presently believed to be dangerous. Unfortunately, it is also true that the SO_2 standards of Tables 6.5 and 6.6 impose a severe constraint on fossil fuel power generation in many densely populated or industrialized areas, because low-sulfur fuels are in short supply and processes for removal of sulfur during or after combustion are still not satisfactorily developed.

Control of particulate emissions from large fossil fuel combustion

[2]Assuming a population of 1 million, with a normal life expectancy of 60 years, whose electricity needs are met by a 1000-MWe power plant, the excess mortality rate (for sulfate concentrations 20 percent above threshold) would be $0.01 \times (1/60) \times 1{,}000{,}000 = 170$ per power plant–year. This is two or three orders of magnitude higher than the total fatality rate for nuclear power shown in Tables 5.7 and 5.8.

Table 6.6
Canadian Federal Air Quality Objectives

Pollutant	Maximum Desirable[a] ($\mu g/m^3$)	Maximum Acceptable[a] ($\mu g/m^3$)
SO_2		
1 hr	450 (0.17)	900 (0.34)
24 hr	150 (0.06)	300 (0.11)
1 year	30 (0.01)	60 (0.02)
Particulates		
24 hr		120
1 year[b]	60	70
CO		
1 hr	15,000 (13)	35,000 (30)
8 hr	6,000 (5)	15,000 (13)
Oxidants		
1 hr	100 (0.05)	160 (0.08)
24 hr	30 (0.015)	50 (0.025)
1 year	20 (0.01)	30 (0.015)
Hydrocarbons		
3 hr		160 (0.24)

[a]volume units (ppm) in brackets.
[b]geometric mean.
Source: Ref. (7).

plants is now a well-established art (17). The EPA has ruled that particulate emissions from new electrical plants must not exceed 0.1 lb/10^6 Btu input, or about 0.1 percent of the coal removed in combustion. For a coal of 10 percent ash content, this would mean 99 percent removal of the ash in the coal. A combination of mechanical and electrostatic precipitators can remove more than 99.5 percent of the particulate matter from the effluent gases. Products of combustion such as smoke and soot range in size from 0.1μ to 100μ or more ($1\mu = 10^{-6}$ m). For particles down to 1μ in size, electrostatic precipitator efficiency can be made very nearly 100 percent. For particle sizes below 0.1μ the collection efficiency drops below 80 percent, though these very minute combustion products have been thought to aggregate fairly rapidly. There is, however, the worry that some of the smallest airborne particles may lodge in the deep recesses of the lung and interact with SO_2. Trace metallic elements from both coal and oil are another concern, and standards for these have not yet been formulated.

General methods of SO_2 control include:

1. burning of low-sulfur fuel
2. use of tall chimneys for widespread dispersion
3. removal of SO_2 from combustion gases
4. removal of sulfur during the combustion process (e.g., with fluidized-bed combustion)

5. sulfur removal from the fuel (e.g., production of sulfur-free gas or oil from coal).

Since low-sulfur fuels are in short supply, development of the technology for sulfur removal is of prime importance at the present time.

Control of SO_2 emissions from small units used in residential, commercial, and industrial heating is feasible only by limiting the sulfur content of the fuel. Most of these small units now use oil or gas. The sulfur content of cleaned natural gas is generally very low, but in North America gas reserves are being rapidly consumed. Low-sulfur oil is also in short supply, though commercial desulfurization processes are available; both Europe and North America are likely to legislate maximum sulfur levels for oil used in the residential, commercial, and industrial sectors. Progressively, then, one may expect natural gas and low-sulfur oils to be denied to electricity generating stations; this is true even in the United Kingdom, which now draws in large measure from North Sea gas fields. In the United Kingdom emissions of SO_2 from electrical power plants are about 40 percent of the total SO_2 emissions; in the United States the corresponding figure is over 60 percent (Table 6.2); in the City of Toronto it is over 80 percent (Table 6.3). Even if (with the use of very small chimneys) SO_2 emissions from electricity generation plants were not the chief contributors to local concentrations of SO_2, one could expect strong pressure to install sulfur removal processes in large plants.

6.3.2 Dispersion by Tall Stacks

In the United Kingdom sulfur removal from fuels used for large power stations is thought to be unnecessary. It is held that with sufficiently tall chimneys atmospheric dispersion of SO_2 will successfully prevent serious augmentation of ground-level concentrations of SO_2 by this source. Figure 6.1 shows the CEGB Didcot coal-fired power station. It has a 2000-MWe generating capacity and a single 650-ft (198-m) chimney for the main plant. As will be shown, atmospheric dispersion is enhanced when the buoyant effects of all of the flue gases are combined in single jet, which typically rises far above the chimney while being blown laterally by the prevailing wind. To achieve the plume rise the gases must leave the stack with enough speed to get clear of the turbulence at the stack top. Otherwise they become entrained in the chimney wake, a phenomenon sometimes called "downwash." With one flue for several units it is not practical to maintain sufficient velocity at low load. Modern practice is to provide separate flues within one chimney structure and to terminate each in a smooth extension to minimize turbulence.

To appreciate the extent to which a tall stack can dilute the SO_2 emitted by a power plant, it is helpful to review a simple approximate

Figure 6.1
Didcot power station. An aerial view of the CEGB's 2000-MW coal-fired power station at Didcot, Berkshire, United Kingdom, showing the main station building with coal bunker, boiler house, turbine hall, and 650-ft multiflue chimney. Instead of a cluster of eight, as is usual on a 2000-MW station, the cooling towers at Didcot are grouped in two sets of three separated by half a mile. Also shown is the 400-kV outdoor substation, the gas turbine house with 325-ft chimney, and the coal stock area, which includes a rail system that carries the 14,000 tons required by the station each day. Courtesy of the Central Electricity Generating Board, London.

model of stack gas dispersion outlined by A. J. Clarke, D. H. Lucas, and F. F. Ross (18). This analysis is discussed here to illustrate the main features of stack gas dispersion, not to present accurate results. For more accurate but more complex analyses, see Refs. (19), (20), and (21).

Figures 6.2 and 6.3 show an idealized view of an exhaust plume from a power plant stack. The gases are assumed to rise above the stack to an effective height H and then disperse in the wind direction in the form of a conical plume. The difference between H and the chimney height h_c is called the thermal plume rise h_t. For large stacks (above 150 m and considerably taller than nearby buildings or hills) simple rules have been developed for estimating h_t. Clarke, Lucas, and Ross use the formula

$$h_t = (8.55 + 0.06\, h_c)\, \frac{Q_e^{1/4}}{u}, \qquad (1)$$

where h_t and h_c are in meters, u is the wind speed in m/sec, and Q_e

(W) is the sensible energy of the exhaust gases relative to ambient temperature.[3] Taking into account the properties of the flue gases and a temperature difference of about 120°C, it may be shown that Q_e is about one-eighth of the electrical output of the plant. Since Q_e appears in the formula to the power 1/4, a simple approximation is satisfactory for our purposes.

Figure 6.3 shows the plume touching the ground when its cone radius reaches H. At this point the volume flow through the base area of the plume is $\pi H^2 u$ (if the wind velocity is assumed to be

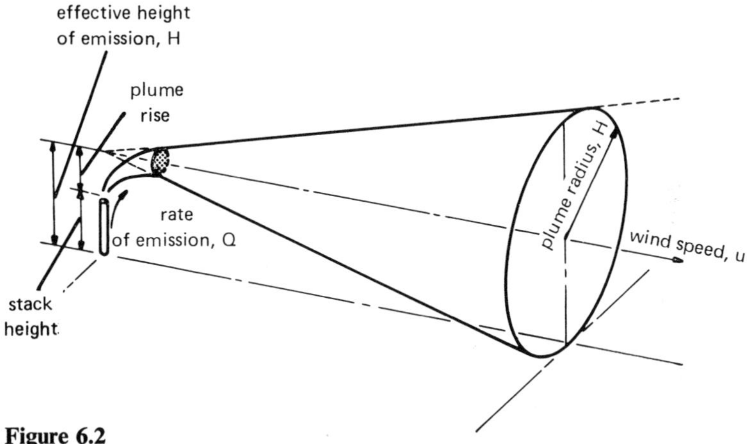

Figure 6.2
Model of plume rise and diffusion. After Ref. (18).

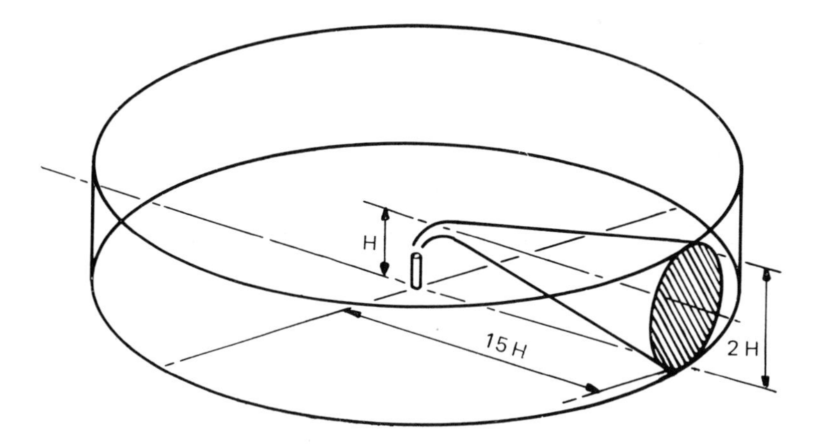

Figure 6.3
Areas for averaging plume dispersion. After Ref. (18).

[3]For large chimneys D. J. Moore (21) confirms an even simpler formula due to Lucas: $h_t = 16.29\ Q_e^{1/4}/u$. For a chimney height of 130 m the two formulas are in agreement. For larger heights the Lucas formula is a more conservative estimate of plume rise.

uniform), and the average pollutant concentration in the plume at that plane is

$$C_0 = \frac{Q_s}{\pi H^2 u}, \tag{2}$$

where the units of H and u are m and m/sec, respectively. If the total flow of pollutant Q_s is expressed in kg/sec, the reference concentration has the units $10^9 \mu g/m^3$; if Q_s is in m^3/sec, then C_0 will be in 10^6 ppm. The concentration C_0 may be taken as a rough measure of short-term ground-level concentration of pollutant.

For long-term average concentration, and assuming a wind of random direction, the cylindrical surface (Figure 6.3) is the appropriate flow area and the indication of long-term average concentration will be

$$\frac{Q_s}{2\pi(15H)2Hu} = \frac{Q_s}{60\pi H^2 u} = \frac{C_0}{60}.$$

United Kingdom measurements cited by Clarke, Lucas, and Ross indicate the results shown in Table 6.7.

The measurements in Table 6.7 suggest that C_0 and $C_0/60$ are reasonable and conservative indicators of short- and long-term ground-level concentrations, respectively. As an example, consider a 2000-MWe power plant burning 2 percent sulfur coal. Suppose the stack is 200 m high, and that wind velocity u is 7 m/sec.
By equation (1),

$$h_t = (8.55 + 0.06 \times 200)\frac{[(2000 \times 10^6)/8]^{1/4}}{7} = 369 \text{ m}.$$

Then the total effective stack height is

$$H = h_t + h_c = 369 + 200 = 569 \text{ m}.$$

Table 6.7
Observations of Pollutant Concentration in the Plume

Sampling Period	Mean Value of Observed Concentrations[a]	Normal Maximum Concentration	Highest Recorded Concentration
Short-Term Peak (3-min average)	$C_0/7$	C_0	C_0
1 hr	$C_0/18$	$C_0/2$	C_0
1 day	$C_0/80$	$C_0/10$	$C_0/4$
1 month	$C_0/320$	$C_0/60$	$C_0/40$

[a] C_0 is calculated with equations (1) and (2), and the measurements were made in the plume-affected area at the distance of maximum concentration.
Source: Ref. (18).

If the power plant operates at 38 percent efficiency and the coal fuel has a heating value of 25×10^6 kJ/tonne, the rate of emission of SO_2 will be

$$Q_s = \left(\frac{2000 \times 10^6 \text{J/sec}}{0.38}\right)\left(\frac{\text{kg coal}}{25 \times 10^6 \text{J}}\right)$$

$$\times \left(0.02 \frac{\text{kg S}}{\text{kg coal}}\right)\left(\frac{76 \text{ kg SO}_2}{44 \text{ kg S}}\right)$$

$$= 7.3 \text{ kg/sec.}$$

Then the concentration C_0 will be

$$C_0 = \frac{Q_s}{\pi H^2 u} = \frac{7.3 \times 10^9}{\pi (569)^2 7} = 1025 \ \mu\text{g/m}^3$$

and

$$C_0/60 = 1025/60 = 17 \ \mu\text{g/m}^3.$$

If, as in Table 6.7, the highest daily concentration recorded is $C_0/4 = 256 \ \mu g/m^3$, the tall stack with a wind of 7 m/sec would meet the standards outlined in Table 6.5.

Equations (1) and (2) show that using one stack for the total emissions of a power plant can result in significantly less ground-level concentration than two or three stacks of the same height.

This model does not allow for special atmospheric conditions such as temperature inversions or hilly terrain. With very low winds the formula predicts a very high plume rise, but in practice a layer of temperature inversion (temperature increasing with height) at some height will restrict this and cause an accumulation of pollutant. Breakup brings pockets of diluted gas to the ground, creating short-term peaks which may be somewhat higher than that given by equation (2) with $u = 7$ m/sec. Inversions at less than H in any wind cause all the plume to remain below, and an inversion at H will cause a doubling of the concentration. This is a rare, short-lived condition. If the inversion is lower still, the whole plume penetrates it and disperses above, giving almost zero concentration at the ground (22).

Figure 6.4 shows a sample result of calculations that use more exact methods of calculation than the one just described. It pertains to a hypothetical 6000-MWe plant exhausting through two 1000-ft (304-m) chimneys, and indicates the possibility of exceeding the hourly SO_2 standard of Table 6.6. For comparison, equations (1) and (2) would predict $C_0 = 0.74$ ppm assuming that the thermal plume rise of the two stacks is independent and that the plumes coalesce as they become horizontal. As mentioned earlier, United Kingdom experience suggests a maximum hourly concentration of $C_0/2 = 0.37$ ppm, which is in quite close agreement with Figure 6.4. However, this kind of agreement should not lead us to overlook the many uncertainties in all such calculations.

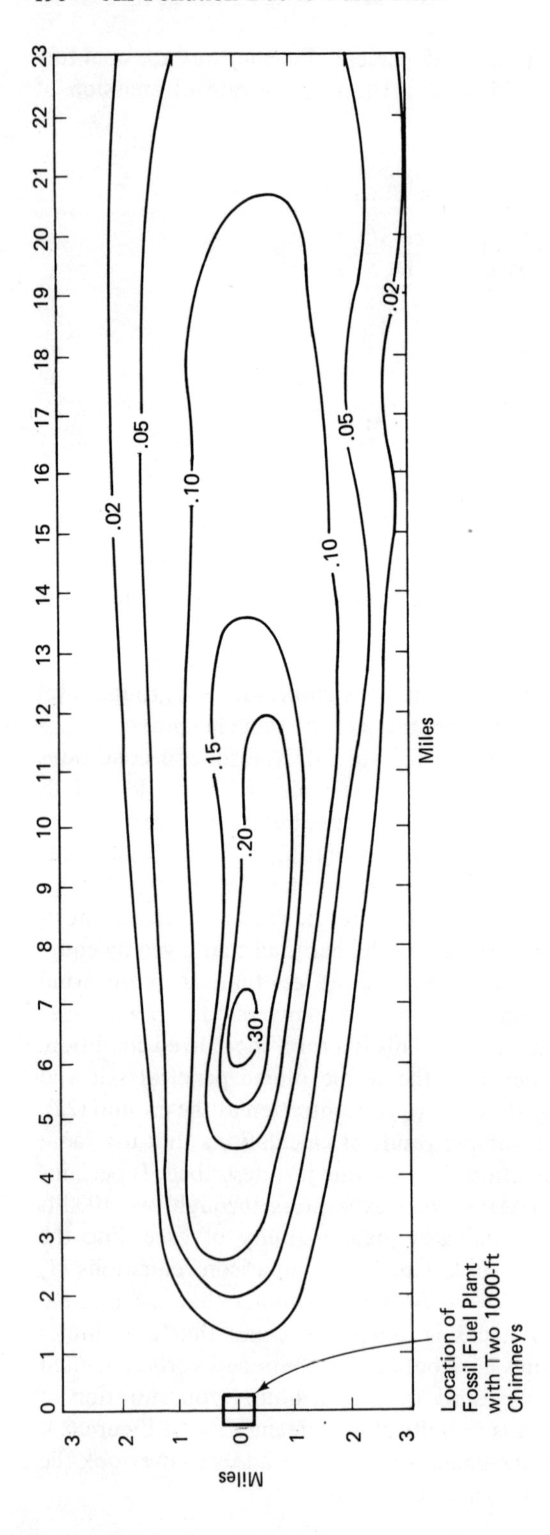

Figure 6.4
Predicted downwind SO_2 concentrations (hourly average ppm) for a hypothetical 6000-MW power station, 2.5 percent sulfur coal; typical weather conditions. After Ref. (7).

The conclusion has been reached in the United Kingdom that tall stacks are negligible contributors to ground-level concentration; the main sources are low-level industrial and residential chimneys (23). At the same time, in Sweden, it is recognized that long-range transport of SO_2, converted to sulfate, may have very serious effects on the soils and lakes. (References (23) to (26) deal with measurements, dispersion, and possible long-range transport of sulfates.) These effects are currently under intensive study. The EPA does not regard dilution of sulfur oxides by tall stacks as necessarily satisfactory. According to Barry R. Korb and Joseph Kivel, "it can be demonstrated that although dispersion can achieve desired regional air quality, it may lead to unacceptable air quality in adjoining regions" (27). They assert that control of ground-level concentration of SO_2 may be insufficient:

this does little to control sulfates, a much more toxic pollutant. Sulfates, which can be dispersed over great distances, are created in the atmosphere through conversion processes about which we know relatively little. There are sulfate levels already extant in large areas of the Lower Great Lakes and Northeast that exceed levels known to be harmful to health.

Environmental protection policy in the United States is based on the assumption that tall stacks are not a satisfactory solution to the SO_2 problem. The EPA has accordingly continued to emphasize the need for absolute control of sulfur emissions. In Canada it has been recommended that future fossil fuel plants (burning high-sulfur fuel) should not be located within 30 miles of major population centers, or of each other (7). In the United States a large expansion in the rate of consumption of high-sulfur fuels appears almost inevitable; there is little question that sulfur removal processes will be required.

6.3.3 Sulfur Removal from Fuels or Stack Gases

Numerous methods of removing sulfur from power plant fuels are being developed. They include gasification of high-sulfur fuels (with sulfur removal before combustion), fluidized-bed combustion (with sulfur removal during combustion), and stack gas cleaning (by liquid scrubbing or by dry catalytic processes). All of these processes have yet to be developed to the point of proven feasibility in large-scale operation, and in general they are expensive.

Removing SO_2 from stack gases is a particularly difficult practical problem, given the great volume of combustion gases and the dilution of SO_2 therein. A 1000-MWe coal-burning plant will burn about 0.1 tons of coal per second. Each ton of coal produces around 15,000 m^3 of combustion gases. The combustion gas volume flow is therefore 1500 m^3/sec, of which the SO_2 volume is only around 0.1 percent.

6.4 CARBON MONOXIDE

As Table 6.2 indicates, the main contributor to CO emissions from fossil fuel combustion is the internal combustion engine.

Carbon monoxide is dangerous because of its strong affinity for hemoglobin, the oxygen-carrying red blood cells. Absorption of CO through the lungs reduces the oxygen supply to the vital tissues. In increasing concentrations it will induce headache, reduced mental acuity, vomiting, collapse, coma, and death. Table 6.8 shows the levels of CO concentration in air that are believed to be associated with the formation of carboxyhemoglobin in the blood and degrees of physical impairment.

In cities, CO levels frequently reach 50 ppm (60 $\mu g/m^3$) and can go as high as 370 ppm (430 $\mu g/m^3$) in traffic jams. Averages of 20 to 30 ppm (25 to 35 $\mu g/m^3$) for eight hours and peaks of several hundred parts per million have been reported from various cities in Europe and North America (2, 28, 29). Table 6.8 refers to nonsmokers; cigarette smokers who average one pack a day are exposed to the equivalent of an ambient CO concentration of 50 ppm.

In Europe these levels of CO contamination have not been taken as seriously as in the United States. M. W. Holdgate and L. Reed of the U.K. Department of the Environment state:

> the only adverse effects known of CO concentration occur in urban areas (especially road tunnels and confined spaces in heavy traffic) where levels can rise sufficiently to block a small proportion of the oxygen-carrying capacity of the blood. Provided that discharges are arranged in such a way that these local concentrations do not exceed those at which a health risk might be incurred, there seems no need to adopt heroic and expensive measures to control emissions. (23)

The response in the United States and Canada has been different. The 1970 U.S. Clean Air Act amendments include the most stringent legislation on automotive emissions that have ever been passed (30). The law required that emissions of hydrocarbons and CO from cars produced in 1975 be reduced 90 percent below 1971 vehicle emission levels. By 1976 (now 1977 since a 1-year delay has been granted) NO_x emissions are to be reduced 90 percent from the average levels measured on 1971 automobiles. Table 6.8 shows the relatively rapid rate at which these standards have been enforced. Canadian

Table 6.8
U.S. Federal Motor Vehicle Emissions Standards (g/mile)

Substance	Uncontrolled (pre-1968)	1971	1972	1973	1974	1975	1976
Hydrocarbons	17	4.1	3.0	3.0	3.0	0.41	0.41
CO	125	34.0	28.0	28.0	28.0	3.4	3.4
NO_x	6	—	—	3.1	3.1	3.1	0.4

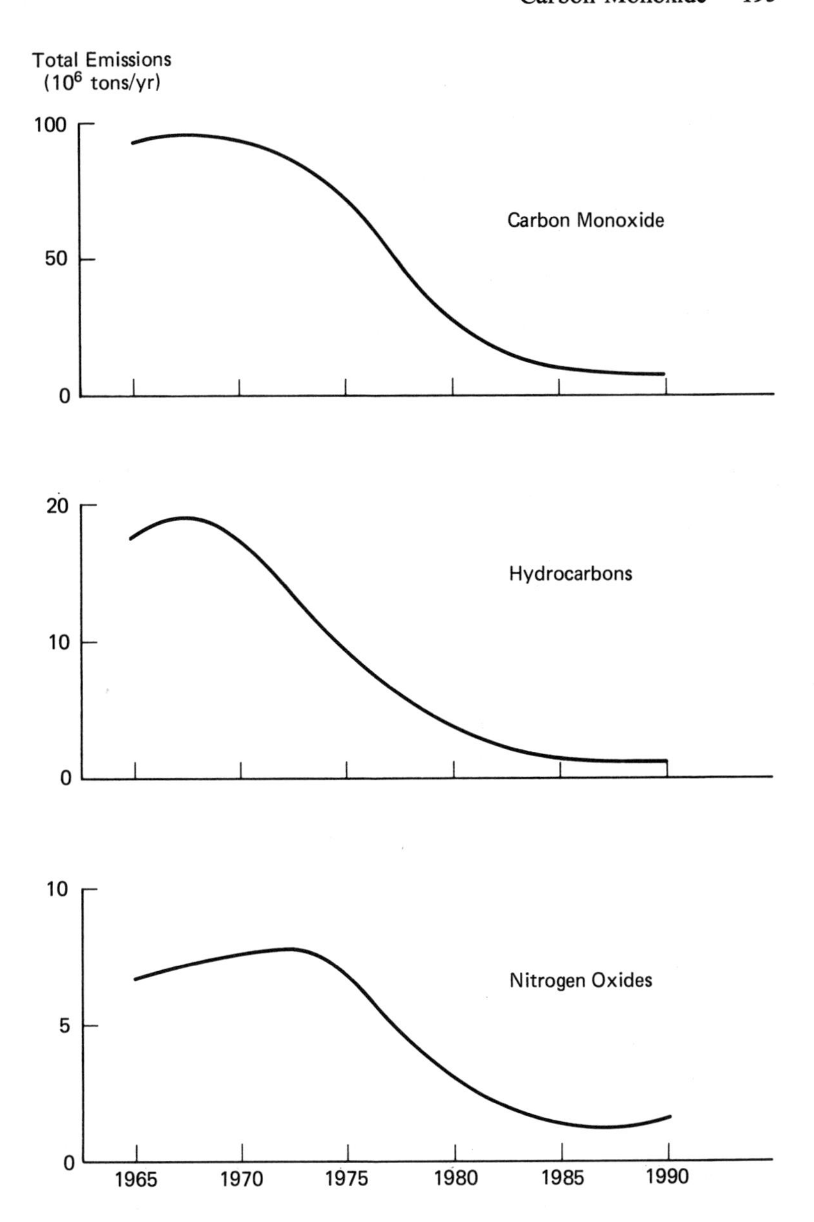

Figure 6.5
Estimated total automotive pollutant emissions in the United States (28).

standards are to be very similar, though more tolerant of NO_x emissions. The Clean Air Act requires that manufacturers guarantee that the vehicle is capable of conforming to the emissions standards for a period of either 5 years or 50,000 miles.

Figure 6.5 shows how total automotive emissions of CO, HC, and NO_x in the United States should decline as a result of the new standards, provided that cars are satisfactorily maintained and continue to provide low emissions as they age.

Control of CO and hydrocarbon emissions in engine exhausts is basically a matter of assuring completeness of combustion. Since passage of the U.S. Clean Air Act of 1970, major improvements have been made in control of fuel–air mixture, fuel evaporation, choking, inlet manifold design, and control of ignition and timing. In addition, new developments in stratified charge and precombustion chamber engines would make the targets set for both CO and HC realistically attainable on production engines within a few years.

The difficulty is that improving combustion to reduce the emissions of these constituents will raise combustion temperature, and so increase the NO_x emissions. Some measures taken to lower combustion temperature, such as recirculation of the exhaust gas into the combustion chamber, tend to lower combustion efficiency and raise the emissions of CO and hydrocarbons. This has called for methods of oxidizing these products outside the engine, either in exhaust burners or catalytic reactors. Preservation of catalysts requires removal of lead from gasoline fuels, which in turn requires reduction of the engine compression ratio (to prevent knocking combustion) and loss of fuel economy.

6.5 NITROGEN OXIDES AND HYDROCARBONS

As shown in the early 1950s by Professor A. J. Haagen-Smit, large quantities of hydrocarbons and nitrogen oxides mixed in the atmosphere, and acted upon by sunlight, lead to the release of photochemical oxidants including ozone and other compounds. These reactions lead to smog formation, and the photochemical oxidants may adversely affect materials, vegetation, and human health. The EPA reports that plants have been injured by 4-hr exposure to 0.05 ppm of photochemical oxidants (31). Eye irritation has been reported at oxidant levels of 0.1 ppm (200 $\mu g/m^3$); this level is exceeded in major California cities as well as in St. Louis, Denver, Philadelphia, Cincinnati, and Washington, D.C., on about half the days of each year. Impairment of athletic performance and increases in attacks among asthmatics have been observed when oxidant levels rise to 0.5 ppm for 1 hr. Ozone levels of 0.9 ppm (1800 $\mu g/m^3$) have been measured in Los Angeles. Table 6.5 shows the general human sensitivity to atmospheric oxidants. It indicates a maximum annual average NO_x concentration of 0.05 ppm (100 $\mu g/m^3$) for protection of health. Although in California only the Los Angeles and San Francisco regions are affected by eye irritation, large areas of the state are susceptible to plant damage and reduced visibility (23).

The significance of NO_x lies mainly in the fact that they are essential to photooxidation of hydrocarbons and the development of photochemical oxidants. Thus, sufficiently reduced levels of either NO_x or hydrocarbons in the air tend to alleviate formation of photochemical oxidants, but the exact relationships are extremely complex. The direct health effects of NO_2 at sufficiently high concentrations derive from its affinity for hemoglobin. It also forms nitric acid, which attacks lung tissue. However, these direct effects of NO_2 would require concentrations much higher than have been measured even in Los Angeles.

The approximate mechanism of photochemical smog formation is shown in Figure 6.6. The conditions in and around Los Angeles which have been so favorable to smog formation include strong sunlight, long periods of nearly stagnant air, and heavy hydrocarbon and

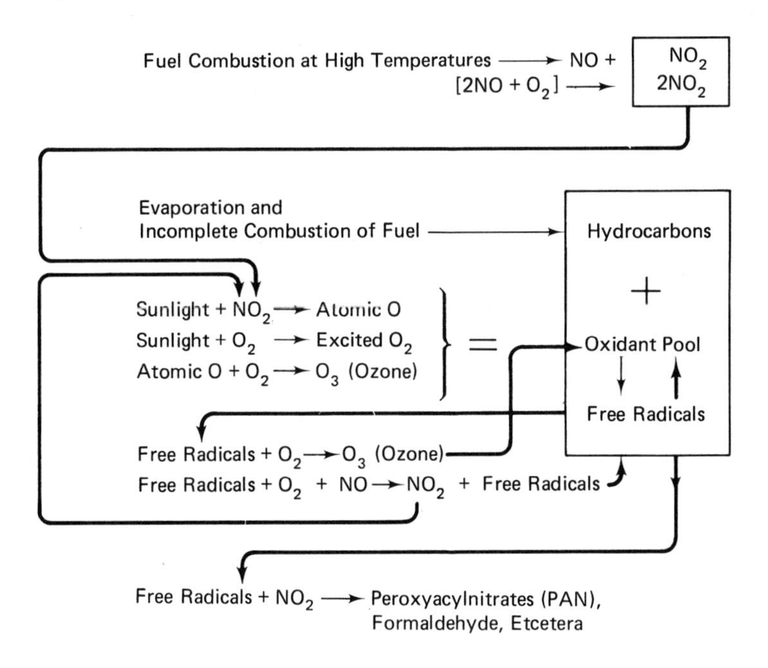

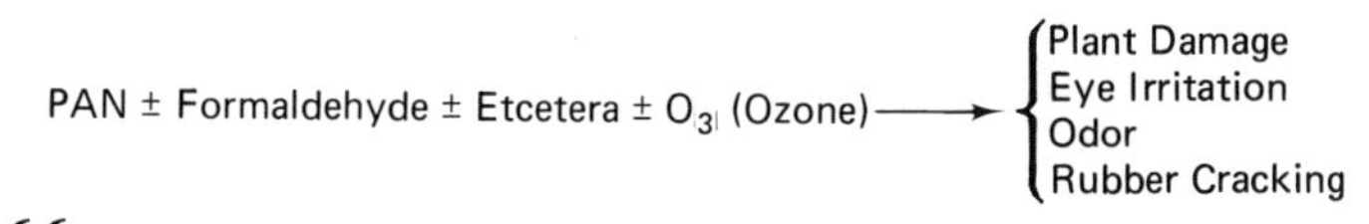

Figure 6.6
A simplified reaction scheme for photochemical smog formation. After Ref. (32). Reprinted by permission of *International Science and Technology* (June 1965).

NO_x pollution from motor vehicles. In brief, sunlight acting on NO_2 in the presence of atmospheric oxide produces ozone, which partially oxidizes the unburned hydrocarbons from combustion exhausts to form what are called free radicals. These in turn interact with NO_2 to form some of the main irritants of photochemical smog, peroxyacyl nitrates (PAN). Other constituents, including ozone and formaldehyde, also lead to effects such as plant damage, eye irritation, odor, and deterioration of rubber compounds. The chemical reactions are extremely complex, and include processes of cyclical regeneration.

Table 6.2 shows the relative contributions of stationary and mobile combustion sources of NO_x; these are equally important. Legislation controlling automotive emissions of NO_x has already been discussed (Table 6.8) as well as the difficulty of reaching these low emission levels. For new electricity generation plants the EPA has ruled that NO_x emissions must be reduced to 0.2 lb/10^6 Btu input for gas, 0.3 lb/10^6 Btu input for oil, and 0.7 lb/10^6 Btu input for coal.

Nitrogen oxide, NO, forms at high temperature during fossil fuel combustion. This remains "frozen" as the combustion gases cool and is oxidized to NO_2 in the atmosphere. The sources of nitrogen are the air and the chemically combined nitrogen in coal and oil fuels. Nitrogen oxide formation is inhibited by reducing combustion temperature or by decreasing time in the hot combustion zone. Peak temperatures in automotive engines can be reduced by exhaust gas recirculation or, in stationary power plants, by two-stage combustion. In the first stage the combustion is fuel-rich and temperatures are fairly low, so little NO is formed; this is followed by a heat exchange process, and then a second low-temperature combustion completes the process. Water injection is another method of reducing combustion temperature. Fluidized-bed combustion can run cool since the bed itself may be cooled by mechanisms such as steam pipes.

6.6 SUMMARY

Air pollution due to fossil fuels has led to specific disasters in the past, and to widespread deterioration in air quality in certain areas. Legislation designed to preserve minimum standards of air quality has brought major improvements. However, the question of what standards are necessary to protect human health (e.g., for ambient concentration of SO_2) is still not clearly settled.

The primary problem in fossil fuel pollution control appears to be the insufficiency of reliable scientific data on the effect of low-level pollution on human health. Standards must be established—even though the data base is and may continue to be unsatisfactory—if disastrous episodes are to be avoided. Even in the absence of specific disasters for which causes can be definitely identified, there is the possibility of general or chronic deterioration of the health of large populations. Such health effects can be identified in a statistical sense, but with uncertainty, since causality is difficult to establish.

The second major problem is the cost and time required for technical developments to reduce pollutant emissions to a level that clearly offers a wide margin of safety from adverse health effects. Given the uncertainty about mechanisms of health damage and threshold pollutant concentrations, there is naturally wide disagreement in the method of technical control required. The classic case is pollution due to compounds of sulfur. If atmospheric dispersion by means of tall stacks is not considered adequate, and it is judged that sulfur must be removed from stack gas before emission, the capital cost of the plant may rise by $100 million. The requisite technology for sulfur removal is still under development. Prototype processes have been unreliable as well as expensive, and some introduce problems of solid and liquid waste disposal. Given a clear and continuing need, however, there seems little doubt that one or more of these processes could be made to work satisfactorily in time. In the absence of adequate data, the cost–benefit or human risk estimate associated with various degrees of pollution control simply cannot be done adequately. The timetable for introduction of new controls is extremely important: if the rules change too rapidly, there may not be time to develop optimum technology, and the rules may then become discredited. Since huge costs are involved, the case for the new rules must be clear; but can it be made sufficiently clear before the arrival of the situation the rules are designed to prevent?

Tremendous reductions in automotive emissions have already been shown to be feasible. Whether the many millions of internal combustion engines that will be "required" by the end of the century can be made to operate with sufficiently low emissions is not yet clear; much will depend on the degree of NO_x control that will be needed, and on the extent to which the low pollutant emissions of older cars can be maintained. Even if the present internal combustion engine ultimately proves inadequate, however, there are other fuels, other engines, and other transport modes which, at a cost, can be developed to meet transportation needs.

With a fairly rich variety of technical options available, there seems no basis for assuming that air pollution should set a rigid limit to the level of energy use; it may, however, force great changes in fuel, engine, and power plant technology, and in energy cost.

References

1. H. S. D. Cole, Christopher Freeman, Marie Jahjoda, and K. L. R. Pavitt, eds. *Thinking about the Future: A Critique of "The Limits to Growth."* Sussex, England: Sussex University Press, 1973.

2. Emil T. Chanlett. *Environmental Protection.* New York: McGraw-Hill, 1973.

3. C. R. Lowe, "Clean·air: The health balance sheet." *Clean Air Conference, Southport, 20–23 October, 1970.* Part I: *Reprint of Papers*, p. 83.

4. Sir Eric Ashby, Chairman. *Royal Commission on Environmental Pollution. First Report.* London: H. M. S. O., February 1971 (reprinted 1973).

5. J. N. Beer and A. B. Hedley. "Air pollution research: Reduction of combustion generated pollution." *Proceedings. Institute of Fuel Conference "Fuel and the Environment,"* Eastbourne, 26–29 November 1973, vol. 1.

6. *The Federal R & D Plan for Air Pollution Control by Combustion Modification.* Report prepared for EPA under contract CPA 22–69–147. Columbus, Ohio: Battelle Memorial Institute, 1971.

7. K. E. Templemeyer, Chairman. *Impact of Energy use on the Environment.* Final Report of the Energy and the Environment Subcommittee, Advisory Committee on Energy, Government of the Province of Ontario, November 1972.

8. William H. Matthews, William H. Kellogg, and G. D. Robinson, eds. *Man's Impact on the Climate.* Cambridge, Mass.: MIT Press, 1971.

9. William H. Matthews, ed. *Man's Impact on the Global Environment.* Report of the Study of Man's Impact on the Climate. Cambridge, Mass.: MIT Press, 1971.

10. Eric Eriksson. "The fate of SO_2 and NO_x in the atmosphere." Chapter 16 in *Power Generation and Environmental Change*, David A. Berkowitz and Arthur M. Squires, eds. Cambridge, Mass.: MIT Press, 1971.

11. U.S. Public Health Service. *Air Quality Criteria for Particulates,* Publication AP-49, 1969; *Air Quality Criteria for Sulfur Dioxide,* AP-50, 1969; *Air Quality Criteria for Carbon Monoxide*, AP-62, 1970; *Air Quality Criteria for Chemical Oxidants*, AP-63, 1970; *Air Quality Criteria for Hydrocarbons*, AP-64, 1970. See also Environmental Protection Agency. *Air Quality Criteria for Nitrogen Oxides*, AP-84, 1971.

12. P. J. Lawther and J. A. Bonnell. "On recent trends on pollution and health in London—and some current thoughts." Paper presented at the Second International Clean Air Conference, Washington, D.C., December 1970.

13. A. D. Bradshaw. "The ecological effects of pollutants." *Proceedings. Institute of Fuel Conference: Fuel and the Environment.* Eastbourne, 26–29 November 1973.

14. U.S. Department of Health, Education, and Welfare. *Air Quality Data from National Air Sampling Networks and Contributory State and Local Networks.* National Air Pollution Administration Publication No. APTD–68–9 (1966 edition).

15. Environmental Protection Agency. "National ambient air quality standards." *Federal Register*, vol. 36, no. 17, April 17, 1971.

16. *Air Quality and Stationary Source Emission Control.* Report prepared for the Committee on Public Works, U.S. Senate, March 1975.

17. A. T. Rosano, ed. *Air Pollution Control.* New York: EKA, Inc., Environmental Science Service Division, 1962, p. 143.

18. A. J. Clarke, D. H. Lucas, and F. F. Ross. "Tall stacks: How effective

are they?" Paper presented at the Second International Clean Air Conference, Washington, D.C., December 1970.

19. F. Pasquill. *Atmospheric Diffusion*, 2nd edition. New York: Halsted Press, 1975.

20. S. J. Williamson. *Fundamentals of Air Pollution.* Reading, Mass.: Addison-Wesley, 1973.

21. D. J. Moore and A. G. Robins. "Experimental and theoretical investigations in chimney emissions." *Proceedings of the Institution of Mechanical Engineers*, vol. 189 (1975), pp. 33–54.

22. F. F. Ross. Personal communication, March 1976.

23. M. W. Holdgate and L. Reed. "The fate of pollutants." *Proceedings. Institute of Fuel Conference: Fuel and the Environment*, Eastbourne, 26–29 November 1973, vol. 1.

24. D. H. Lucas. "The effect of emission height with a multiplicity of pollution sources in very large areas." Paper 2.1–2, *Transactions of the 9th World Energy Conference*, Detroit, Michigan, September 22–25, 1974.

25. G. Persson. "Problems, Policy and Practice in the International Field." *Proceedings. Institute of Fuel Conference: Fuel and the Environment*, Eastbourne, 26–29 November 1973, vol. 1.

26. J. S. S. Reay. "Monitoring of the environment." *Proceedings. Institute of Fuel Conference: Fuel and the Environment*, Eastbourne, 26–29 November 1973, vol. 1.

27. Barry R. Korb and Joseph Kivel. "Air pollution control under the Clean Air Act and its energy implications." Paper No. 2.1–9, *Transactions of the 9th World Energy Conference*, Detroit, Michigan, September 22–25, 1974.

28. P. E. Trott. *Continuous Measurement of Carbon Monoxide in Streets.* Ministry of Technology, Warren Springs Laboratory, 1967–1969.

29. P. Chauvin, "Carbon monoxide: Analysis of exhaust gas investigations in Paris." *Environmental Research*, July 15, 1967.

30. J. J. Brogan. "Recent automotive air pollution control legislation in the U.S. and a new approach to achieve control: Alternative engine systems." Paper C141/71 in the Automobile Division and Combustion Engines Group conference volume, *Air Pollution Control in Transport Engines.* London: Institution of Mechanical Engineers, October 1971.

31. Environmental Protection Agency. *Annual Report of the Environmental Protection Agency to the Congress of the United States in Compliance with Section 202 (6)(4), Public Law 90–148—The Clean Air Act as Amended.* July 1, 1971.

32. Seymour Tilson. "Air pollution." *International Science and Technology*, June 1965.

33. Eric Hirst. "Pollution control energy costs." Paper 73/WA/Ener-7 presented at the Annual Winter Meeting of the American Society of Mechanical Engineers, Detroit, Michigan, December 6, 1973.

34. *Sulfur Dioxide (SO_2)—An Air Pollutant.* Report by the Technical Committee, National Society for Clean Air, 134/136 North Street, Brighton, U.K., 1973.

35. Jack Nord Forsker. "Meso-scale and large-scale transport of air

pollutants." Paper presented at the Third International Clean Air Conference, Düsseldorf, October 8–12, 1973.

Problems

6.1 This problem involves estimation of the SO_2 emissions from the plant shown in Figure 6.1, using the simplified method presented by Clarke, Lucas, and Ross (18). To determine the maximum ground-level concentration of SO_2, use the following assumptions:

1. The plant operates with 2.5 percent sulfur coal and a heat rate of 10,000 Btu/kWh. The energy content of the coal is 25×10^6 Btu/ton.
2. The thermal plume rise is given by equation (1).
3. The plumes of each stack rise independently to a height H, but coalesce thereafter into a single cone of horizontal axis and uniform concentration across any cross section. The wind speed is 7 m/sec.
4. The volume flow rate of SO_2 can be determined from the mass of sulfur consumed in the coal, assuming the SO_2 is at atmospheric pressure and temperature far downstream of the stack. (The atomic weight of sulfur is 32.)
5. The plume touches the ground at a distance $15H$ from the stacks.

Estimate:

a. The effective stack height H.
b. The volume flow rate of SO_2.
c. The typical short-term ground-level concentration $C_0/2$.
d. The typical long-term concentration $C_0/60$.

Compare results with those of Table 6.5, using 1 ppm SO_2 = 2620 $\mu g/m^3$.

6.2 Clarke, Lucas, and Ross (18) present required stack heights for coals of various sulfur contents as shown in Table 6.A.

Table 6.A
Required Stack Height for Coal of Various Sulfur Contents

Sulfur Content (percent)	Required Emission Height (m)	Plume Rise (m)	Required Stack Height (m)
2	550	350	200
3	675	425	260
4	775	460	315
5	865	505	360

The plant is 2000 MWe, and the wind velocity is assumed to be 7 m/sec. Assume that the plant heat rate is 9000 Btu/kWh and the heating value of the coal is 25×10^6 Btu/ton.

a. Determine the value of C_0 ($\mu g/m^3$) corresponding to use of equation (2) to determine the emission height H in Table 6.A.

b. Use equation (1) to verify (approximately) the total emission height, using the stack heights in Table 6.A.

c. Determine, for the total emission heights cited in Table 6.A, the total emission height using the formula

$$h_t = 16.29\, Q_e^{1/4}/u.$$

Again take Q_e as one-eighth of the electrical power output.

6.3 Using the simplified dispersion model of Section 6.3.2, determine the percentage changes in the estimated long-term average ground-level concentration of SO_2 under the following conditions:

a. The wind velocity is 3 m/sec or 10 m/sec instead of 7 m/sec.

b. The thermal rise of the plume is only half as great as in equation (1).

The thermal emission Q_e is 125×10^6 W and the stack height h_c is 250 m.

Note: For refined methods of estimating atmospheric dispersion of pollutants see Refs. (19) and (21).

6.4 A 2000-MWe coal-fired plant emits 3000 m^3/sec of exhaust gases (at standard atmospheric pressure and temperature). The concentration of NO_x (mostly NO) in the stack is 800 ppm, midway in the range indicated in Table 9.8. Assuming the stack height is 240 m and the wind velocity is 7 m/sec, estimate the maximum short-term and long-term ground-level concentrations of NO_x. Compare these results with the NO_x standards of Tables 6.5 and 6.8. Use equation (1) for estimating thermal plume height.

6.5 Given the data in Table 6.B and approximate carbon–hydrogen ratios for gas, oil, and coal fuels, determine the stack gas concentrations of NO_x (calculated as NO_2) in stoichiometric combustion.

Table 6.B
Characteristics of Various Fuels

	Gas	Oil	Coal
Approximate Composition	CH_4	$CH_{1.5}$	$CH_{0.8}$
Heating Value (Btu/lb)	25,000	19,000	12,000
NO_x Emission Limit (lb NO_2 per 10^6 Btu)[a]	0.2	0.3	0.7

[a]EPA.

First show that if the fuel is CH_n, and air consists of 3.76 moles of N_2 per mole of O_2, the number of moles of products (CO_2, H_2O, N_2) per unit mass of fuel is $(4.76 + 1.44n)/(12 + n)$. Then calculate NO_x

concentration in $\mu g/m^3$ and compare results with typical data on stack gas NO_x concentrations shown in Table 9.8.

6.6 The SO_2 level will build up in a stagnant atmosphere over a city. Estimate the increase in average concentration level of SO_2 in a windless atmosphere if the products of combustion of residential furnaces are contained for 24 hr and uniformly mixed in a layer of air 1000 m from the surface of the city. Assume the city has 200,000 dwelling units in a space of 1000 km^2 and that each consumes 1 gal/hr of oil with 1.0 percent sulfur fuel. The specific gravity of the oil is 0.8. The air density is 1.18 kg/m^3.

Suppose a 1000-MWe coal-fired power plant burning coal of 3 percent sulfur but with 85 percent removal of SO_2 from stack gases were discharging into the same air volume in the same period and that the SO_2 emitted were uniformly distributed within the volume. What additional increment in SO_2 concentration could conceivably result? Assume the heating value of the coal is 25×10^6 Btu/ton and the plant efficiency is 0.39.

6.7 Lave and Seskin (see Refs. (17) and (18) of Chapter 5) find from a statistical study that an additional microgram per cubic meter of mean annual SO_2 concentration is associated with increased mortality of 0.039 per 10,000 population per year.

In London, the annual mean concentration of SO_2 in the atmosphere is about 115 $\mu g/m^3$. Using the coefficient of Lave and Seskin, calculate the extra death risk per million population corresponding to the difference between 115 $\mu g/m^3$ and the standards of Table 6.5. Relate this to other risks (Table 5.6).

Table 6.C
Central Station Power Plant Emissions

Emission	Coal (lb/ton of coal burned)	Residual Oil (lb/1000 gal of oil burned)	Natural Gas (lb/10^6 ft^3 of gas burned)
Aldehydes	0.005	1.0	3.0
CO	1.0	0.04	0.4
Hydrocarbons	0.3	5.0	40
Other Organics			4.0
NO_x	18	100	390
SO_2[a]	38(S%)	157(S%)	0.6
SO_3[a]		2(S%)	
Particulates		8	15
Fuel Heating Value	25×10^6Btu/ton	160,000 Btu/gal	1030 Btu/ft^3
Plant heat rate (Btu/kWh)	9500	9800	10,300

[a](S%) is the fuel sulfur content in percent by mass.

6.8 Typical pollutant emissions from fossil fuels consumed in central station power plants have been categorized as shown in Table 6.C.

a. Calculate the SO_2 production rate in lb/sec for coal, residual oil, and natural gas plants operating at a 2000-MWe power level with 2.5 percent sulfur in the coal and oil fuels.

b. Calculate the NO_x production rate in lb/sec for coal, residual oil, and natural gas plants operating at a 2000-MWe power level.

c. Determine the relative production rates of particulates for coal, oil, and gas plants of the same power level. For the coal plant assume 10 percent ash content in fuel and 99.5 percent particulate removal from stack gases.

7
The Cost of Electrical Power Generation

7.1 INTRODUCTION

The decade of the 1960s appears in retrospect to have been a period of remarkably stable costs for electrical power generation. Table 7.1 shows the results of a biennial survey, from 1962 to 1974, of power station costs in the United States. The average production cost increase in the 1960s was only about 1.4 percent per year. Table 7.1 shows that plant construction costs were quite steady over the period and that fuel costs even declined slightly. Also, average heat rates (the ratio of fuel energy consumed in Btu, to the electrical energy produced, in kWh) fell slightly over the period. There are a number of minor variations from year to year in average capital, fuel, and other operational expenses. However, in 10 years the total increase in the unit cost of electrical energy was only about 15 percent. Allowing for the average effect of inflation on the value of the dollar over the same period, the cost of electricity actually declined about 25 percent (in constant dollar terms).

The 1970s have been a period of very different power generation economics. Fossil fuel prices have increased drastically (see the 1974 value shown in Table 7.1); so have the capital costs of both fossil fuel and nuclear plants. The single most shocking factor (since it took place so rapidly) has been the rise in world price of oil. Imported crude oil now costs \$11 to \$12/bbl, or about \$2/10^6 Btu. (In

Table 7.1
Survey of Power Costs in the United States, 1962–1974

Year	1974	1972	1970	1968	1966	1964	1962
Total Busbar Energy Cost (mills/net kWh)	15.09	7.99	7.20	6.04	5.84	6.82	6.78
Construction Cost Excluding Switchyard ($/kW)	193.21	144.33	125.57	117.94	118.81	126.80	145.95
Fixed Charges (mills/net kWh)	7.58	3.78	3.72	3.12	2.59	3.32	3.42
Operating Costs (mills/net kWh)	7.51	4.21	3.48	2.92	3.25	3.50	3.36
Fuel Cost (mills/net kWh)	6.05	3.49	2.83	2.40	2.73	2.93	2.78
Fuel Cost Portion of Operating Costs (percent)	82.2	82.6	81.8	82.8	84.7	82.6	83.5
Fuel Cost (cents/10^6 Btu)	72.29	35.86	28.36	24.12	26.24	26.26	27.24
Manpower, Operating and Maintenance Employees (per MW)	0.162	0.172	0.144	0.160	0.174	0.206	0.271
Station Net Heat Rate (average Btu/net kWh)	10,379	10,115	9927	9980	9713	10,347	10,098
Annual Plant Factor, Average (percent)	54.56	57.86	62.80	62.65	65.00	62.51	65.38
Utilization Factor, Average (percent)	93.16	98.60	98.02	99.68	98.50	96.67	99.57

Source: Ref.(1). Reprinted from *Electrical World*, Nov. 1, 1973, and Nov. 15, 1975.

early 1973, the price was \$3 to \$3.40/bbl.) At the same time coal prices have risen from about \$6 or \$8/ton to \$25 or \$30/ton for many utilities, or around \$1/10^6 Btu. The relative scarcity of natural gas makes it virtually unobtainable for new electrical power generation plants. In general, environmentally acceptable fossil fuels are either scarce or expensive, or both.

Owing to the high cost and possible shortages of fossil fuels, as well as the problem of satisfying emissions regulations, there has been a strong trend toward nuclear–electric power, despite rapidly increasing construction costs for nuclear stations. The decisive factor is that fueling costs are much less for nuclear than for fossil fuel plants. The cost of uranium ore has been relatively stable. Even if it were not, the total cost of nuclear power is not strongly dependent on the price of uranium ore. However, a switch to nuclear power cannot be made quickly; it takes six to nine years to construct a new plant after the go-ahead decision, and old plants will be phased out gradually. Thus huge increases in electricity generation costs must be expected for the coming decade, even if the switch to nuclear power is made as quickly as possible.

The second major factor that has caused a drastic increase in power production costs involves monetary inflation and high interest rates, which place a severe burden on the capital-intensive plant used for electrical power generation. The current rate of inflation of plant costs is 8 percent or higher, so that inflation alone will double capital costs in eight or nine years (the time needed to put a new nuclear plant into operation).

Environmental protection has also increased power costs. The extent to which government regulations restrict sulfur emissions from power plants will have a strong effect on the competitive position of coal and oil. Limitations on SO_2 emissions from power plants in the United States have led to premium prices being charged for limited supplies of low-sulfur fuel oil and low-sulfur coal. Since these are in such short supply, many utilities now burning coal or contemplating new coal-burning plants are forced to consider installation of SO_2 removal equipment, and this might increase the capital cost of the plant by 25 or 30 percent. This could raise the capital cost per kilowatt of the coal-fired plant to nearly the level of the nuclear plant. A variety of clean substitute fuels can be made from coal, but the technology for these is not fully developed and large-scale production is not yet available. Even the cheapest of these processes could double the price of fueling to the utility.

At the same time, increasing safety precautions required in the design of nuclear plants are adding to the cost of nuclear power. Examples include new requirements for emergency cooling of the reactor core and for emergency shutdown.

Although the relatively high capital costs of a nuclear plant are typically balanced by low fueling costs, the future costs of fuel refining and reprocessing and waste disposal are not certain. Many believe that the United States will not be able to obtain sufficient supplies of low-cost uranium to fuel its light-water reactors over the next three or four decades, and that a new kind of nuclear plant, the breeder reactor, will have to be built in large numbers. This would introduce many new problems of safety, reliability, and capital cost. Others are hopeful that, since the age of uranium exploration is new and uranium is relatively abundant in the earth's crust, relatively large supplies of uranium ore will be found at costs that will not mean an abnormal increase in the price of nuclear power. Moreover, there is the possibility of exploiting other nuclear fuels such as thorium, which is relatively abundant.

No attempt is made here to predict absolute power costs, though possible costs of fossil and nuclear power are compared under various assumptions. Many questions remain to be answered about the cost of producing clean liquid or gaseous fuel from coal. For the far future, questions arise concerning possible competition from fusion or solar power. Any present answers to these questions must be regarded as tentative, but let us draw the picture as well as we can with the information now available.

7.2 FUEL COSTS

Figure 7.1 shows the relative costs to United States utility companies of coal, gas, and oil fuel up to 1974, the last year for which

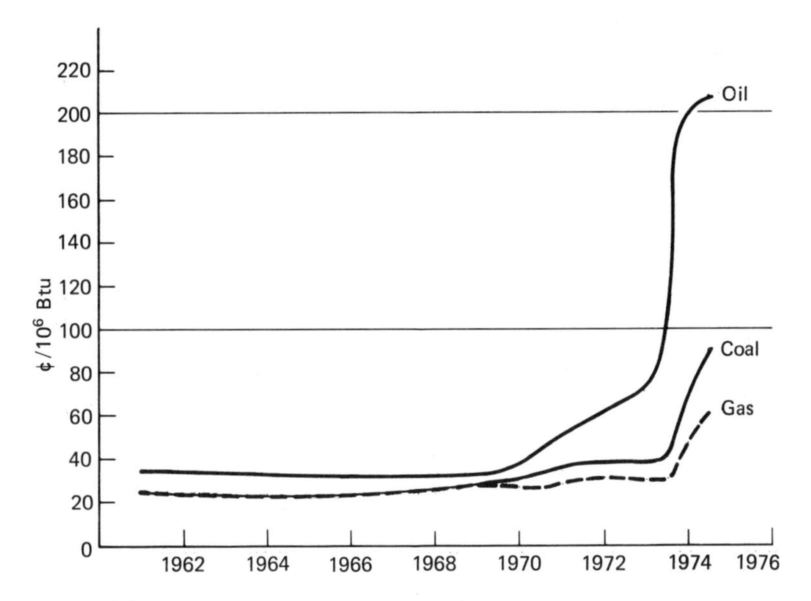

Figure 7.1
Electric utility average fuel costs (2, 3).

the tabulation has been published. The average fuel price for utilities declined in the early 1960s, and then a fairly rapid rise in oil prices occurred between 1970 and 1972, even before the large increases set by OPEC in 1973 and early 1974. The cost of imported oil, currently five to six times the 1970 value, may not decrease significantly in the future, given the serious possibility of early depletion of worldwide reserves of oil.

Other fossil fuels have also become rapidly more expensive. Recently, the price of coal has been moving quickly upward from its fairly stable level in many areas of $6 to $7/ton in the 1960s. The rise is due in part to increases in wages and increasingly stringent requirements for mine safety equipment. But the basic reason is possibly that new oil prices make coal seem a much more valuable market commodity. By 1975, metallurgical coking coal already cost about $40/ton, and coal for electrical power plants cost $25 or $30/ton. Gas prices are also rising rapidly, now that the early depletion of continental gas reserves is a matter of general agreement.

Nuclear fuel is still relatively stable in price, though reserves are limited. Unlike the cost of fossil fuel power, nuclear power cost depends relatively weakly on fuel costs. Taking the present total generation cost as 8 to 10 mills/kWh, only 1.5 to 2 mills are due to fueling cost for light-water reactors, and less than 1 mill for the natural-uranium-fueled heavy-water reactor. For light-water reactors only about 30 percent of the total fueling cost (1.5 to 2 mills/kWh) is due to the cost of the uranium ore; the rest is due to enrichment and fabrication. Thus, tripling the cost of the uranium ore may only increase the total nuclear power generation cost by 10 to 15 percent. In contrast, tripling the cost of fossil fuel may double the cost of power generation from a fossil fuel plant.

The relative independence of nuclear generation costs from fuel price is a great advantage, but it means that the capital cost is all the more important. Unfortunately, the capital costs of nuclear power plants have undergone drastic upward revisions in recent years.

7.3 CAPITAL COSTS

A review of capital costs of light-water nuclear reactors built in the United States has been compiled by F. C. Olds (4) and is presented in Table 7.2. The average cost figures in Table 7.2 are the actual costs or best estimates available by 31 March 1974; these may be considerably higher than estimates of average plant cost made when the plants were contracted. For example, plants contracted in 1967 were estimated, on the average, to cost $146/kWe and to require six years for completion. Actual construction experience for these same plants led to average figures of $346/kWe and 7.5 years for completion.

Table 7.2
Nuclear Plant Costs and Schedules, 1974

Year Contract Let	Average Date Entering Service	Number of Plants	Cost Range ($/kWe)		Average Cost ($/kWe)	Average Size (MWe)	Schedule Range (years)	Average Schedule (years)
			Low	High				
1965	1971	6	124	330	199	645	5–8	6.3
1966	1973	20	112	482	260	821	4–9	6.7
1967	1975	30	109	652	354	852	5–13	7.5
1968	1977	14	197	720	413	925	6–13	8.7
1969	1978	7	187	530	395	1030	6–13	9.2
1970	1979	14	240	577	370	1020	6–10	8.4
1971	1980	20	296	572	475	983	6–12	8.7
1972	1981	36	290	645	458	1105	7–11	8.5
1973	1982	38	313	650	456	1120	6–13	8.8
1974	1983	8	425	655	558	1215	8–12	9.4

Source: Ref.(4). Reprinted by permission of *Power Engineering*.

Table 7.2 points out how rapidly the capital costs of light-water nuclear plants have been increasing. Since 1970, the average annual increase has been about 26 percent. The time required to build plants has also increased markedly, 9 or 10 years not being unusual. As will be shown later, these two factors are closely related, and capital costs could be substantially decreased if the period of construction could be shortened.

Table 7.2 shows that average unit sizes have roughly doubled since 1965. This, by itself, could have reduced cost per unit power by 10 to 20 percent. However, this effect has been overwhelmed by increases due to inflation in the costs of labor, materials, and money. New regulations on plant design for health and environmental protection have also added to cost. The wide discrepancy between high and low cost figures is significant in this relatively early period of commercial nuclear power.

Table 7.2 indicates that nuclear light-water plants commissioned in 1974 will cost about $558/kWe. Olds suggests that, for a plant entering service in 1981, the average cost will be $600 to $700/kWe, or $800 to $900/kWe for a plant entering service in 1984 (4). The trend toward large-scale units of at least 1000 MWe is now firmly established.

Figure 7.2 shows the results of a series of estimates of capital costs for light-water nuclear plants. The estimates increased fivefold, from $134/kWe in 1967 (for 1973 operation) to over $700/kWe in 1974 (for 1983 operation). Such a large increase is extraordinary in the history of plant cost estimates; the reasons for it are worth considering.

The total capital cost of the plant is the sum of direct, indirect, and time-related costs:

1. Direct costs include nuclear steam supply system, turbine generator, construction material and equipment, and craft labor.
2. Indirect costs include professional services, construction tools and materials, and contingency funds.
3. Time-related costs include escalation during construction and interest during construction.

It is the third of these categories that accounts for most of the total increase in estimated plant cost. Interest costs have increased not only because of high interest rates but also because of lengthened construction periods.

For the 1967 estimate no allowance was made for escalation of costs during the construction period. For the 1973 estimate, escalation of costs during construction was estimated at the following rates:

Turbine generator: 6 percent (except for first three years)
Craft labor: 8 percent
Material and equipment: 5 percent

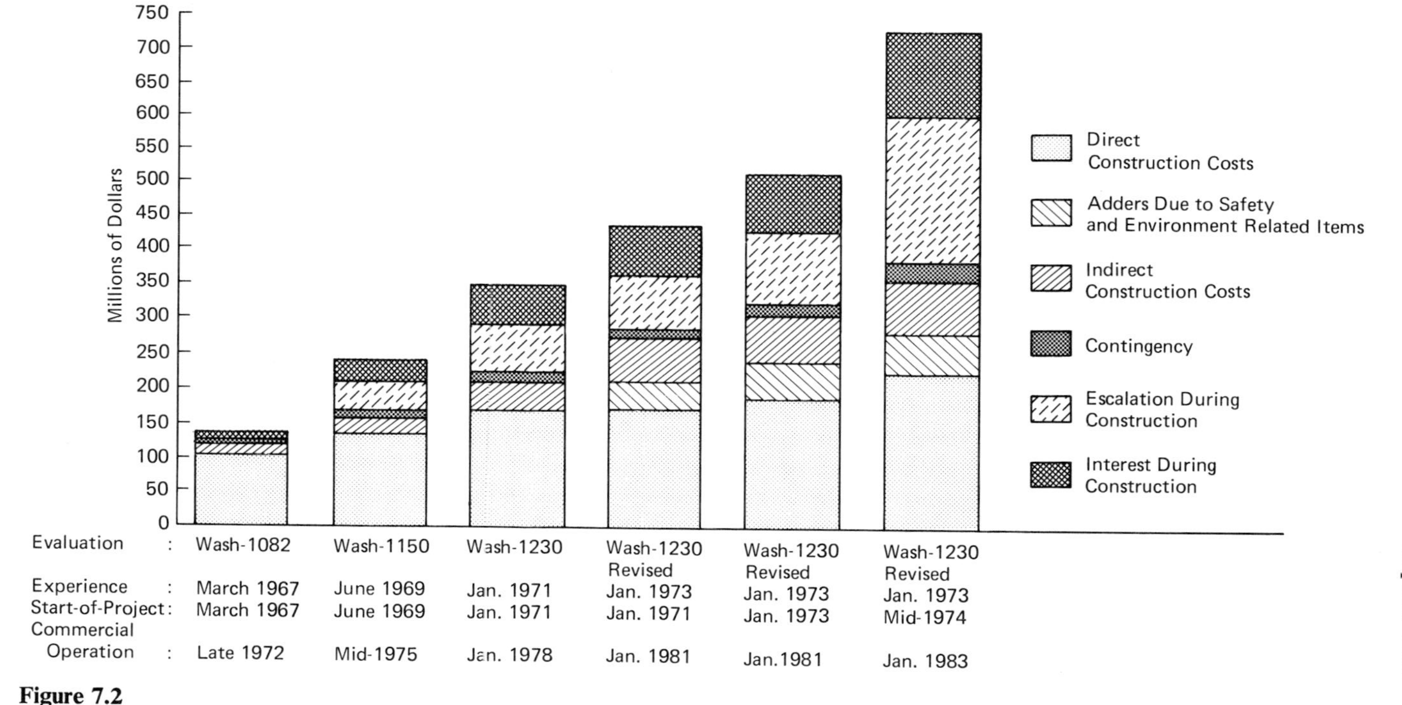

Figure 7.2
Comparison of nuclear plant cost estimates (total investment cost for 1000-MWe units). Compiled from United Engineering and Constructors, Inc., reports WASH-1082, -1150, and -1230 for the USAEC. From (5).

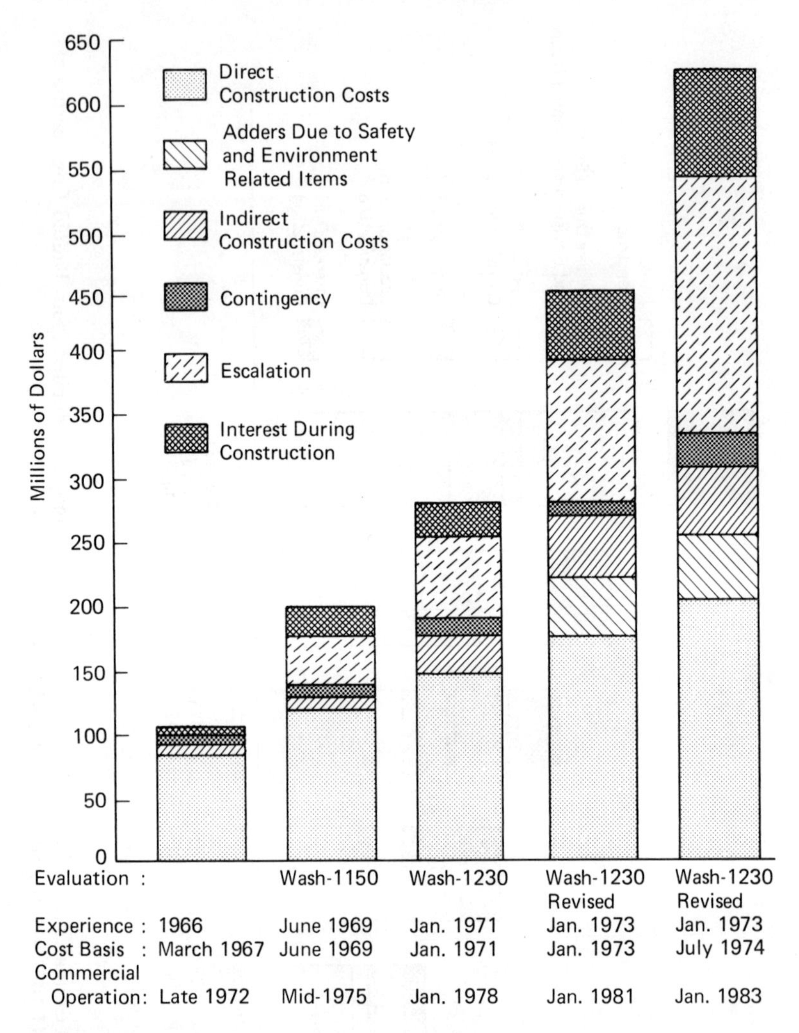

Figure 7.3
Comparison of coal-fired plant cost estimates (total investment cost for 1000-MWe units) (5).

Professional services: 6 percent.
Also, interest during construction was taken to be 7.5 percent with a construction period of 8.5 years. Escalation and interest during construction were estimated to be nearly half the total plant cost (for the 1983 date of commercial operation).

Figure 7.2 indicates a significant increase in total cost due to safety and environmental provisions. These include the effects of many small measures to reduce radiation dose, to increase reliability of reactor control, and to reduce the environmental effect of the plant cooling water (by reducing the temperature rise of the cooling water from 20 or 25°F to 15°F and reducing cooling water inlet velocity from 2.5 ft/sec to 0.5 ft/sec). The costs of these measures, with associated indirect costs, total nearly $60/kWe.

For the 1967 estimate, the nuclear steam supply system and turbine generator accounted for nearly 40 percent of total cost; for the 1973 estimate, the corresponding figure is only 15 percent. The nuclear steam supply system is more important than this last figure might suggest: it may have a considerable effect on construction time, and therefore on both interest and escalation costs, as well as on professional labor.

With Figure 7.2 in mind, one can appreciate why nuclear power cost suffers so severely from even moderate rates of inflation. Unfortunately these same effects also raise capital costs of fossil fuel plants. Figure 7.3 shows successive upward revisions of the capital costs of coal plants, which are similar to the nuclear estimates of Figure 7.2.

For the most recent of these estimates, the coal plant construction period was taken to be 4.5 years. Escalation and interest rates were as cited previously for the nuclear plant. Again, there was a fivefold increase in estimated cost between 1967 and 1974. Also, the interest and escalation costs rose to nearly half the total capital costs. The additional costs for environmental control are mainly due to an allowance of $50/kWe for the provision of SO_2 removal equipment.

Power plant cost estimates are typically quite sensitive to unit size, as demonstrated by Figure 7.4 (pertaining to a December 1981 operation date). However, such a dependence is difficult to discern in historical construction data because of differences in local labor and materials costs and other factors. Figure 7.5 shows capital costs of fossil fuel plants entering service in the United States in the period 1970–1971. It indicates a considerable spread in capital costs. No clear effect of capital cost on unit size is evident, although the data do not disprove a size effect. The average cost figure of about $150/kWe is considerably less than the figure of $199/kWe obtained from Table 7.2 for a nuclear plant entering service in 1971. The difference between the two is of the order of the cost of sulfur re-

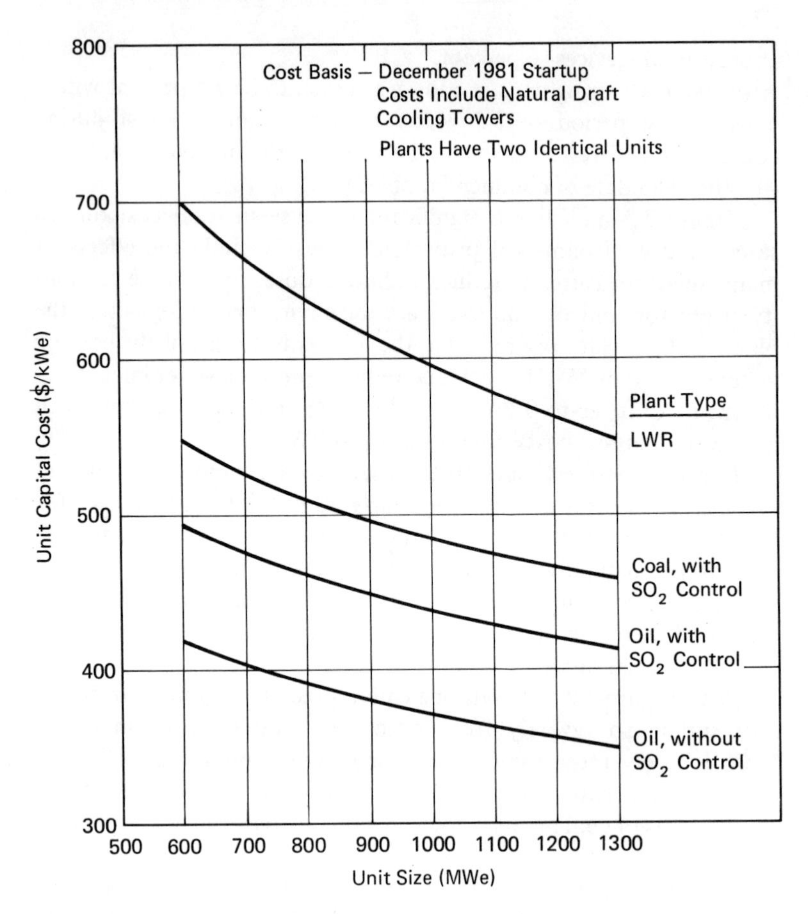

Figure 7.4
Unit capital costs of power plants as a function of size (5).

moval equipment, which would currently be required for fuels high in sulfur.

Capital costs of the most efficient oil, and coal units are shown against year of initial service in Figure 7.6. The gas-fired plants are not quite as efficient as coal or oil plants (for reasons to be discussed in Chapter 9), so only three of them are indicated in Figure 7.6. These show much lower capital costs than for oil or coal, since the gas-fired plant does not need equipment for fuel storage, and fuel handling is much simpler than for coal or crude oil. The capital costs of these most-efficient coal and oil plants vary over a considerable range; the high point is for the Eddystone plant in Philadelphia, designed for exceptionally high steam pressures and temperatures (see Table 9.4).

Capital costs of both nuclear and fossil fuel plants continue to increase year by year, beyond the levels shown in Figures 7.2 and 7.3. Table 7.3 provides an example of recent capital cost estimates for the Long Island Lighting Company. The alternatives considered were: two 1150-MWe nuclear units or three 800-MWe fossil fuel

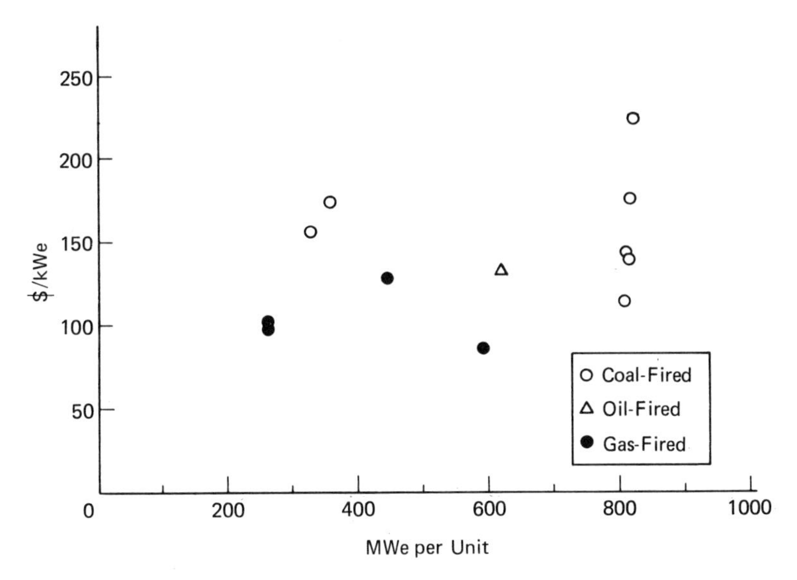

Figure 7.5
Capital costs of fossil fuel plants entering service in the United States, 1970–1971, with heat rates 10,000 Btu/kWh or less (2).

units (using either high-sulfur coal or oil). The average capital costs were $805/kWe (nuclear), $740/kWe (coal), and $637/kWe (oil). For the same time period gas turbine power for peak loads is estimated at $168/kWe.

With provision for sulfur removal, the capital cost of the coal plant is estimated at 9 percent of the nuclear plant costs. Other estimates have placed coal plant capital costs in the range 75 to 85 percent of nuclear costs (4, 6).

Given the great cost of nuclear plants, finding the capital to finance new plants has become a serious problem. It has been suggested that the ability to attract the necessary capital may be the major governing factor for the rate of expansion of nuclear power (7). Unfortunately, the capital cost of coal units, with provision for sulfur removal, are almost as high, and the high fueling costs make that option quite unattractive. As will be shown, this argument is much less compelling in western North America, where low-sulfur coal can be strip-mined at low cost. In this case both the capital and fuel costs of the coal option can be greatly reduced.

The capital costs of nuclear plants will probably always exceed those of fossil fuel plants of the same size. The former require extra systems for radiation monitoring, fuel failure detection, and coolant processing, as well as secondary containment and complex control of radioactivity inside and outside the reactor.

According to the IAEA the capital costs of all water-cooled reactors are about the same (excluding the heavy-water cost, where

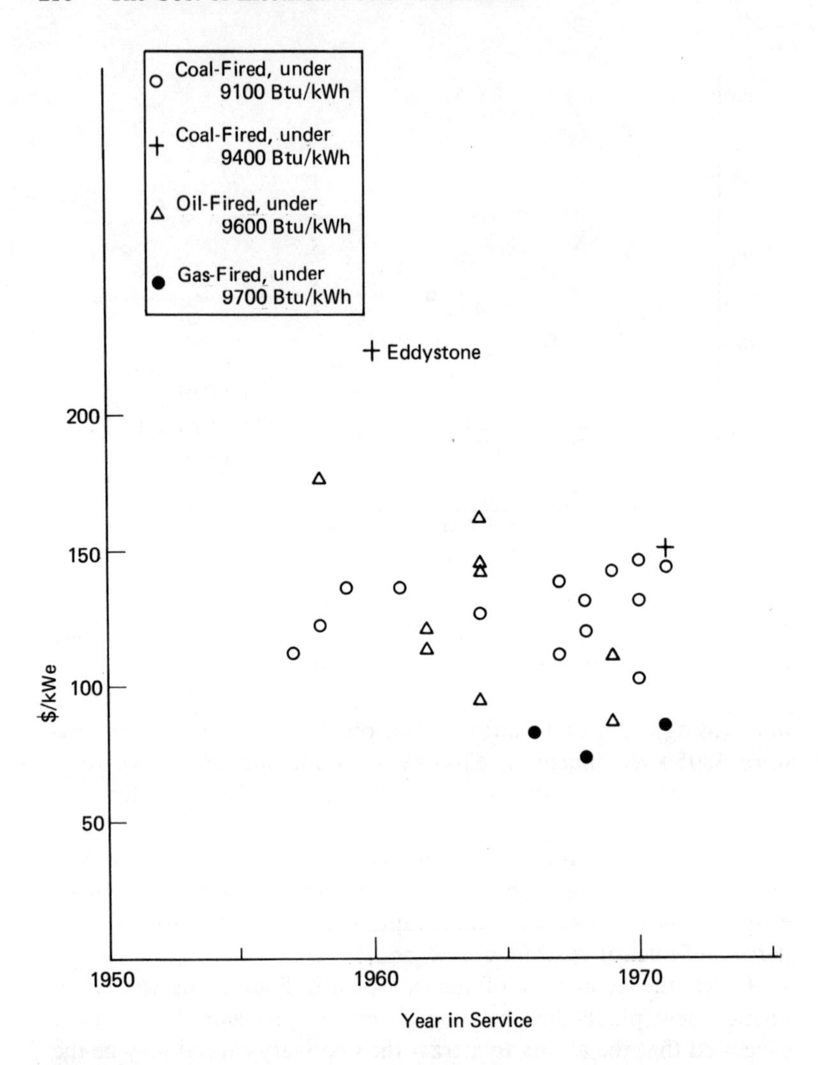

Figure 7.6
Capital costs of U.S. fossil fuel plants with lowest annual heat rates, 1972, by year of initial operation.

Table 7.3
Alternative Power Plant Capital Cost Estimates, September 1975 (10^6 $)

	Present-Day Cost[a]							
	Labor	Materials and Equipment	Escalation on Present-Day Costs[b]	Allowance for Funds during Construction	Contingency[c]	Additional[d]	Total	Average Capital Cost
1150-MWe Nuclear LWR Units								
Unit 1 (1982)	240	240	162	163	87	77	970	$805/kWe
Unit 2 (1984)	200	200	195	147	80	59	881	
800-MWe High-Sulfur Coal Units								
Unit 1 (1982)	159	194	141	85	62	44	685	$740/kWe
Unit 2 (1983)	112	137	135	60	47	27	518	
Unit 3 (1984)	112	137	175	66	52	30	572	
800-MWe High-Sulfur Oil Units								
Unit 1 (1982)	140	152	118	71	52	44	577	$637/kWe
Unit 2 (1984)	103	111	152	57	45	31	499	
Unit 3 (1986)	103	111	152	57	45	31	499	

[a]As of January 1975.
[b]Escalation rates: for materials and equipment, 10 percent in 1975 and 4 percent in following years; for labor, 12 percent in 1975 and 6 percent in following years.
[c]Contingency allowance: 10 percent.
[d]Includes taxes during construction, training, and startup; land; utility engineering; miscellaneous.
Source: Long Island Lighting Company.

needed). The costs of the high-temperature gas-cooled reactor (HTGR) may be higher due to the helium inventory and the great size of the pressure vessel. Owing to operation at high temperatures and high neutron fluxes, and the need for special safety systems to deal with sodium coolant, the liquid-metal-cooled fast breeder reactor (LMFBR) may be more costly than the light-water reactor.

To indicate the capital cost penalty of the heavy-water inventory in a heavy-water reactor, Table 7.4 shows the cost breakdown of the Pickering Canadian deuterium (CANDU) unit. The heavy water accounts for 25 percent of the direct cost of construction, but only about 16 percent of the total capital cost. The average capital cost of the unit ($370/kWe) is substantially higher than what Table 7.1 would indicate for an average light-water reactor entering service in 1971. However, in view of the variations of light-water reactor costs reported in Table 7.1 and the large increases in costs of all power plants in recent years, one cannot draw a firm conclusion from this comparison. The use of heavy water for moderation has two important advantages (discussed in Chapter 8): lower fueling cost since fuel enrichment is not needed (as in the light-water reactor), and greater utilization of the fission energy per unit mass of uranium. Precise comparisons of heavy- and light-water reactors are difficult to make, although the extra cost of the heavy water in the one concept is partly compensated by the cost of fuel enrichment in the other.

Until recently, fuel enrichment capital costs were not a matter of

Table 7.4
Capital Cost of Four 500-MWe Pickering CANDU Reactors

	Cost	
	10^6 1971 dollars	percentage of total
Direct Costs		
Reactor, Boiler and Auxiliary	102.102	13.7
Fuel	8.470	1.2
Heavy Water	119.260	16.0
Turbine Generator and Auxiliary	66.172	8.9
Other	168.562	22.6
Total Direct Costs	464.566	62.4
Indirect Costs		
Professional Services	112.365	15.0
Escalation	6.793	0.9
Interest during Construction	101.788	13.6
Other	60.498	8.1
Total Indirect Costs	281.434	37.6
Total Costs	746.000[a]	100

[a]Average capital cost: $373/kWe.
Source: Ref. (8).

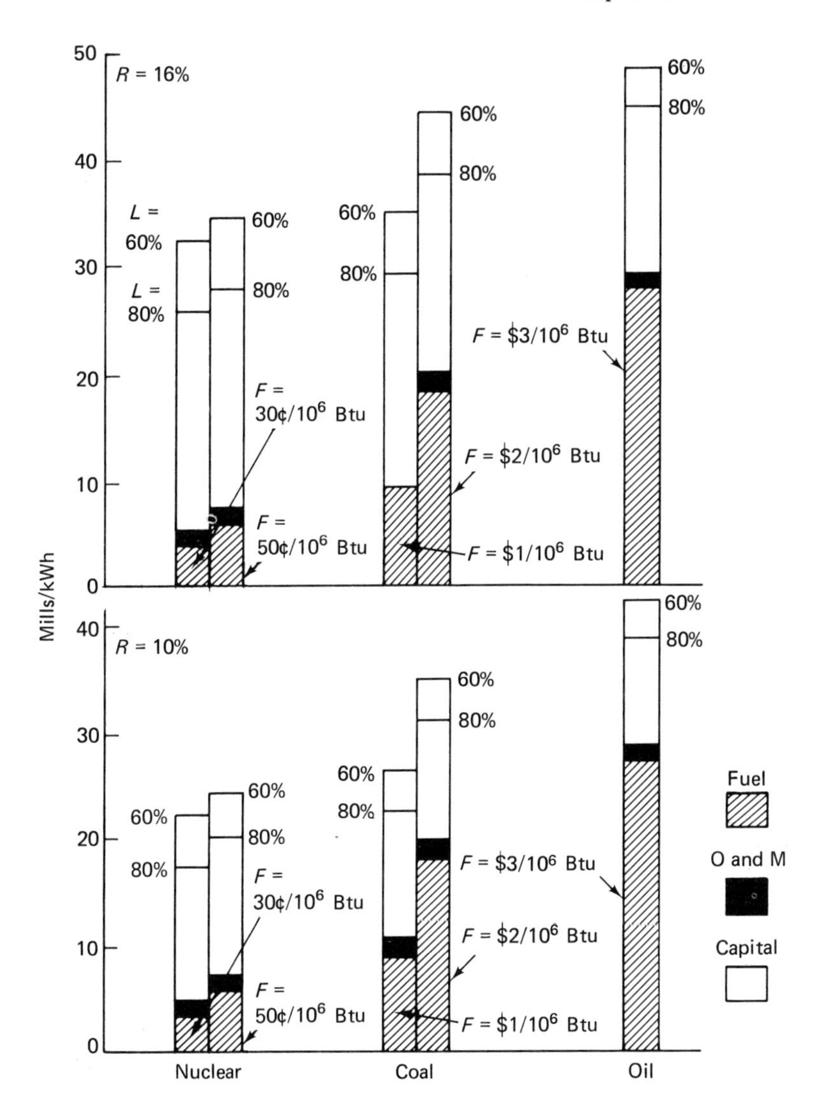

Figure 7.7
Electricity generation costs for 1000-MWe plants entering service in 1984. See Table 7.5.

concern in the United States, or the United Kingdom, since spare capacity was available in facilities provided at government expense for other purposes. Now, however, there is a developing need for substantial new enrichment facilities, both in Europe and the United States. It is expected that fuel enrichment by gaseous diffusion will effectively increase the capital cost of the reactors served by less than 7 percent (9). For the newer centrifuge process the figure may be nearer 5 percent. Both processes are heavy consumers of electricity, and the capital required for an additional electricity generation plant to serve the enrichment plant is not included in the above figures.

7.4 GENERATION COSTS

Figure 7.7 and Table 7.5 show projections of possible electricity generation costs for plants coming into commercial operation in 1984. These projections incorporate various assumptions on how capital, fuel, and financing costs will increase over the next eight years. Figure 7.7 must not be regarded as indicating predictable costs, but rather, as outlining reasonable possibilities. What is quite clear is that, compared with the 1960s, very large costs are in store (see Table 7.1).

Table 7.5
Possible Costs of Electricity Generated by 1000-MWe Plants Ready for Operation in 1984

	Nuclear[a]	Coal[b]	Oil[b]
Total Capital Cost C (\$/kWe)	900	800	700
Load Factor L (percent)	60(80)	60(80)	60(80)
Fuel Cost F (¢/10^6 Btu)	30(50)	100(200)	300
Heat Rate H (Btu/kWh)	11,400	9000	9200
Plant Efficiency $3413/H$ (percent)	30	38	37
Capital Charge Rate R (percent)	10(16)	10(16)	10(16)
Generation Costs (mills/kWh)			
Fuel Costs ($FH/10^5$)	3.42(5.70)	9.00(18.00)	27.00
Operation and Maintenance	1.50	2.00	2.00
Capital Charges ($RC/8.760L$)			
$L=60$ percent $R=10$ percent	17.12	15.22	13.32
$L=60$ percent $R=16$ percent	27.40	24.35	21.31
$L=80$ percent $R=10$ percent	12.84	11.42	9.99
$L=80$ percent $R=16$ percent	20.55	18.26	15.98
Total Generation Costs (mills/kWh)[c]			
$L=60$ percent $R=10$ percent	22.04(24.32)	26.22(35.22)	42.32
$L=60$ percent $R=16$ percent	32.32(34.60)	35.35(44.35)	48.31
$L=80$ percent $R=10$ percent	17.76(20.04)	22.42(31.42)	39.00
$L=80$ percent $R=16$ percent	25.47(27.75)	29.26(38.26)	44.98

[a]LWR.
[b]With sulfur-removal equipment.
[c]1976 fuel prices escalated at 5 percent per year to 1984. The two figures are for the alternative values of F being considered for nuclear and coals.

Table 7.6
Nuclear Reactor Load Factors

	Annual Load Factor, June 1974 (percent)	Cumulative Load Factor, June 1974 (percent)	Million MWh Generated (gross cumulative)
Light-Water Reactors			
PWR	62.19	49.35	248
BWR	54.26	49.89	189
Heavy-Water Reactors	60.42	56.62	37
Magnox	60.59	58.02	538

Source: Ref. (10). Reprinted by permission of *Nuclear Engineering International.*

As emphasized already, plant capital costs have been subject to large increases in recent years. The capital costs for 1984 shown in Table 7.5 are roughly in accord with the trends indicated in Refs. (4) and (5), and are very sensitive to interest and inflation rates, since the period of construction is long.

Nuclear fuel cost is taken as either 30 or 50¢/10^6 Btu for 1984. The first figure escalates the 1976 value (20¢/10^6 Btu or 0.4 mill/kWh) at 5 percent per year; the second allows for uranium price rising with demand. Two figures are also used for coal costs. The first—\$1/10^6 Btu—might represent the 1984 price in the West of strip-mined western coal (\$16 and 16×10^6 Btu per ton); the second coal price could correspond to eastern coal from underground mines. The oil price used (\$3/10^6 Btu) corresponds to imported oil at \$11.60/bbl in 1976, escalated at 5 percent per year to 1984.

The plant load factor L in Table 7.5 is the ratio of the annual average electrical power output to the gross station capacity (which includes the electrical power consumed in running the station itself).[1]

Table 7.6 shows load factors reported for June 1974 for various types of nuclear reactors in Europe, Japan, India, Canada, and the United States. The reactor types (described in Chapter 8) are the light-water-cooled PWR (pressurized-water-cooled reactor), the BWR (boiling-water reactor), the heavy-water reactors, and the Magnox (carbon-dioxide-cooled reactor). As of 1974, the Magnox type had generated the largest number of megawatt-hours, but the annual load factor up to June 1974 was only 60 percent. Effectively this means that the plant capacity is idle for about 40 percent of the year. Since nuclear plants are so capital-intensive they should be utilized year-round for best system economy.

A load factor of 60 percent rather than 100 percent has a large

[1]Elsewhere this factor is called the capacity factor, to distinguish it from the ratio of average annual power level to maximum power during a given period.

effect on the unit cost of the electricity generated. Annual load factors of other reactor types are also around 60 percent or less. The difference between the annual load factor up to June 1974 and the average cumulative load factors up to that date (from the date of station operation) is significant. In the early days of large-scale nuclear power plants there have been many problems such as licensing delays, steam generator water chemistry problems, heavy water shortage, etc. Particular units have had load factors as low as 10 percent. On the other hand, as an example, the Pickering Unit 1 had a load factor of 93 percent in 1973. Variability of system demand is another reason load factors are less than 100 percent.

The availability factor, in contrast to the load factor, is the fraction of the year that the unit is ready for service. According to J. R. Dietrich (11) the average availability of light-water reactors in the United States in 1972 was 73.4 percent, and the average availability of fossil fuel plants in the year 1971 in the United States was 81 percent. As commercial nuclear technology matures (with the development of reliable standardized designs and improved quality control in manufacturing), availability factors should rise. Further, with the low fuel costs of nuclear plants, it is reasonable to expect system planning to make maximum use of nuclear stations when they are available. This may well call for the development of energy storage systems (12, 13). There seems no clear reason why nuclear plant load factors as high as 80 or 90 percent could not be developed in the future. Table 7.5 and Figure 7.7 show the results of calculations for load factors of 60 percent (present experience) and 80 percent (expected future experience).

The heat rates shown in Table 7.5 are nominal values for each plant type; reasons for these differences are presented in Chapters 8 and 9. The fueling cost in mills per kilowatt-hour is simply (fuel cost in ¢/10^6 Btu) × (10 mills/¢) × (heat rate in Btu/kWh).

Capital charge rates depend not only on interest rates but also on whether the utility is public or private. In the latter case it must generate enough income to provide a return on investment attractive to investors and must also allow for tax payments. This is why two quite different values of capital charge rate are shown in Table 7.5. A capital charge rate of 10 percent would be equivalent to an annual interest rate of 6.7 percent and straight-line depreciation (equal annual increments of depreciation) over 30 years. A capital charge rate of 16 percent could be equivalent to an annual total of interest and tax rate of 12.7 percent and straight-line depreciation over 30 years (14). The effect of charge rate on the generation cost of power plant is calculated from

$$\frac{(\text{capital charge rate in }\%) \times (\text{capital cost in \$/kWe}) \times (1000\text{ mills/\$})}{(8760\text{ hr/yr}) \times (\text{load factor in }\%)}.$$

Despite all the uncertainties in fuel and capital costs and in capital charges, the following tentative conclusions can be drawn from Figure 7.7.

1. Assuming the requirement for sulfur removal will raise the cost of coal-burning plants, nuclear generation costs will be considerably lower than costs for plants burning eastern coal, particularly if both kinds of plants can operate at the same load factor. Plants burning western coal in the West may remain competitive with nuclear plants.

2. A 67 percent increase in the cost of nuclear fuel will raise the cost of nuclear power generation by around 10 percent. The fuel prices in Figure 7.7 (except for the high nuclear value) correspond to 1976 fuel prices escalated at 5 percent per year. A greater fuel escalation rate would show a more decisive advantage for nuclear power.

3. If nuclear plants continue to have load factors around 60 percent (while coal plants can operate at 80 percent and higher), coal power could be cheaper than nuclear power, but probably only for low-cost surface-mined coal. If load factors are taken to be the same for nuclear and fossil fuel plants, nuclear plants will generally have the cost advantage.

4. Reducing the capital charge rate from 16 percent improves the competitive position of nuclear power, but this is not a strong effect. If environmental regulations were relaxed so that 1984 capital costs for coal plants could be reduced to, say, \$600/kWe, the cost advantage would not be restored to coal-fired generation with eastern coal prices and equal fossil fuel and nuclear plant load factors.

Thus, although large increases in fossil fuel prices have tended to make nuclear power appear cheaper than fossil fuel power, the conclusion is not clear-cut. Many local factors affecting fuel and capital costs could swing the decision of a utility one way or the other. In this situation environmental considerations may often play an important role in the final decision. Other important factors might be the availability of capital and the market interest rate at the time the decision is being made.

Despite their much lower capital cost, many existing fossil fuel plants are now much less economical to operate than nuclear plants. Table 7.7 shows current generation costs of two plants that have been operating for several years. Though the nuclear plant had nearly three times the capital cost, its generation cost is less than that of the coal plant, and would be even less if the capital charge rate were raised to 15 percent. This is the case even with coal costs at \$19/ton (the price is now considerably higher) and without the high capital cost of sulfur-removal equipment.

Tables 7.1 and 7.5 suggest that over the 12-year period 1972–1984 a fourfold increase in electricity generation cost could be realized. This would be equivalent to an average annual growth rate of 12 per-

Table 7.7
Example of Comparative Nuclear and Coal Generation Costs

	Pickering (nuclear)	Lambton (coal)[a]
Parameters		
Capacity (MW)	4 × 514	4 × 495
Life (years)	30	30
Interest Rate (percent)	8	8
Capital Cost ($/kWe)	370	133
Station Capacity Factor (percent)	80	80
Unit Energy Cost (mills/kWh)		
Capital Charge Rate (R=8.7%)	4.6	1.70
Operation and Maintenance	0.64	0.61
Heavy Water Upkeep	0.20	
Fuel	0.90	6.57
Total	6.34	8.89

[a]Coal at $12/ton.
Source: Ref. (15), Ontario Hydro comparison (1975).

cent, nearly 10 times as high as the average rate of increase during the 1960s. Consistent with safety needs and environmental protection, there is much incentive for technological development to reduce future costs of power generation. The trend toward nuclear power generation could continue, provided that satisfactory nuclear power safety records can be maintained. This could lead to a shortage of natural uranium fuel and to a strong demand for breeder reactors. However, intensified uranium exploration may yet provide uranium at prices that would not severely affect the competitive advantages of nuclear power generation even if commercial use of the breeder were long delayed.

The present cost disadvantage of fossil fuel power generation does not provide incentive for development of clean synthetic fuels from coal. It would not be desirable, in principle, to have to depend exclusively on nuclear reactors for electrical power, but the question is how much we are willing to pay for an environmentally acceptable option.

The IAEA concluded after surveys of the market that the large increase in oil prices in 1973 would cause a substantial shift in demand for new nuclear plants (9). Table 7.8 compares their projections of world demand for nuclear power before and after the sudden jump in oil prices ("early 1973" and "accelerated demand").

The IAEA estimated that if the capacities in Table 7.8 are reached, present assured reserves of uranium would be exhausted by 1990 for the "early 1973" projection of capacity and by 1988 for the "accelerated" case. The problem could be solved by increasing the reserves finding rate from what it has been recently, around 65,000 tons/yr, to about 230,000 tons/yr. If the price of U_3O_8 rises

Table 7.8
Projections of World Nuclear Plant Demand

Year	Early 1973 Projection (GWe)	Accelerated Demand Projection (GWe)
1975	104	103.5
1980	302	316
1985	693	888
1990	1390	1900
1995	2355	3365
2000	3580	5330

Source: Ref. (9).

$20 or $30/1b, the IAEA believes it likely that the additional reserves will be found and that provision of enrichment facilities will not be a limiting factor. They are worried, however, about safety: adequate testing and inspection of pressure vessels, piping, valves, etc., could be a problem. Lack of trained people to administer and control licensing may also be a problem in a number of countries, particularly if a number of plants are built at the same time.

7.5 ENERGY TRANSMISSION COSTS

Up to this point, only the costs of generating electrical power have been discussed. The costs of transmitting and distributing electricity to consumers is relatively large, and highly dependent on population density. Sample relative costs of generation and delivery of electricity are shown in Table 7.9.

"Transmission" in Table 7.9 refers to high-voltage lines that are used over long distances. These proportions can vary widely from utility to utility, depending on whether the area served is urban or rural. In any case it can be seen that the price paid by the consumer must be substantially higher than the generation cost calculated at the plant.

Table 7.9
Distribution of Costs in the Provision of Electrical Power

	United States (percent)	United Kingdom (percent)
Generation	35	60
Transmission	25	7
Distribution	40	33
Total	100	100

As shown earlier, there is considerable economic incentive to build large plants. A utility will typically build the largest plant it can keep fully loaded. Interconnecting utility networks to distribute load more widely will mean that each utility can plan for steadier load and hence for larger plants. The resulting economy of scale for the plant may mean increased transmission costs. Siting of large plants close to major population centers is becoming more and more difficult. Shortage of land and cooling water and local fears of air pollution and nuclear radiation tend to force plants into increasingly remote locations. Serious thought has been given in recent years to floating power plants on large offshore platforms (e.g., off the New Jersey coast). Although this would be expensive, high-voltage transmission for 100 to 200 miles on land does not usually mean a severe cost penalty. The principal difficulty may be in securing rights-of-way for the transmission line.

Increasingly, the environmental effects of energy transportation must be considered alongside the direct monetary costs. Overhead transmission lines are generally considered ugly. In the United States more than 7 million acres are now set aside for transmission line rights-of-way. This is only about 0.2 percent of the total land area, but the lines spoil the view from a much larger fraction of land area. What are the relative costs and environmental effects of electrical and fuel energy transmission lines? How can existing rights-of-way best be utilized to increase transmission capacity? These questions are expected to be of increasing importance as the national demand for electricity continues to grow.

The "simple" solution—underground electrical transmission lines—is very expensive. The cost of a 500-kV underground cable may approach $1 million/mile; the cost of the right-of-way is a negligible part of this since the required trench may be fairly narrow. The heat generated by the electrical resistance of such a high-voltage cable is far larger than the earth can absorb, so intensive cooling must be provided. Several possible systems are under development, including use of compressed gas coolant and cryoresistive or superconducting systems. Compressed gas cables, using sulfur hexafluoride (a good insulator) have been built in lengths much less than a mile. Cryoresistive systems, in which the cable is cooled to the temperature of liquid nitrogen (77°K) to lower its resistance by a factor of 10, would need to be supplied with refrigerators every few miles. Superconducting cables would need to operate with a suitable material, such as niobium or niobium–tin, that is superconducting at temperatures below 18°K.

Since most of these systems are only in the research stage, costs are quite uncertain. It would appear, however, that costs of high-voltage underground cable are 5 to 20 times the cost of overhead transmission, not allowing for land costs. In cities, where land costs

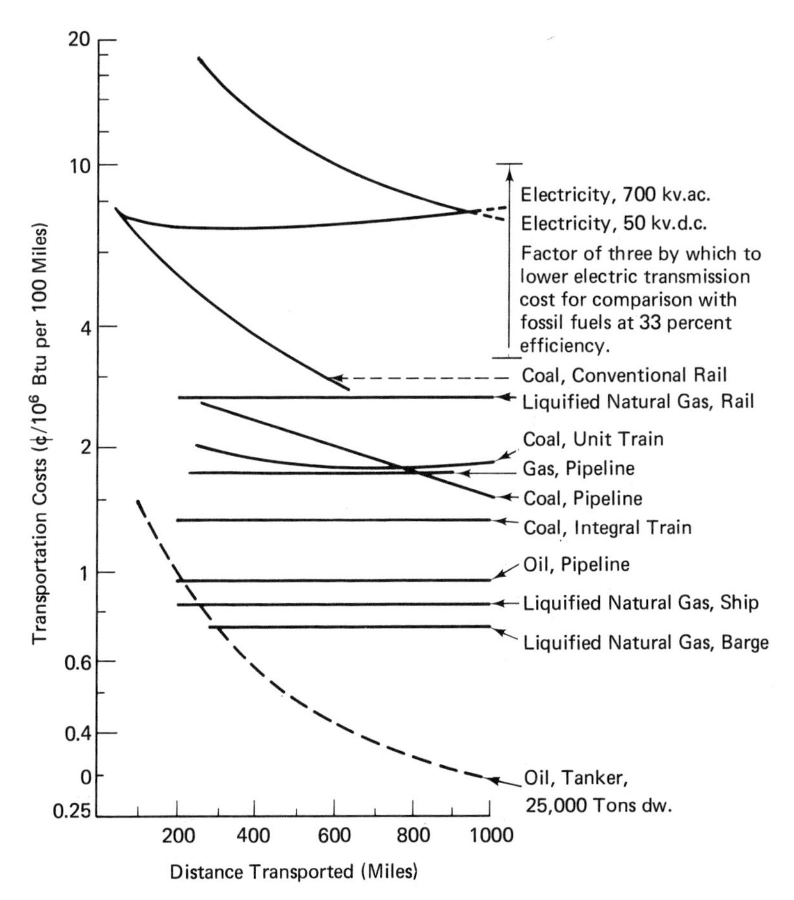

Figure 7.8
Energy transportation costs, 1972 (16).

are high or the land may be unavailable, the land requirement of the overhead line (up to 12 acres/mile) may make local use of underground transmission economical or even unavoidable. Underground cables are being used increasingly in and around large cities.

Figure 7.8 shows that high-voltage energy transmission may be up to three times as expensive as rail transportation of coal. This would refer to high-voltage electricity transmissions (500 to 1000 kV). Taking into account an electricity generation efficiency of about 0.33 means that (per unit of electrical energy) these modes of long-distance energy transport are about equally expensive. Pipeline transport of oil and natural gas is, however, much cheaper per kilowatt-hour of electricity produced, perhaps by another factor of three. Transportation cost for nuclear fuel would be negligible on the scale of Figure 7.8.

B. I. Spinrad has pointed out that with nuclear fuel able to produce 1000 to 10,000 MW-days/ton of uranium, even a transport charge of $1000/ton would add an increment of only 0.04 mill to total generation cost (17). In contrast, oil shipped by tanker for as low as 25¢/bbl would be equivalent to 0.25 to 0.30 mill/kWh. The relative costs of coal and electrical transmission over 1000 miles are shown in Figure 7.9 as a function of power plant capacity. For very large plants the two costs are nearly equal if the coal is shipped by unit train. Transport of coal in a slurry in pipelines could be much cheaper than high-voltage electrical transmission. One 280-mile length pipeline carrying 5 million tons of coal per year is said to be operating very satisfactorily in the southwestern United States (19). Another 1000-mile length pipeline is being designed.

In the past, the low cost of transporting oil and natural gas by pipeline (relative to electrical transmission cost) suggested that electrical plants should be near load centers to minimize transmission costs. For nuclear plants this consideration is still important; for coal plants and rail transport it may be equally beneficial to have the

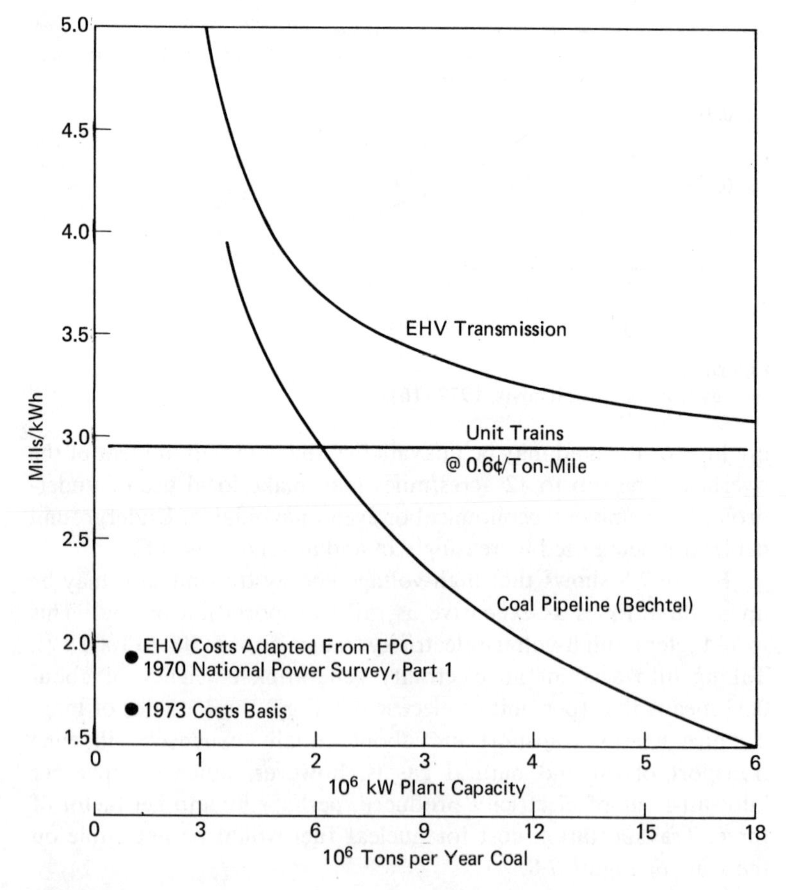

Figure 7.9
Cost comparison of alternative modes of coal energy transmission (1000-mile transport distances), 1974 (18).

Table 7.10
Relative Generation and Transportation Costs for Various Gases

	Density (Btu/scf)	Cost (¢/10^6 Btu) Generation or Wellhead	Transport	Total
Natural Gas	1000	60	20	80
Low-Btu Gas	160	140	120	260
High-Btu Gas	1000	250	20	270
Hydrogen	300	510	30	540

plant near the coal mine. In both cases, however, plant siting will be strongly dependent on the availability of cooling water and on the difficulty of obtaining site and transmission line approval from authorities who are increasingly concerned about the environment.

Although natural gas is too scarce to consider for fueling new power plants, clean synthetic gas made from coal is a possibility (see Chapter 9). Hydrogen has also been proposed as a medium for long-distance energy transmission. Both of these will be expensive to generate, but transmission costs, which depend on energy content and fluid properties, could be reasonable.

Natural gas has an energy density of around 1000 Btu/scf, hydrogen around 300 Btu/scf, and low-Btu synthetic gas around 100 to 160 Btu/scf. Transported in the same pipeline at the same pressure, temperature, and velocity, each gas would have pipeline costs varying inversely as energy density. This means that low-Btu gas would be 6 to 10 times as expensive to transport (per unit of energy delivered) as natural gas. For 1000-mile transmission, relative costs might be as shown in Table 7.10.

These costs are approximate; natural gas has cost as little as 16 ¢ /10^6 Btu, but the price is rising fairly rapidly. Costs of low- and high-Btu gas production are still uncertain. The cost quoted for hydrogen would correspond to electrolysis, with electricity costing 12 mills/kWh, in a plant costing $100/kW and operating at 75 percent efficiency. Even though the hydrogen energy content per unit volume is only about one-third that of natural gas, it can travel nearly three times faster for the same percentage pressure drop per unit length. Pumping costs and safety measures could of course cost more for hydrogen than natural gas. Thus transmission costs can be low while generation cost is quite high. Efforts are now being made to develop other and less costly methods of hydrogen production, but these have not yet proved feasible.

7.6 SUMMARY

Between 1970 and 1985 the costs of electrical power generation will increase by a factor of four or five. A part of this is due to worldwide inflation, which has greatly increased the capital cost of both nuclear and fossil fuel plants. Part is also due to increasing concern with the environment, which has both raised the price of low-sulfur fossil fuels and increased the cost of plants designed to operate with high-sulfur fuel and low SO_2 emissions. The main cause of the increase in power cost has been the large increase in the cost of fossil fuels, especially the rapid jump in imported oil price in the latter part of 1973. The present high cost of coal could discourage the substantial efforts needed to develop large supplies of clean synthetic fuels from coal and to provide an environmentally acceptable alternative to nuclear power.

In projecting costs for plants that will come into operation in the 1980s, it appears that nuclear power will generally be cheaper than fossil fuel power for steady base loads. Since uranium and thorium are relatively abundant in the earth's crust, it is hoped that future exploration will assure a large supply of nuclear fuel at stable prices. However, even if the cost of uranium ore doubles or triples in price, this will not remove the competitive advantage of nuclear power generation, since the cost of the uranium ore is at present only a small fraction of the total cost of nuclear power generation. Again, if reasonably low-cost reserves of uranium are found to be in very short supply even after much further exploration, the nuclear breeder or near-breeder reactors are potentially available to greatly extend the life of those reserves.

The main concerns affecting the future expansion of nuclear power appear to be safety and the availability of capital. The availability of capital will depend greatly on whether future demand growth continues at an average of about 7 percent per year (as it has done for many decades), or whether a more modest rate of 4 or 5 percent characterizes growth in the next few decades.

Fossil fuels will still supply the energy for much of our future electricity. Oil and gas appear to be generally uneconomical (or simply unavailable) for new plants. In special circumstances coal could be more economical to use than nuclear fuel (e.g., if supplies of low-cost, low-sulfur coal are located near the power plant, or if government regulations on sulfur emissions are relaxed). The social costs of air pollution from fossil fuel plants are still so uncertain that pollutant emission limits must be regarded as being the result of current official judgment; they could be revised upward or downward in the future.

Whether coal or nuclear power is more economical depends quite strongly on load factors and relative capital costs. For low load factors, as in intermediate or peak load units, the lower capital costs of fossil fuel plants make them more economical than nuclear plants. There is, however, considerable economic incentive to couple nuclear plants with energy storage systems so that the nuclear plant can run at a high load factor, with the energy storage unit used for peak or intermediate loads.

References

1. "18th [and 19th] steam station cost surveys." *Electrical World*, November 1, 1973, and November 15, 1975.

2. Federal Power Commission. *Steam–Electric Plant Construction Cost and Annual Production Expenses. Twenty-Fifth Annual Supplement, 1972.* Washington, D.C., April 1974.

3. Federal Power Commission, Bureau of Power. *Annual Summary of Cost and Quality of Steam–Electric Plant Fuels, 1973 and 1974.* Washington, D.C., May 1975.

4. F. C. Olds. "Power plant costs going out of sight." *Power Engineering*, August 1974.

5. U.S. Atomic Energy Commission, Division of Reactor Research and Development. *Power Plant Capital Costs: Current Trends and Sensitivity to Economic Parameters.* WASH-1345. Washington, D.C., October 1974.

6. Arthur. D. Little Inc. *A Study of Base Load Alternatives for the Northeast Utilities System.* Cambridge, Mass., July 5, 1973.

7. Denis M. Slavich and Charles W. Snyder. "Meeting the financial needs of the nuclear power industry." *Nuclear Engineering International*, March 1975, pp. 161–164.

8. W. G. Morrison, C. E. Beynon, E. K. Keane, and R. D. Wardell. "Pickering Generating Station." Paper 4.1–2, *Transactions of the 9th Conference*, Detroit, Michigan, September 22–27, 1974.

9. J. A. Lane, R. Krymm, N. Raisic, and J. T. Roberts. "The role of nuclear power in the future energy supply of the world." Paper 4.1–22, *Transactions of the 9th World Energy Conference*, Detroit, Michigan, September 22–27, 1974.

10. G. Greenhalgh. "World nuclear programme." *Nuclear Engineering International*, March 1975, pp. 164–169.

11. J. R. Dietrich. "The introduction of new power generation systems: Implications of LWR experience." Paper 4.1–15, *Transactions of the 9th World Energy Conference*, Detroit, Michigan, September 22–27, 1974.

12. J. L. Haydock. "Energy storage and its role in electric power systems." Paper 6.1–21, *Transactions of the 9th World Energy Conference*, Detroit, Michigan, September 22–27, 1974.

13. R. Fernandez, O. D. Gildersleeve, and T. R. Schneider. "Assessment of concepts in energy storage and their application on electric utility systems." Paper 6.1–17, *Transactions of the 9th World Energy Conference*, Detroit, Michigan, September 22–27, 1974.

14. W. F. Stoecker. *Design of Thermal Systems.* New York: McGraw-Hill, 1971.

15. J. C. Gray. *Why CANDU? Its Achievements and Prospects.* Report AECL 4709, Atomic Energy of Canada, January 1975.

16. H. J. Hottel and J. B. Howard. "An agenda for energy." *Technology Review*, vol. 74 (1972), p. 38.

17. B. I. Spinrad. *The Role of Nuclear Energy in Meeting World Energy Need.* Report IAEA-3M-16412, International Atomic Energy Agency, Vienna, 1971.

18. National Academy of Engineering. *U.S. Energy Prospects: An Engineering Viewpoint.* Washington, D.C., 1974.

19. Seymour Baron. "Cost–benefit analysis of advanced power-generation methods." *Energy Sources*, vol. 1, no. 2 (1974), pp. 201–221.

Supplementary Reference

A. J. Surrey. "The future growth of nuclear power. Part 1. Demand and supply. Part 2. Choices and obstacles." *Energy Policy*, vol. 1, nos. 2, 3 (September and December 1973), pp. 208–224.

Problems

7.1 Show that a sum of money P invested now at interest rate i (compounded annually) will have a value S in n years of

$$S = P(1 + i)^n$$

and that the present value P of a sum S which must be spent n years from now is

$$P = \frac{S}{(1+i)^n}.$$

The construction period of a 1000-MWe power plant is estimated to be eight years, with the cash expenditures distributed as follows:

Year 1 : \$ 50 million
Year 2 : \$100 million
Year 3 : \$150 million
Year 4 : \$175 million
Year 5 : \$150 million
Year 6 : \$125 million
Year 7 : \$100 million
Year 8 : \$ 50 million.

If the annual compound interest rate is 10 percent, estimate the present value of these payments.

7.2 Show that regular payments R made over n regular periods at interest rate i per period will accumulate to a sum

$$S = R\frac{(1 + i)^n - 1}{i},$$

that is,

$$R\sum_{j=0}^{n-1} (1 + i)^j = R\frac{(1 + i)^n - 1}{i},$$

the sum of n terms in the series

$$S = R(1 + (1 + i) + (1 + i)^2 + \cdots).$$

A so-called "sinking fund" is to be established by regular equal payments which will accumulate over n periods to a sum S to be used for replacement of a power plant. Compare the annual sinking fund depreciation cost per kilowatt,

$$R = \frac{Si}{(1 + i)^n - 1},$$

with the "straight-line" depreciation method,

$$R = \frac{S}{n},$$

for a \$1000/kWe power plant depreciated over 30 years, with interest at 10 percent, compounded (a) annually and (b) semiannually.

7.3. To make allowance for the inflation of costs during the period of construction, suppose that average construction cost escalation is 8 percent per year. Adjust the cash flows of Problem 7.1 accordingly and estimate the new present value of the cash flow during the period of construction. Calculate the percentage increase in plant cost due to escalation, still assuming a 10 percent interest rate.

Now estimate the percentage decrease in plant cost (calculated as the value at the end of the construction period) due to shortening of the construction period from eight years to four years. Suppose the same schedule of cash flows as in Problem 7.1, but let each period be reduced to six months instead of one year.

7.4 An electrical utility is considering the purchase of a 1000-MWe nuclear plant, and is weighing up the costs of light-water (LWR) and heavy-water (HWR) nuclear plants under the parameters presented in Table 7.A. Estimate the generation costs for both public and private utilities.

Table 7.A
Comparative Nuclear Plant Parameters

	LWR	HWR
Total Capital Cost (\$/kWe)	1000	1300
Load Factor	0.7	0.8
Fueling Cost (mills/kWh)	2.4	1.0
Interest Rate, compounded annually (percent)	10	10
Depreciation Period[a] (years)	30	30
Tax Rate (percent)		
Private Utility	5	5
Public Utility	0	0
Operation and Maintenance Cost (mills/kWh)	2	2

[a]Assume straight-line depreciation (see Problem 7.2).

7.5 Compare the sensitivity of power generation cost to fuel price for coal and nuclear plants. Use the assumptions in Table 7.B.

Table 7.B
Comparative Nuclear and Coal Plant Parameters

	Coal	Nuclear
Capital Cost ($/kWe)	500	850
Load Factor	0.8	0.6
Fuel Cost ($/$10^6$ Btu)	2.00	0.25
Heat Rate (Btu/kWh)	8750	11,000
Capital Charge Rate (interest, depreciation, and taxes) (percent)	0.17	0.17
Operation and Maintenance Cost (mills/kWe)	2.0	1.5

Estimate the percentage increase in power generation cost in both cases if fuel price doubles.

7.6 Estimate power generation cost as a function of load factor for nuclear, coal-fired, and gas turbine power units under the assumptions of Table 7.C. Consider load factors of 0.1, 0.3, and 0.8, corresponding to peaking, intermediate, and base load service.

Table 7.C
Comparative Parameters for Three Types of Power Units

	Coal	Nuclear	Gas Turbine
Capital Cost ($/kWe)	500	850	200
Fuel Cost ($/$10^6$ Btu)	1.50	0.30	3.00
Annual Capital Charge Rate (interest, depreciation, and taxes) (percent)	16	16	16
Operation and Maintenance Cost ($/yr/kWe of capacity)	12	9	12
Plant Heat Rate (Btu/kWh)	9500	11,400	13,650

7.7 A helium-cooled high-temperature reactor (HTGR) is capable of operation at temperatures such that calculated thermal efficiency (using a combined helium gas turbine and steam cycle) is 45 percent. The fuel is highly enriched and so is more expensive than light-water reactor (LWR) fuel. How much capital cost would be justified for this plant, in competition with a light-water reactor, if the parameters in Table 7.D hold?

7.8 A 2000-MWe power plant is to be fueled with coal from a mine located 1100 miles from its load center, a large city. Consider the two options of locating the mine either 100 miles or 1000 miles from the city on a straight line between the city and the mine. Adequate land area and cooling water are available at both sites.

Table 7.D
Comparative Parameters for LWR and HTGR

	LWR	HTGR
Capital Cost ($/kWe)	850	?
Fuel Cost ($/$10^6$ Btu)	0.3	0.5
Thermal Efficiency	0.31	0.45
Operation and Maintenance Cost (mills/kWh)	1.5	1.5
Capital Charge Rate (percent)	16	16
Load Factor	0.75	0.75

Using the data of Figure 7.8, estimate the total fuel plus electricity transmission costs (mills/kWh) for each site. The plant efficiency is 39 percent.

Check the costs for 1000-mile transmission of coal and electricity with the data of Figure 7.9.

7.9 Coal prices paid by utilities in 1974 varied greatly with region as shown by the Federal Power Commission data in Table 7.E.

Table 7.E
Coal Price and Sulfur Content in Various Regions

	Coal Price (¢/10^6 Btu)	Sulfur Content (percent)
New England	114.5	1.4
Middle Atlantic	86.1	2.1
East North–Central	70.1	2.8
West North–Central	44.8	2.3
South Atlantic	97.3	1.8
East South–Central	59.5	2.9
West South–Central	16.9	0.6
Mountain	26.0	0.5
Pacific	36.7	0.7
National Average	71.0	2.2

Consider the alternatives faced by a coal plant located in the New England region of

a. installing stack gas sulfur-removal apparatus costing $75/kWe and introducing a unit operating cost of 1 mill/kWh so that high-sulfur coal can be burned; or

b. transporting low-sulfur coal 2000 miles by unit train from the West South–Central Region, and avoiding the need for stack gas sulfur removal.

Suppose that the plant heat rate is 9500 Btu/kWh and the capital charge rate is 17 percent. The plant load factor is 0.8.

7.10 Suppose that maximum steam temperature could be raised 100°F and thermal efficiency raised from 0.390 to 0.395 in a coal-fired steam plant by using an alloy for the boiler and superheater tubes whose extra expense would be \$7.5/kWe. Would the extra expense be worthwhile?

Coal costs, which were 25¢/10^6 Btu in the late 1960s, have risen to \$1/10^6 Btu in the mid-1970s.

Assume an annual capital charge (including interest, depreciation, and taxes) of 15 percent. The plant load factor is 0.8.

8
Nuclear Power

8.1 INTRODUCTION

The most important issues today in nuclear power are safety, costs, fuel utilization rate, and efficiency.

Public anxiety over the safety of large light-water reactors has been high (1). Some experts are worried that so many reactors are being built before all their safety features have been fully tested. Fears have also been expressed that the plutonium hazard may not, in fact, be controllable. It may be some years before the full costs are known, not only of providing an acceptable standard of safety but also of demonstrating this convincingly. This problem is acute in the United States, where full public disclosure of all possible environmental effects of new power plants must be made before construction and operating permits can be obtained.

Even though commercial reactors have not yet had accidents serious enough to harm people nearby, it must be remembered that the age of nuclear power is only beginning. The total level of reactor power could increase by 100 times in the next few decades. Society could become extremely dependent on nuclear energy. It would be difficult to argue that there has been excessive caution against nuclear radiation hazards, though it might well be true that not nearly enough attention has been paid to the effects on human health of air pollution from fossil fuel power plants.

Second only to the safety issue is the cost of nuclear power. As shown in Chapter 7, the capital cost of nuclear power falls significantly as reactor size increases. A. J. Surrey reports that the average size of reactors commissioned in the United States will double from 361 MWe in 1969 to 777 MWe in 1977 (2). Reactors in the 950- to 1200-MWe class will account for half the nuclear capacity entering service in the next five or six years. Reactors of 2000-MWe capacity are not out of the question, although many utility companies are not large enough to use them. This relatively rapid growth in reactor size means that designs are far from standardized and that little direct experience is available for estimating reliability and lifetime costs. For the time being, the USAEC has limited approvals of all new reactors to those of not more than 3800 MW thermal (or about 1200 MWe for light-water reactors, whose efficiencies are around 30 percent).

Rapid growth of the nuclear supply industry means that manufacturing costs have not had a chance to settle into a rational pattern. Some firms have entered the field by building equipment at a loss. Various kinds of government subsidy have been significant. Expensive delays have stretched construction times to eight or nine years.

Because nuclear power plants are heavily capital-intensive, their costs have risen substantially with generally increasing world interest rates. Indeed, it has been suggested that a world shortage of capital could be one of the major constraints limiting the future development of nuclear or any other kind of power plant. Estimation of nuclear plant costs is strongly dependent on assumed capital charge rates. Public utility charge rates are sometimes taken as low as 10 percent, whereas for private utilities, which must pay taxes and dividends, capital charge rates are as high as 20 percent or more.

In view of all the local variations and future uncertainties, it is not realistic to try to establish a single narrow range of absolute costs of nuclear power generation. About the best that can be done is to estimate relative costs of fossil fuel and nuclear plants under stated assumptions of capital cost, fuel cost, interest rates, and plant load factors (see Chapter 7).

Although fuel costs are relatively small for nuclear plants (compared with coal or oil plants), overall efficiency is nonetheless very important.

Among the reasons for seeking higher efficiency of reactor systems are

1. lower fuel cost per kilowatt-hour
2. reduced reactor cost per kilowatt
3. reduced heat rejection per kilowatt
4. improved fuel utilization.

Since fuel cost is a relatively small fraction of nuclear–electric generation cost, it might seem that reactor plant thermal efficiency would have relatively little economic importance. The following argument will show that this is not necessarily so.

The cost of producing electrical power (\$/kWh) may be written as

$$\frac{R\, C_{\text{plant}}}{8760\, L} + \frac{3412\, C_{\text{fuel}}}{10^6\, \eta},$$

where

C_{plant} is the plant cost (\$/kWe),
C_{fuel} is the fuel cost (¢/10^6 Btu),
η is the plant thermal efficiency,
R is the annual capital charge (percent),
L is the annual load factor (percent),
8760 is the number of hours per year,
3412 is the number of Btu per kilowatt-hour.

For a given type of plant, the reactor size per unit electrical output will be proportional to the nuclear heat output. This will vary inversely as the thermal efficiency. As an approximation one might assume that reactor cost is proportional to size, so that

$$C_{\text{plant}} \propto \frac{1}{\eta}.$$

If this is so, the total production cost will be inversely proportional to plant efficiency, regardless of the fraction of the total cost directly attributable to fuel.

It cannot be concluded from this that production costs for gas-cooled reactors with plant efficiency around 40 percent would be less than for water-cooled reactors with efficiency around 30 percent. These are plants of entirely different design.

Fuel utilization is important since nuclear resources are limited. The estimate of uranium resources in Chapter 1 suggested that uranium resources could be exhausted in a few decades. This estimate referred to relatively high-grade ore, and may not adequately allow for future discoveries. Uranium exploration is a relatively recent art, and was not very active during much of the 1960s. If future uranium finds are small, there will be increasing need for the breeder reactor, which should allow extraction of 60 to 70 times as much energy per gram of uranium as present types, but the breeder reactor may have severe safety problems and cost more to run than present reactors.

The future may see substantial innovations in the reactor plant aimed at lowering capital costs and improving fuel utilization and efficiency. Of the many possible reactor types, the pressurized light-water reactor has captured over 70 percent of the market so far. But there are many other practical possibilities. Heavy-water reactors improve fuel utilization by a factor of two; gas-cooled reactors could raise thermal efficiency from 30 to 40 percent or more. One or more

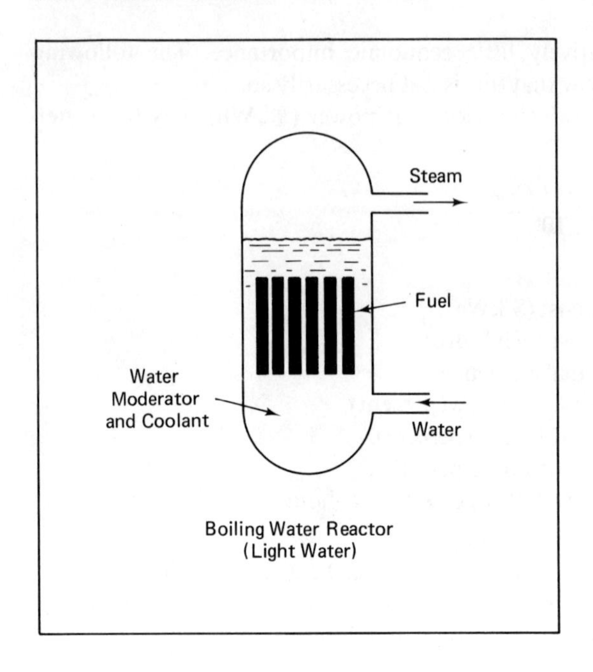

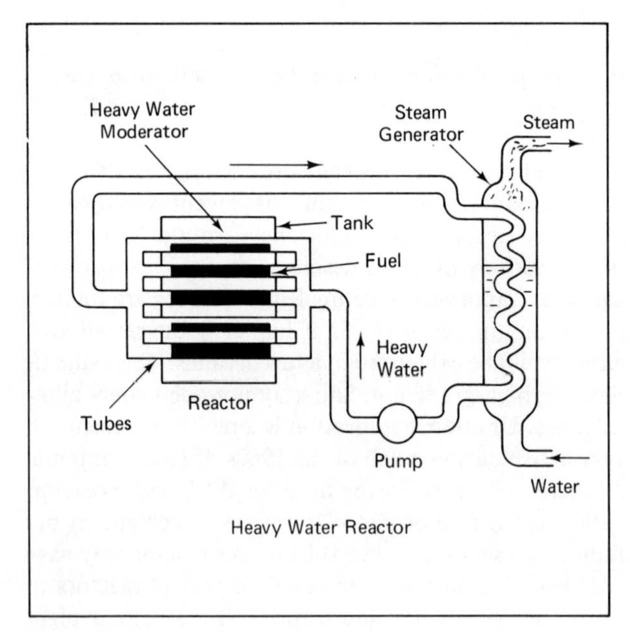

Figure 8.1
Thermal reactor types.

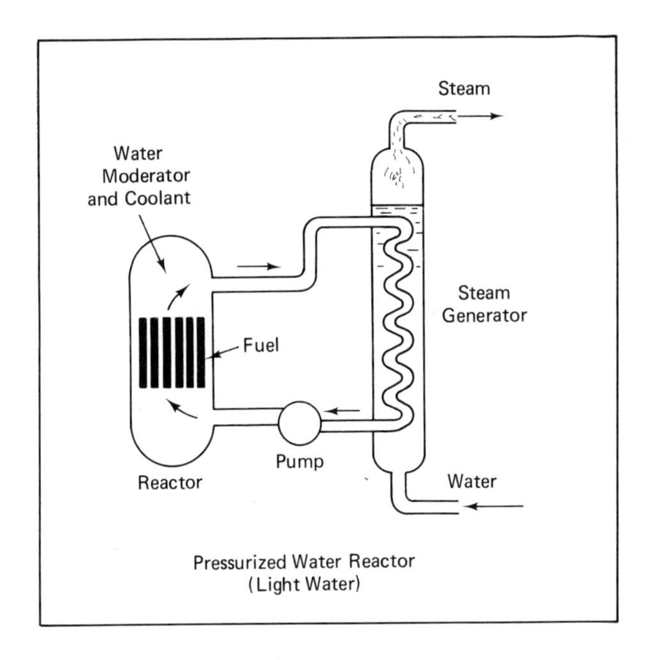

Pressurized Water Reactor
(Light Water)

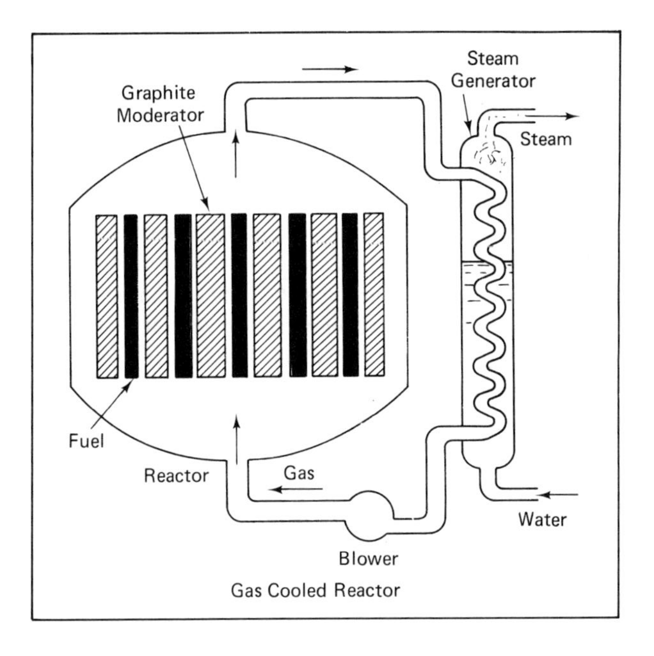

Gas Cooled Reactor

types of breeder reactor could effectively generate new fuel and make useful the very large quantities of U-238 and thorium that are not fissionable in present reactors.

The next few decades may broaden the use of nuclear energy, possibly by integrating power generation with heat generation, desalination, or the manufacture of synthetic fuels and feedstocks. Depending mainly on economics, there is a rich variety of possibilities for maximizing the utility of nuclear energy to man. Our concern here is primarily with power generation and with those technical and economic factors that will constrain reactor development and application.

8.2 REACTOR CHARACTERISTICS

8.2.1 General Features

Figure 8.1 shows schematically the kinds of reactors that have been of greatest commercial interest. All of them derive their energy largely from the fission of the uranium isotope U-235; they also extract energy from the fission of plutonium created by neutron capture in U-238. The kinetic energy of the fission fragments, captured in the fuel, is transferred as heat to the working fluid of the power cycle. Each reactor type is named after the particular coolant used to transfer heat from the fuel.

The two principal types of light-water-cooled reactors are the pressurized light-water reactor (PWR) and the boiling light-water reactor (BWR).

The two main kinds of heavy-water-cooled reactors are the Canadian deuterium–uranium reactor (CANDU) and the steam-generating heavy-water reactor (SGHWR).

Three kinds of gas-cooled reactor have become prominent: the Magnox reactor, the advanced gas-cooled reactor (AGR), and the high-temperature gas-cooled reactor (HTGR). The first two of these are cooled by CO_2, the third by helium.

The relative commercial popularity of the different kinds of reactors is indicated by Table 8.1. Light-water reactors, particularly the PWR, hold the field and will do so for some time. However, safety and cost issues and progress toward better fuel utilization may mean a different pattern of reactor ordering in coming decades.

Natural uranium contains an average 0.72 percent U-235; the rest is U-238, so that the ratio of U-235 to U-238 atoms is about 1:139. Both of these are fissionable, but the probability of U-235 fission is much higher. The important reaction is

$$\text{U-235} + 1 \text{ neutron} \rightarrow \text{fission fragments} + 200 \text{ MeV energy} + 2.5 \text{ neutrons (average)}.$$

Most of the energy released is in the form of the kinetic energy

Table 8.1
Commercial Reactors Entering Service in the Noncommunist World, 1972–1977

Reactor Type	Number
Light-Water Reactors	
PWR	101[a]
BWR	51
Heavy-Water Reactors	6
Gas-Cooled Reactors	
AGR	14
HGTR	2
Other	3
Other (including prototype fast breeder reactors)	10
Total	187

[a]Approximate.

of the fission fragments, which is transformed to thermal energy as the fission fragments are captured in the fuel matrix. The released neutrons have an average kinetic energy of about 2 MeV per neutron, which is much more than they would have in thermal equilibrium at the fuel temperature (the neutron thermal energy is 0.025 eV at room temperature). Thus the released neutrons also undergo a slowing-down process.

At their highest speeds some neutrons cause fission with U-238. However, it is much more probable that high-speed neutrons interacting with U-238 will simply be slowed down without being captured or causing fission. This slowing-down process is called "inelastic scattering." At low speeds the likelihood of neutrons causing fission in U-238 is negligible, but the probability of fission capture in U-235 is appreciable, and varies approximately inversely with neutron speed. Reactors that rely on low-speed neutrons for most fissions are called "thermal reactors."

One design task for thermal reactors is to provide a means for slowing neutrons to thermal velocities without losing too many by nonfission capture in the fuel or supporting structure or by escape from the reactor. For the nuclear chain reaction to continue, an average of 1 of the 2.5 neutrons released must be slowed down to the thermal speed range to cause a subsequent fission.

To slow the neutrons without nonfission capture, thermal reactors use substances whose atoms absorb a fraction of the neutron momentum on each collision. These substances are called moderators. Light water, heavy water, and graphite have been extensively used in reactors as moderators, since they perform the slowing-down function well without absorbing too many neutrons in nonfission capture. Light water slows neutrons more rapidly (in a smaller volume) than heavy water, but absorbs many more neutrons after doing so.

If the moderator is light water, it is always necessary to use enriched fuel in which the fraction of U-235 is raised to perhaps 2 or 3 percent. This is three or four times the concentration in natural uranium, and is necessary in order to sustain a chain reaction. The enrichment process is costly, and the need for it can be avoided by using heavy water as the moderator. However, heavy water is expensive and requires an increase in reactor size per unit power output.

The absorption of neutrons by U-238 produces plutonium (Pu-239), which itself is a nuclear fuel. Perhaps half of the plutonium fissions in place after it is produced. The remainder could be extracted from spent fuel, concentrated, and used in thermal reactors, but so far this process has not been economical. Inventories of plutonium in spent fuel from thermal reactors could serve as an important part of the fuel supply of the breeder reactor, which will be discussed later. Over the lifetime of the fuel as much as 30 percent of the energy produced in a light-water reactor may be due to plutonium fission.

The coolant is the fluid medium used to transfer thermal energy from the fuel elements to the working fluid of the power plant. The coolant and moderator may be one and the same, as in light-water reactors, or they may be two distinct materials. The reactor coolant may also be used as the working fluid of the turbine, but this is not usual.

Thermal reactors differ mainly in:

1. fuel (natural or enriched uranium)
2. coolant (light water, heavy water, CO_2, or helium)
3. moderator (light or heavy water or graphite)
4. pressure vessel (single large cylinder or multiple pressure tubes).

Figure 8.1 shows in a simplified way how the nuclear fuel, moderator, and coolant are arranged; it is vital that the three be intimately close in each reactor type.

The BWR is a large pressure vessel containing light water (which serves as both moderator and coolant) and enriched fuel. The coolant boils as it passes through the reactor, and the steam produced is taken directly to a steam turbine. In other nuclear plants, steam is generated indirectly by heat exchange with the coolant. The fuel for the BWR is assembled in arrays of rods made up of pellets of UO_2 clad with an alloy of zirconium to withstand corrosion by H_2O. The reactor must be shut down and the lid of the pressure vessel removed for refueling. The pressure in the reactor is about 1000 psia (7 MPa) and the wall thickness is large.

The PWR operates at even higher pressure than the BWR. The higher pressure suppresses boiling; this permits more intensive heat transfer, so that for the same power level the PWR can be somewhat smaller than the BWR. The higher pressure permits the PWR to operate at higher temperature and thus somewhat higher efficiency, but this effect is small. The BWR and PWR operate with nearly the same efficiencies. Their capital costs also are virtually the same, and have been somewhat lower than those of reactors cooled by gas or heavy water.

In contrast to the light-water reactors, the heavy-water reactor shown in Figure 8.1 contains the high-pressure coolant in tubes that pass through the unpressurized heavy-water tank. Again the fuel typically consists of UO_2 pellets clad with zirconium alloy. The coolant and moderator are both heavy water. The pressurized tubes are made of zirconium alloy. The pressurized tube concept has the advantage that as the reactor is scaled up to large size, there is no need for a pressure vessel with extraordinarily thick walls. The most prominent of the heavy-water reactors is the CANDU, which does not have quite so high a cycle efficiency as the PWR or BWR, but has considerably better overall utilization of the fission energy available in natural uranium. This is because it can use natural uranium directly and consumes a large fraction of the U-235 contained therein. It also converts a larger fraction of U-238 into Pu-239.

For gas-cooled reactors the pressure vessel is typically made of prestressed concrete. The coolant is either CO_2, used in the United Kingdom for the AGR, or helium, used in the HTGR. Both of these reactors use graphite as moderator. They require enriched fuel and have high capital costs, but they have the merit of high cycle efficiency. Helium (though not CO_2) is capable of operation with graphite at very high temperatures provided it can be kept sufficiently pure. Graphite has excellent high-temperature behavior, so that the gas-cooled reactor may be capable of high cycle efficiency, as compared with water-cooled reactors.

Table 8.2 shows fuel, coolant, and moderator materials for these reactors, as well as for a typical fast breeder reactor. Typical sizes and power densities are also indicated, as well as operating pressures, temperatures, and efficiencies for reactors of largest typical size.

The performance figures shown in Table 8.2 point to a potential advantage in efficiency for the gas-cooled reactor, but at the same time they indicate a penalty in reactor size. The power density for the reactor core is 30 times as high for the PWR as for the AGR. Even with special techniques of construction, the reinforced concrete pressure vessel used in the AGR is expensive. On the other hand, the steel pressure vessel of the PWR has such large wall thickness that fabrication to required quality and safety is difficult.

Table 8.2
Typical Reactor Characteristics

	PWR	BWR	CANDU	SGHWR	Magnox	AGR	HTGR	LMFBR
Net Electrical Power Output (MWe)	1100	1100	540	500	500	560	1200	350
Fuel								
Fuel Material	UO_2	UO_2	UO_2	UO_2	U	UO_2	UO_2, ThC	PO_2, UO_2
Can Material	Zr	Zr	Zr	Zr	Mg	SS[a]		SS[a]
Feed Enrichment (percent U-235)	3.0	2.5	natural	2.0	natural	2.0–2.6		
Coolant								
Composition	H_2O	H_2O	D_2O	H_2O	CO_2	CO_2	He	Na
Pressure (psia/MPa)	2200/15.2	980/6.8	1350/9.3	1000/6.9	385/2.7	490/3.4	700/4.8	147/1.0
Inlet Temperature (°C)	283	275	250	274	247	300	340	380
Outlet Temperature (°C)	316	286	294	283	414	670	780	540
Core								
Diameter × Height (m)	3.7 × 3.4	4.8 × 3.7	6.4 × 6	7 × 3.7	17.4 × 9.1	9.6 × 8.3	8.4 × 6	1.8 × 0.91
Average Power Density (MW thermal/m³)	95	51	9.2	12	0.9	2.6	8.4	400
Moderator	H_2O	H_2O	D_2O	D_2O	graphite	graphite	graphite	none
Pressure Vessel	steel	steel	Zr tubes	Zr tubes	concrete	concrete	concrete	SS[a] tank
Inside Diameter × Height (m)	4.4 × 13	6.4 × 22	0.1 m dia.	0.13 m dia.	29 × 29	20 × 17.7	30.6 × 27.8	6.2 × 16
Wall Thickness (mm)	215	160	5	5	3300	6400	4700	25

Power Plant								
Steam Pressure (psia/MPa)	860/5.9	970/6.7	570/3.9	900/6.3	665/4.7	2400/16.7	2400/16.7	2400/16.7
Temperature (°C)	274	291	250	280	400	566	566	566
Station Net Thermal Efficiency (percent)	33	33	31	33	31.4	43	43	43

[a]Stainless Steel.
Source: Data from Refs. (3) (4).

8.2.2 Fuels

Except for the Magnox reactor, which uses natural uranium clad with magnesium, these reactors use UO_2 fuel elements. Both the CANDU heavy-water reactor and the Magnox CO_2-cooled reactor use natural uranium, 0.72 percent of which is the fissionable isotope U-235. The other reactors use enriched fuel (up to 3 percent U-235) to sustain fission. To maintain the fission chain reaction with natural uranium fuel it is necessary to use heavy water as moderator. The heavy water does not absorb nearly as many neutrons as ordinary water. It therefore permits better fuel utilization. The U-235 ratio in fuel rejected from the CANDU heavy-water reactor is as low as 0.18 percent. The enrichment percentage of the fuel rejected from the light-water reactors is as high as, if not higher than, the natural percentage. Thus the light-water uranium reactor makes use of perhaps half as much of the energy in the natural ore as does the heavy-water reactor. Heavy water is expensive to produce (around $15/kg) and forms an appreciable part of the capital cost of a heavy-water reactor (see Table 7.4). It also greatly increases the size of the reactor core, and this too contributes to capital cost.

On the other hand, enrichment is a costly process and adds 60 percent or more to the cost of the fuel. It requires elaborate apparatus to separate the two isotopes U-238 and U-235, since these differ so little in atomic weight. First, natural uranium ore is processed to uranium oxide U_3O_8. For enrichment, this is converted to uranium hexafluoride gas UF_6, which is separated into heavy and light constituents (corresponding to U-235 and U-238) either in diffusion or centrifuge plants. The established process—gaseous diffusion—uses a vast number of cells through which the U-235, in gaseous form as UF_6, diffuses slightly faster than the U-238, making progressive enrichment possible. A competitive process (now being developed for commercial application) uses a long series of rotating cylinders to separate the gaseous compounds of U-235 and U-238 by means of a centrifugal force field. Gaseous diffusion plants must be built on an enormous scale for low-cost enrichment. Only five of these are now operating in the noncommunist world: three in the United States, one in the United Kingdom, and one in France. They are heavy consumers of electricity; the three United States plants together consume a total of nearly 6000 MWe. Enriched fuel for commercial nuclear power has been processed in plants built at government expense for military use of nuclear energy. However, the commercial demand for enriched fuel is outgrowing the capacity of these plants, so large new capital investment will be required.

8.2.3 Efficiencies

Table 8.2 shows the relative efficiencies of various reactor plants. Water-cooled reactors are considerably less efficient than fossil fuel steam plants, owing to limitation of steam temperature to 570°F (300°C) or less. Water coolant will dissociate to some extent under irradiation, so that corrosion of the fuel cladding is a serious problem. Zirconium alloy, selected because it absorbs very few neutrons, is said to have good corrosion resistance in pure water at temperatures up to 660°F (350°C), but radiation and water impurities reduce the allowable temperature limit.

Water-cooled reactors are capable of the high power densities shown in Table 8.2 only if the water stays in the liquid state so that heat transfer rates are high. This brings in the pressure problem. At 560°F (294°C) the vapor pressure is 1133 psia (7.88 MPa); at 660°F (350°C) it increases to 2365 psia (165 MPa). Raising the pressure to these levels places a tremendous mechanical load on large reactor pressure vessels, so that fabrication is difficult and cost rises rapidly. As Figure 8.1 suggests, the light-water reactor is particularly subject to this problem since the large coolant–moderator vessel is subject to maximum pressure. The heavy-water reactor deals with this problem by confining the hot coolant to many small-diameter tubes that surround the fuel; these are easier to design to safe working stresses. Since a large number are required, the reactor becomes mechanically complex; a compensating feature is that the reactor may be refueled, one tube at a time, while operating at full power.

The great advantage of the gas-cooled reactor is that coolant pressure is not directly linked to temperature. Thus if corrosion can be avoided, high temperatures and high efficiencies are possible without unduly high mechanical pressure loads. A temperature increase need not imply a pressure increase, as with liquid coolant. Also, high-temperature corrosion can be avoided by use of an inert gas such as helium. With CO_2, dissociation and corrosion are potential problems; for the AGR to operate successfully at 566°C, the fuel must be clad with stainless steel.

8.2.4 Development Possibilities

First, there may be some gains in temperatures and efficiencies. Pressure vessel reactors with light-water cooling are already a difficult design problem: higher pressures are unlikely in these large-size vessels. Pressure tube reactors will be helped by zirconium alloys of higher strength, which will permit water cooling with somewhat higher pressure. Switching to an organic coolant with low vapor pressure could permit higher temperatures. Organic coolants have been tested under simulated reactor conditions, and may be usable at top temperatures as high as 750°F (400°C). Inert gas coolant should ultimately permit very high temperatures. The HTGR uses UO_2 fuel encased in graphite; temperatures of 850°C (1600°F) or more are considered quite feasible as development targets. Combining the HGTR with a helium-and-steam binary cycle should permit efficiency to rise significantly above best present levels for fossil fuel plants.

Second, there is scope for improved fuel utilization. In present thermal reactors, significant quantities of the U-238 are converted into Pu-239 by fission capture (0.3 to 0.6 g of Pu-239 per gram of U-235 consumed). This plutonium could be extracted from the reject fuel and processed as relatively concentrated PuO_2 for recycling through a thermal reactor, although this has not yet been economical· The main interest today is in using the plutonium as feed for the fast breeder reactor. Breeding of fissionable fuel (Pu-239) from U-238 is feasible in a fast breeder reactor, and Th–U-233 breeding is a practical possibility in a thermal reactor. Thorium, like uranium, is an abundant element.

Several fast breeder reactor prototypes are under development. Demonstration LMFBR plants are operational in France (Phénix) and in the United Kingdom (Dounreay). A demonstration plant is under construction in the United States (Clinch River). The LMFBR uses liquid sodium as coolant. For fast neutrons to survive, such moderating materials as hydrogen, deuterium, or carbon must be absent; this and other factors lead to the choice of sodium. Sodium has high thermal conductivity and so permits high power density. The typical thermal energy release rate is 600 MW thermal/m^3; this is 10 times as high as in the light-water reactor, and perhaps 100 times as high as in the gas-cooled reactor.

8.3 LIGHT-WATER REACTORS

8.3.1 The PWR and the BWR

The PWR has a history of successful operation, first in submarine propulsion and then over the last decade in a number of electrical power stations. The engineering feasibility of light-water reactors was proven in the 1950s, but it was not until 1960 that the first demonstration light-water power reactor began to operate. It was not until 1967 that the first commercial-sized station came into operation. At present many 1100- or 1200-MWe units are operating. Thus the pace of development of light-water reactors has been rapid indeed and has necessitated advances in the arts of pressure vessel construction, water quality control, and radiation control, as well as in the design of plant components.

Figure 8.2 shows a typical steam cycle for a PWR. Steam is generated at 925 psia (6.35 MPa) and superheated to 570°F (298°C) by the reactor heat exchanger. At the entry to the turbine the steam conditions are 700 psia (4.82 MPa) and 566°F (296°C). In the high-pressure turbine, the steam expands to about 250 psia (1.72 MPa) and 7 percent moisture. It is then reheated to about 500°F (258°C) and sent to the low-pressure turbine. The moisture content at the

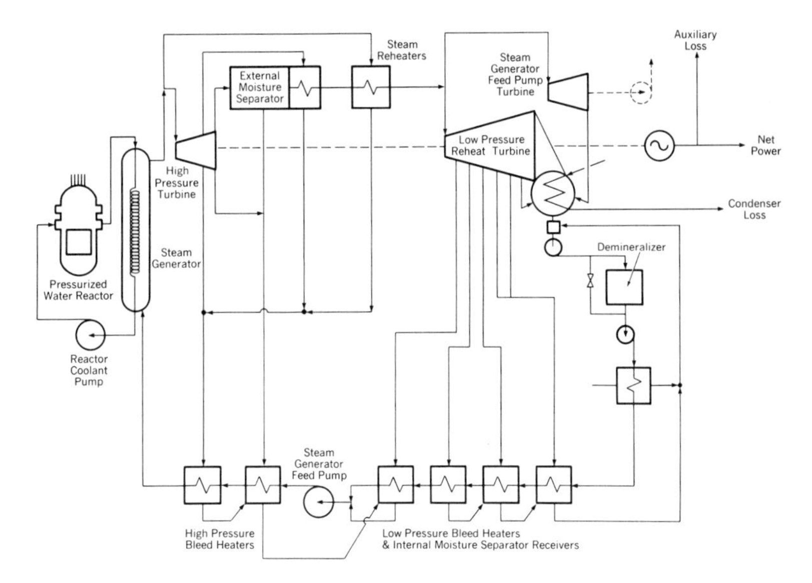

Figure 8.2
Power cycle diagram, nuclear fuel—reheat by bleed and high-pressure steam, moisture separation, and six-stage regenerative feed heating—900 psia, 566°F/503°F steam. Courtesy of Babcock and Wilcox (5).

end of this expansion is nearly 14 percent, despite ample provision for moisture separation. Six feedwater heaters are used.

Figure 8.3 shows the cross section of the BWR manufactured by General Electric (6). Fuel is arranged in the cylindrical reactor core in the form of square lattice fuel assemblies. Each fuel assembly contains a seven-by-seven lattice of fuel rods, each lattice located within its own channel; the channel coolant flow is adjusted (by what is termed "orificing") to correspond with variations in local heat release rate across the reactor.

The fuel material in the BWR is enriched uranium in the form of sintered and ground UO_2 pellets. The pellets are contained in rods enclosed in zirconium alloy cladding. Grids are used to space the fuel rods and also to prevent flow-induced vibrations.

The required enrichment of the fuel depends upon its location within the reactor core. At the outer part of the core more highly enriched fuel is used. Annual replacement of fuel rods is typical, and this requires the shutdown of the reactor and removal of the fuel rods through the top of the reactor vessel. Approximately one-fourth to one-third of the fuel core is removed at each refueling.

The rate of fission within the reactor is controlled by the disposition of substances that absorb neutrons and also (in the case of the BWR) by changes in the coolant flow rate. Neutrons are absorbed by cruciform-shaped control rods that contain boron carbide powder in stainless steel tubes. These rods can be moved into or out of the

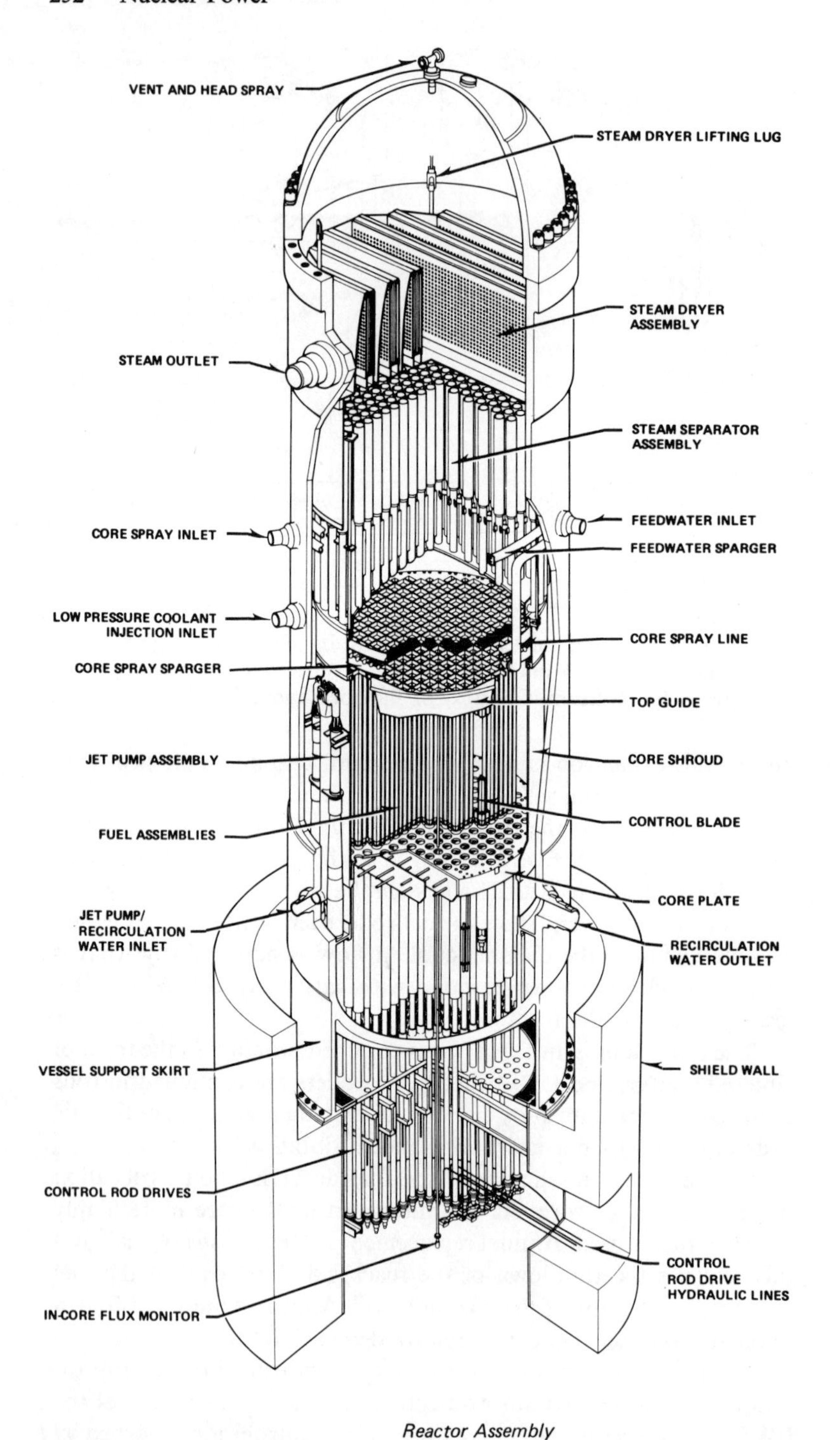

Figure 8.3
Boiling water reactor (BWR). Courtesy of General Electric Co.

reactor core. During emergency shutdown of the reactor (SCRAM), all control rods can be inserted simultaneously into the reactor core with hydraulic actuation from a stored high-pressure gas accumulator. The coolant flow through the reactor core is adjusted by altering the flow rate through the ejector jet pumps located around the side of the control. As the flow rate through the reactor channels goes up or down, the liquid–vapor volume fraction changes, thus altering the average concentration of moderator within the core. This changes the neutron concentration and so enables the reactor to increase or decrease energy production rate, depending on the electrical load demanded by the system.

The core coolant is driven downward in the outer annular space between the reactor core and the inside of the pressure vessel. Then the flow rises vertically through the reactor core channels as a steam–water mixture. The circulation is driven by jet pumps supplied by two or more pumps external to the reactor. The use of jet pumps minimizes the quantity of coolant that must recirculate outside the reactor vessel itself. Leaving the core, steam rises through moisture separators and driers and emerges nearly dry (moisture content less than 0.3 percent maximum) through the steam outlet port shown in Figure 8.3. Emergency cooling of the core is done by sprays in the top head and near the core, and by flooding of the core itself. The reactor has a control system to cool the reactor during shutdown and to remove the "decay heat" after shutdown. Control of water quality to prevent corrosion and eliminate impurities is done with filters and demineralizers.

Whereas the liquid H_2O coolant in the PWR is used indirectly to generate steam, the coolant in the BWR pressure vessel is used directly as steam in the turbogenerator. The hazard of excessive radioactivity in the turbine is avoided in normal operation by venting gaseous activation products (noncondensables) from this loop. Contamination of the turbine by solid fission products escaping from the fuel is slight since these remain in the liquid phase of the coolant, from which they are fairly easily removed.

In the BWR vapor voids in the coolant can lead to oscillations of the flow field, vibrations, and possibly even to burnout. Prevention of such adverse flow conditions requires that the BWR power densities be lower than those for the PWR.

Figure 8.4 shows a typical PWR pressure vessel manufactured by Westinghouse. Again, the reactor core is cylindrical and is built up with square lattice fuel assemblies. Each fuel assembly consists typically of a 14 × 14 lattice of fuel rods. Separate flow channels are not used in the PWR core, so the coolant can mix radially between adjacent fuel assemblies. Again, the fuel is enriched UO_2 pellets clad in zirconium alloy. The degree of enrichment varies with location in the core.

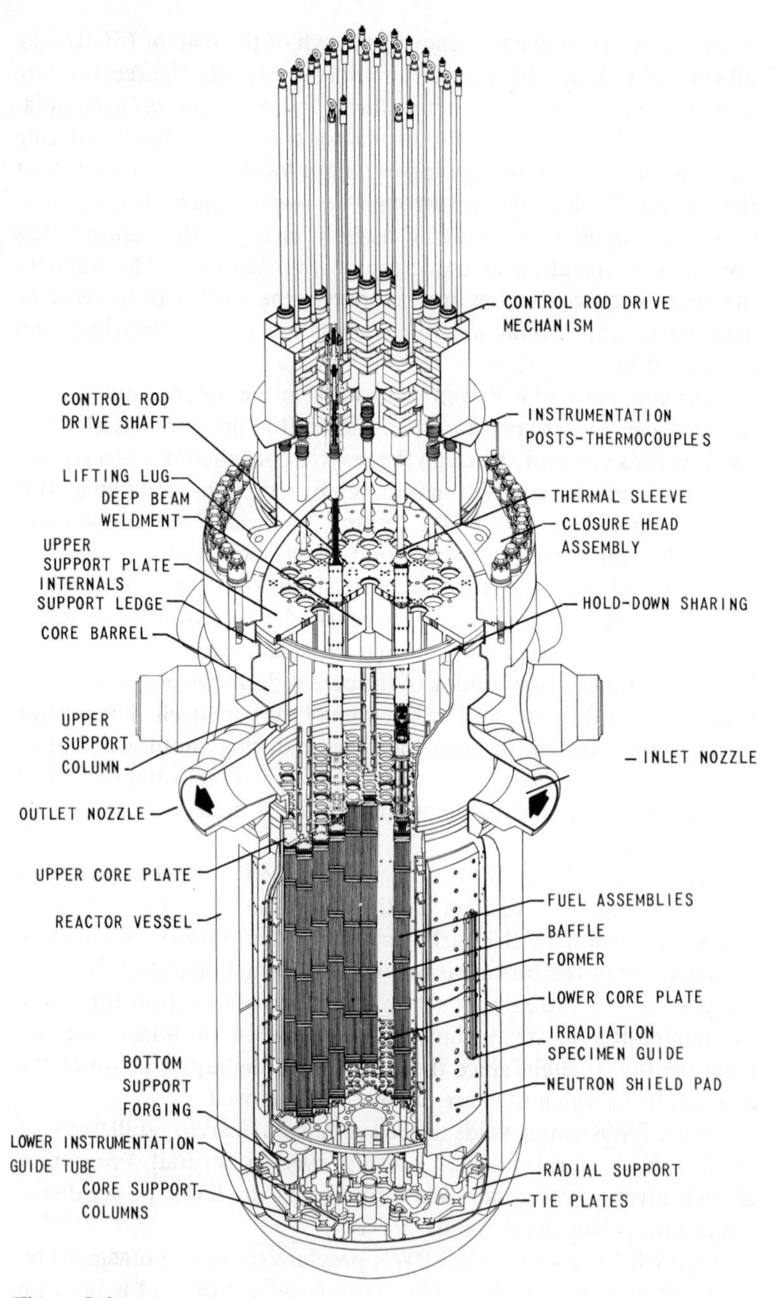

Figure 8.4a
Pressurized water reactor (PWR). Courtesy of Westinghouse Electric Corp. (6).

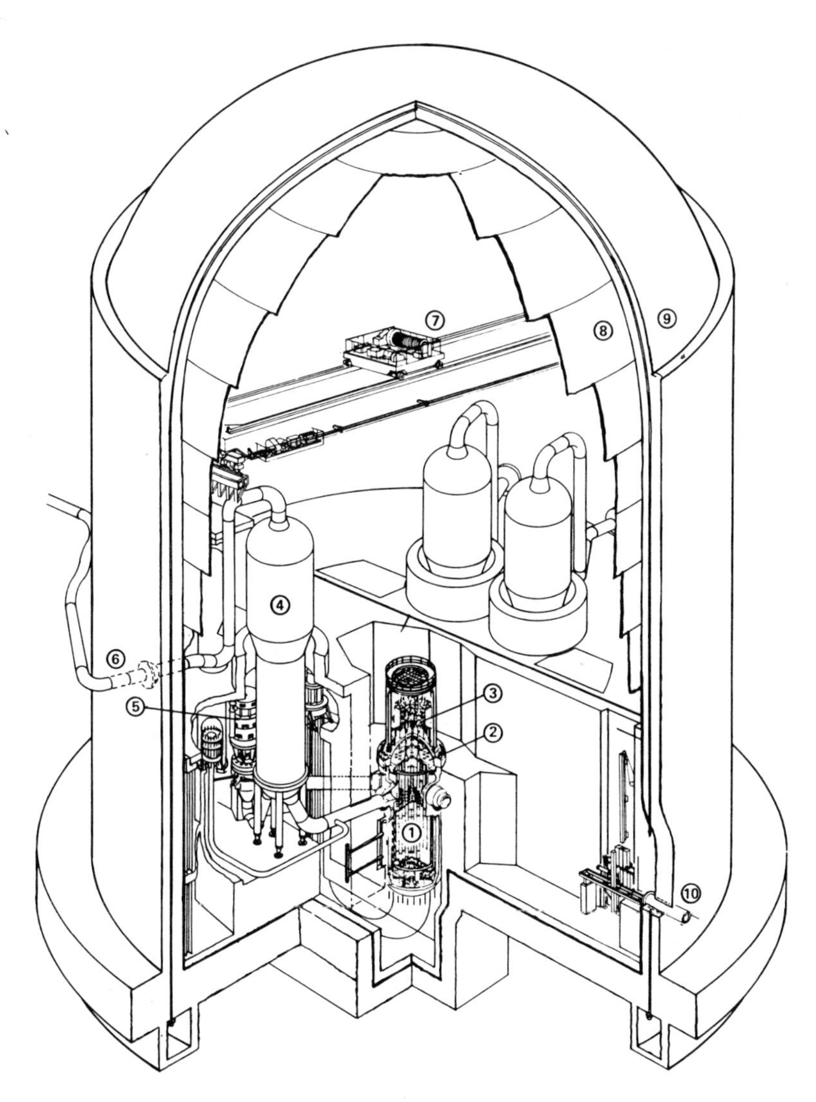

1 Reactor Core
2 Upper Internals
3 Control Rod Drive Mechanisms
4 U-Tube Steam Generator
5 Main Coolant Pump
6 Main Steam Line
7 Polar Crane
8 Steel Containment Liner
9 Reinforced Concrete Containment
10 Fuel Transfer Canal

Figure 8.4b
PWR and containment vessel. Courtesy of Westinghouse Electric Corp. (6).

Fission rate is controlled in the PWR with control rods. The control rod material is typically boron carbide or silver–indium–cadmium clad with stainless steel and inserted into control rod tubes within the fuel assemblies. The level of the control rods within the core is adjusted mechanically; during emergency shutdown the control rods fall by gravity into the reactor core.

Since vapor is absent in the PWR core, flow control cannot be used to adjust the level of reactivity. Another control method is provided by using boric acid dissolved in the primary coolant. The boric acid concentration can be changed to compensate for slow changes in the characteristics of the core due to fuel burnup and the accumulation of fission products. In the PWR, the coolant enters the reactor vessel through the inlet nozzle (shown in Figure 8.4) and passes vertically downward in the annular space between the reactor core and the inside of the pressure vessel. It then flows upward through the core fuel assemblies and leaves through the outlet nozzle, passing to a heat exchanger that generates steam for the turbine. In one type of heat exchanger, subcooled feedwater is heated to saturation temperature and then evaporated, leaving the heat exchanger with a moisture content of not more than 0.25 percent. In the second type of heat exchanger, steam is superheated by as much as 28°C. As with the BWR, several methods of emergency cooling are provided. In the PWR, borated coolant is injected directly into the reactor vessel if the coolant pressure decreases below a specified level.

8.3.2 Emergency Core Cooling

The single most serious source of danger with nuclear reactors is melting of the core fuel and release of radioactivity for dispersion in the atmosphere. Nuclear safety systems are designed to prevent overheating of the fuel, and in the event that the fuel does melt, to prevent release of the radioactivity.

Fuel overheating can be caused by loss of coolant due to a failure in the cooling system, such as a pipe rupture. Alternatively, the reactor fuel can overheat if, following a shutdown of the reactor, the heat removal systems fail to operate. Immediately after shutdown the decay heat is about 7 percent of the full thermal power of the reactor; this dies rapidly away to about 1 percent, which lasts for a long period and for which a heat removal system is required. The first of these failures is called a "loss-of-coolant" accident. Should one of the coolant ducts leading to or from the reactor rupture, the pressure of the coolant would drop very quickly. Vapor voids would then form in the coolant channels. The first effect would be to reduce the supply of neutrons through removal of a portion of the liquid moderator. This would tend to decrease radioactive heat release, but not immediately, due to fissions caused by the "delayed neutrons." In this transient the local loss of liquid coolant could allow the surface of the fuel element cladding to melt, followed by massive melting of the fuel and reactor core structure.

The Reactor Safety Study set up by the USAEC estimated the risks due to such accidents and compared them with nonnuclear risks to which the population is exposed (7). The Study Group reviewed a large number of ways in which core-melt could conceivably occur following a loss-of-coolant accident. In nearly all cases it was found that multiple failures of the emergency core cooling system would be required for the core to melt. One exceptional case was the failure of the pressure vessel itself; however, the Study Group concluded that "accumulated experience with pressure vessels indicates that the chance of such a failure is indeed very small," one in about 10 million vessel-years. This would make vessel failure an insignificant contributor to population risk. Not very different is the finding of the Advisory Committee on Reactor Safeguards that the disruptive failure probability of reactor pressure vessels constructed and operated in accordance with the ASME *Boiler and Pressure Vessel Code* is less than 10^{-6} per vessel-year (8).

The Study Group found that the most probable causes of loss-of-coolant accidents were pipe-rupture in the reactor coolant system and failure of the decay heat removal system that should come into operation during an unplanned reactor shutdown.

In the large reactors under consideration by the Study Group, the decay heat levels following a loss-of-coolant accident were so large that if the emergency core cooling system did not work, the heat could not be dissipated through the reactor containment vessel wall. The level of radioactivity that could be released from these large reactors would require a fairly thick concrete shielding around the outside of the containment vessel to prevent overexposure of persons nearby in the event that reactor core failure had occurred. In some cases the pressure could rise sufficiently for the containment vessel to rupture, allowing gaseous radioactivity to diffuse over the countryside.

The containment building is designed to prevent any escape of radioactivity initially released, but would not guarantee containment of all radioactivity should the core melt. According to the Study Group, "Although there are features provided to keep the containment building from being damaged for some time after the core melts, the containment will ultimately fail, causing a release of radioactivity" (7). An essentially leak-tight containment building is provided to prevent the initial dispersion of the airborne radioactivity into the environment. The containment would not likely fail for a number of hours after the core melts; during that time the radioactivity released from the fuel would be deposited by natural processes on the surfaces inside the containment. In addition, water sprays and pools are provided to wash radioactivity out of the building atmosphere, and filters are installed to trap radioactive particles. Since the containment buildings are made essentially leak-tight, the radioactivity would be contained as long as the building remained intact. Even if the building were to have sizable leaks, most of the airborne radioactivity would probably be trapped and not released.

Following core-melt, the molten mass could melt through the thick concrete base into the ground. But this would take time—about half a day or so. In the meantime, radioactive decay and washing of radioactive gaseous products on the internal walls could occur. According to the Study Group, melt-through would not be as catastrophic as is often imagined:

> Most of the gaseous and particulate radioactivity that might be released would be discharged into the ground, which acts as an efficient filter, thus sufficiently reducing the radioactivity released to the above-ground environment. Accidents which follow this path are thus characterized by relatively low releases and consequences. (7)

In plants that have relatively large volumes of containment, the melt-through path described apparently represents the most likely course of the accident.

The safety systems used in current light-water reactors aim to protect against the consequences of a loss-of-coolant accident, in the following sequence:

1. After the reactor coolant system ruptures, and high-pressure, high-temperature water discharges into the containment vessel, the emergency core cooling system operates to drench the core and keep it adequately cool.
2. The reactor containment vessel is designed to be leak-tight and to contain essentially all of the radioactivity released from the core.
3. Following the accident, radioactivity is removed from the interior of the containment vessel. For the PWR this is done by spraying water that contains chemical additives.
4. Iodine and other radioactive materials are washed out of the containment atmosphere. The BWR reactor containment vessel removes radioactivity from its atmosphere by means of a vapor-suppression pool and also by use of a filter.
5. Heat is removed from the interior of the containment vessel to prevent overpressurization, which might lead to the rupture of the vessel. This is done by heat transfer, from warm water within the containment to cold water outside the containment.

The Study Group examined a large number of events and faults associated with possible PWR accidents and concluded that the most probable sequence of events leading to radioactive release from the reactor was a large rupture in the reactor coolant system, followed by failure of the emergency core cooling system, followed in turn by core-melt and melt-through at the base of the reactor, and thus rupture of the containment vessel.

For the BWR the most likely cause of containment failure appears to be by overpressure rather than by melt-through; this has somewhat lower probability of failure but somewhat larger release of radioactivity.

For the PWR the mean probable failure rate for radioactive release from the containment vessel following core-melt is estimated to be 60 per million reactor-years; for the BWR it is 30 per million reactor-years.

Having estimated the probabilities of reactor containment vessel failure and release of radioactivity to the environment, the Study Group estimated the magnitude of the release of various types of radioactivity and also estimated the effects of these releases on health and property damage in the surrounding areas. To estimate radiation health effects they made use of the findings of the National Academy of Sciences (9).

Other causes of reactor failure—earthquakes, tornadoes, floods, aircraft crashes, or missiles ejected by a disintegrating turbine—all led to lesser probabilities of failure than those mentioned above. Failure due to human sabotage was not considered.

Following the important paths for release of the radioactive constituents to nearby populations, the Study Group estimated short-term and long-term health effects. The estimated annual fatalities due to accidents in a population of 100 reactors are shown in Figures 5.2 and 5.3.

The conclusions of the Study Group depend on many assumptions and on estimates of probable component failure rates. A degree of support for their conclusions is to be had from experience up to the present with commercial and military power reactors: 2000 reactor-years with no major accidents. This suggests that the likelihood of even a small accident is less than 1000 per million reactor-years. Since power reactors have not yet had even dangerously high fuel temperatures, this suggests that core-melt should be much less likely than 1000 per million reactor-years. The Study Group therefore finds the computed probability of core-melt of 60 per million reactor-years to be reasonable. Calculations further indicate that 9 out of 10 core-melt accidents cause less than 10 fatalities. The probability-weighted average of the consequences indicates 400 deaths per million reactor-years, essentially the same for both PWR and BWR plants.

Assessment of reactor safety has been shown by the Study Group to be extremely complex, and to involve, in the end, many uncertainties. Nevertheless, they have gone a long way toward systematically tracking down all the possible failure sequences that can rationally be identified by present knowledge. They see no reason for despair over reactor safety provided extreme care is exercised in every stage of construction, maintenance, and operation.

8.4 HEAVY-WATER REACTORS

Heavy water (D_2O) has much the same characteristics as ordinary water. Its boiling point at atmospheric pressure is 214.6°F (101.4°C), and its density at room temperature is only 10 percent above the density of ordinary water. However, its nuclear properties are very different from those of ordinary water. Its slowing-down length (the distance required to slow a neutron through a given reduction in velocity) is about twice as great as for ordinary water. This means that a relatively large volume of moderator is required for the heavy-water reactor, but the probability of neutron capture is much less than in light water. As pointed out earlier, this makes possible the use of natural uranium, a valuable feature for a country like Canada with substantial uranium resources but no enrichment facilities. The inventory of U-235 per unit of electrical power output from the CANDU reactor is about half what it is for the PWR and about a third what it is for the BWR. The production rate of plutonium per gram U-235 consumed in the CANDU reactor is about twice what it is in the PWR and the BWR (10).

8.4.1 The CANDU Reactor

Figure 8.5 shows a cross section of the CANDU heavy-water reactor at Pickering on the shore of Lake Ontario (10, 11). The horizontal cylindrical vessel which contains the reactor fuel, coolant tubes, and moderator is called the "calandria." The reactor fuel is arranged in bundles of small, short tubes; each fuel bundle consists of 28 zirconium alloy tubes 49.5 cm long, containing compacted and sintered pellets of UO_2. Twelve of these fuel bundles are mounted end-to-end in each reactor channel, which is a larger zirconium alloy tube through which the coolant flows. There are 370 coolant tubes, each pressurized separately. The coolant tubes are also made of zirconium alloy and are surrounded by a thin annular space containing nitrogen, which is bounded by yet another zirconium alloy tube called the "calandria tube." This is directly supported by the end shields of the calandria. The insulating nitrogen-filled annulus retards heat transfer from the coolant tubes to the low-pressure moderator surrounding the calandria tubes. In the event of coolant failure leading to pressure tube rupture, the surrounding moderator would act as a heat sink; this is an important safety feature, protecting against the possibility of melting. Another safety feature is that leakage of coolant from a tube can be detected by monitoring moisture level in the surrounding nitrogen atmosphere. Control rods are located within the moderator.

The 500-MWe reactor shown in Figure 8.5 contains about 276 tons of heavy water within the stainless steel calandria vessel. Below the calandria is a dump tank, normally pressurized by helium, into which the heavy water can be quickly transferred. At $65/kg, the heavy water in the reactor represents a capital cost of around $35/kWe. Since it is so expensive and there are so many seals with this configuration, leakage of coolant from the reactor could be a major problem. However, close attention to design and development has reduced leakage of heavy water in the Pickering reactors to a negligibly small quantity.

The heavy water coolant leaves the reactor channel at a temperature of about 250°C (see Table 8.2); it is circulated by pumps through boilers that generate the turbine supply steam.

Fuel can be removed and loaded while the reactor is operating at very nearly full power. The fueling machine takes one channel out of service at a time, removing the coolant tube sealing plug, inserting new fuel at one end, and removing spent fuel from the other. On-power refueling could mean higher availability than is possible with pressure vessel reactors, which require an annual shutdown period for refueling.

The reactor is housed in a large building. Should a loss-of-coolant accident occur and the pressure within the reactor building exceed about 1.5 atm, the contents of the reactor building will discharge into a separate vacuum building that is maintained at about 0.1 atm.

Since the coolant is separated from the moderator by pressure tubes, other coolants could be used, with heavy water retained as moderator. The use of organic coolant, for example, could substantially raise the allowable temperature without exceeding reasonable pressures in the coolant, and would make possible more efficient power generation (up to around 40 percent efficiency with 400°C temperature of the organic coolant).

Table 8.3 shows the potential for uranium conservation in CANDU-type reactors, according to A. J. Mooradian (10). Case 1 corresponds to present technology with the CANDU reactor. Case 2 shows the possible benefit of using uranium silicide fuel, which has not yet been developed for commercial application but has advantages because of its higher density of uranium within the fuel. This benefit could lead either to lower capital costs of the nuclear plant or to lower fuel utilization, and Table 8.3 shows only its possible effect on reducing fuel use per unit of power output. Case 3 indicates a CANDU-type reactor for a boiling indirect cycle optimized for minimum natural uranium consumption. Case 4 shows the possible benefit of using the plutonium generated from the reactor itself, recycled with the fuel in the same reactor; the fuel is then a mixture of natural uranium plus the recycled plutonium. Plutonium recycling could halve the amount of natural uranium used. The required reactor modifications would not be large (to accommodate this new fuel), but a new industry would be needed to process the highly toxic plutonium. Mooradian points out that while plutonium re-

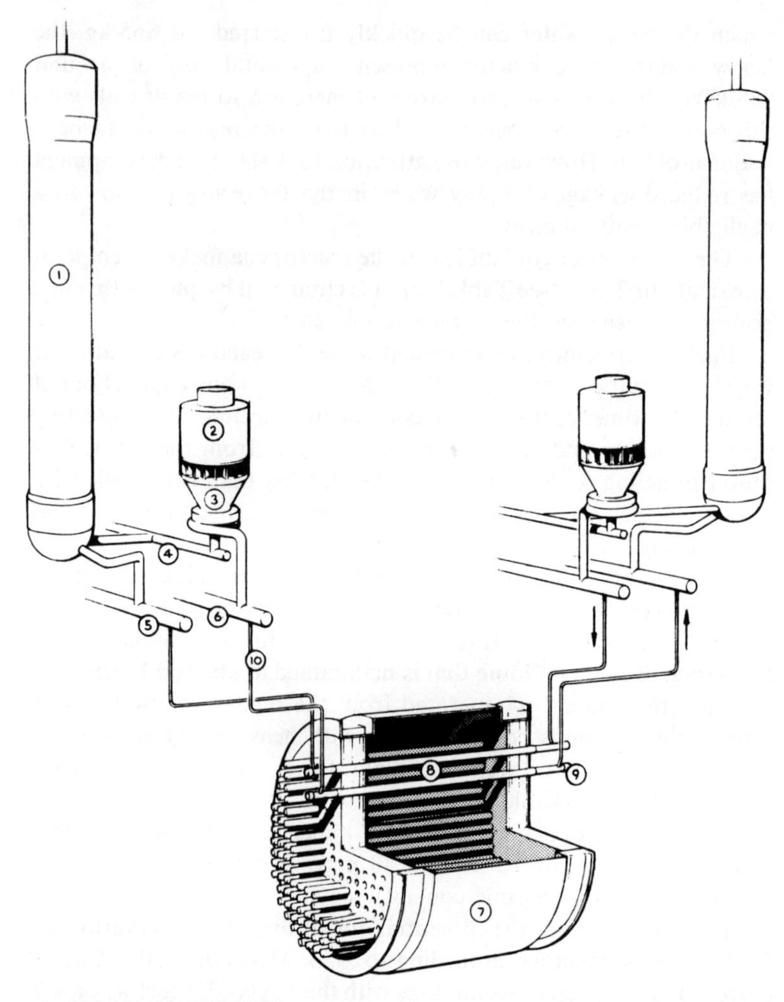

1 STEAM GENERATORS
2 PUMP MOTOR
3 PUMP
4 PUMP SUCTION HEADER
5 REACTOR OUTLET HEADER
6 REACTOR INLET HEADER
7 CALANDRIA
8 COOLANT TUBES
9 END FITTINGS
10 FEEDERS

Figure 8.5a
CANDU reactor and primary circuit. Courtesy of Atomic Energy of Canada.

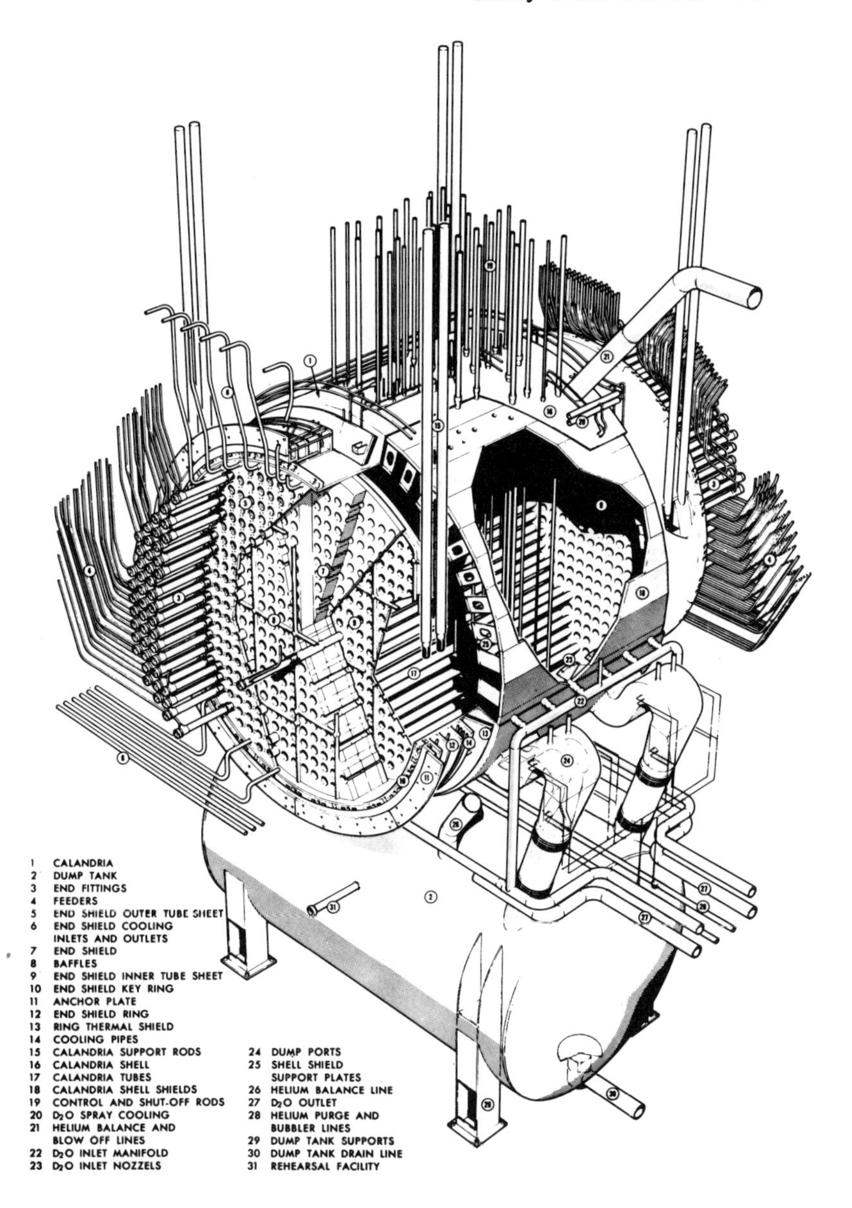

Figure 8.5b
CANDU reactor assembly. Courtesy of Atomic Energy of Canada.

Table 8.3
CANDU–PWR Potential for Uranium Conservation

Case and Technology	Fuel Burnup (MWe/tonne heavy metal)	Running Consumption of Natural Uranium (kg/MWe-yr)
1		
Current Pickering; natural UO_2 fuel:		
a. Absorber rods	7,500	168
b. Booster rods	8,200	154
2		
Natural U_3SiAl fuel; absorber rods	9,000	140
3		
Pickering design for U conservation; natural U_3SiAl fuel; booster rods; boiling coolant	10,500	120
4		
Natural UO_2 fuel plus Pu-recycling; absorber rods	15,000	84
5		
Th–U-233 fuel cycle with Pu from a natural CANDU–PWR	30,000	35
6		
Self-sufficient Th–U-233 cycle	12,000	0

Source: Ref. (10).

cycling is a plausible means of reducing uranium consumption, a combination of the plutonium cycle with the thorium one (as shown in Case 5) could be more effective. Fuel recycling would involve a system of 4.5 reactors: 3.5 of these would be fueled with thorium and plutonium, while one reactor would produce the plutonium required by the other reactors of the system. The technology of the reactor would be similiar to those in commercial operation today; the main problem would be in plutonium processing. The burnup figure quoted in Table 8.3 for Case 5 results from an optimization based on a particular assumption for fuel cost. If uranium became extremely expensive, Case 6 could imply virtual independence of uranium; nuclear power would be produced with thorium fuel, given an initial inventory of plutonium enrichment. Mooradian expects that the capital costs of the reactor plants for Cases 2 through 6 would be no higher than those for the present commercial CANDU system. The major capital uncertainties would be in the fuel processing plants (Cases 4, 5, 6).

The BWR version of the CANDU reactor produces a mixture of water and steam at the reactor exit. The steam is separated and passed directly to the turbine. This version of the CANDU reactor lowers the capital costs of the plant but has poorer fuel utilization than a reactor with heavy-water coolant.

8.4.2 Safety Systems

Safety systems for the standardized 600-MWe CANDU reactor are designed for two basic kinds of malfunction (12). In the first it is assumed that there is a failure in the operational system but that all safety systems function. In the second a failure of the operational system is accompanied by a failure in the safety system. The reactor complex is designed so that each kind of failure will occur with a specified maximum frequency and a specified maximum dose to individuals and populations. These frequencies and doses are given in Table 8.4. A much higher dose is considered tolerable for the second class of failure since its probability is so much lower than the first. The central idea in the design process is to be sure that these probabilities are realistic. According to G. L. Brooks,

> Each safety system must be independent (to the maximum practical extent) of the process systems [i.e., systems that are required to operate the reactor] and of the other safety systems. In addition, each safety system must have a demonstrated unavailability (fraction of time unavailable to function) less than 10^{-3}. This means that the system must be functional . . . 99.9 percent of the time. During design of the system the unavailability is calculated using failure rate estimates for the components. . . . The designer will sometimes use redundancy (valves in parallel for instance) to achieve the required reliability. The actual reliability is demonstrated by testing during the first few years of operation. (12)

Safety systems in the CANDU reactor are provided to shut down the reactor, to remove decay heat, to prevent subsequent failure after an initial failure in the reactor operating system, and, ultimately, to prevent radioactive emissions from exceeding specified limits.

Reactor shutdown is accomplished by dropping control rods into the reactor and, if needed, by injecting concentrated gadolinium nitrate into the moderator.

Table 8.4
Dose Limits for Accident Conditions

Situation	Assumed Maximum Frequency	Maximum Individual Dose Limits	Maximum Integrated Dose Limits
Serious Process Failure (single failure)	1 per 3 years	0.5 rem whole body, 3 rems thyroid	10^4 man-rems, 10^4 thyroid-rems
Serious Process Failure When a Safety System Is Unavailable (dual failure)	1 per 3000 years	25 rems whole body, 250 rems thyroid	10^6 man-rems, 10^6 thyroid-rems

Source: Ref. (12).

Decay heat is removed from the fuel by core coolant, which generates steam for discharge from the boilers. If the coolant supply to a coolant loop fails, the emergency core cooling (ECC) system provides a supply of light-water coolant to the loop from a storage tank in the roof of the containment vessel. The coolant that flows through the core and leaks out of the coolant system to the basement of the reactor buildings is returned by pumps to the reactor core, after passing through heat exchangers. A cooling system is provided to transfer heat from the moderator. An emergency water supply can provide water to the boilers for removing decay heat to any unfailed coolant loop.

The containment building is designed to withstand overpressure due to energy release during a serious failure. To reduce the rate of pressure buildup, a water spray operates automatically to cool the interior of the containment building. Air is passed through a filter system before being discharged.

8.5 GAS-COOLED REACTORS

An inherent advantage of gas cooling is that the maximum temperature and pressure of the working fluid can be selected independently; raising the temperature does not necessarily imply raising the coolant pressure, as with liquid-cooled reactors. The gas-cooled reactor will nevertheless operate at quite high coolant pressure, to keep reactor size and pumping power within acceptable limits.

The material most favored for gas-cooled reactor pressure vessels is prestressed concrete. Figure 8.6 shows a cross section through the HTGR. It will be noted that the structure is essentially a block

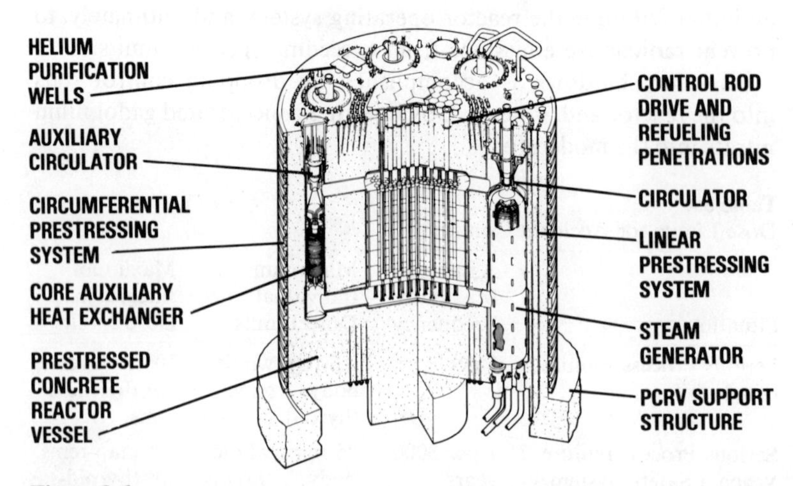

Figure 8.6a
High-temperature gas-cooled reactor (HTGR). Courtesy of General Atomic Company.

of concrete with cavities in which the reactor core and the heat transfer and turbomachinery components are contained.

Gas cooling appears to be fundamentally safer than liquid cooling, in that local overheating will not reduce the heat transfer rate drastically. This is in contrast to burnout in a liquid-cooled reactor, when the liquid film on the fuel element surface vaporizes and the cooling rate decreases so much that there is danger of fuel melting.

The choice of coolant has come down to a choice between CO_2 and helium. Many gaseous coolants are ruled out because they become radioactive and would be unsuitable for use in a gas-cooled reactor coupled, for example, to a closed-cycle gas turbine. Helium does have one isotope that becomes radioactive, but it is very rare; the only serious radioactivity in helium would be caused by impurities.

Carbon 13, which has about 1 percent concentration in natural carbon, becomes C-14 under irradiation, with a half-life of about 5700 years. Because of this long life, the use of CO_2 in nuclear reactors means a continuous buildup of radioactivity. Another disadvantage of CO_2, as British experience has shown, is dissociation at higher temperatures, which can induce corrosion. Corrosion can be avoided by using stainless steel, but corrosion of graphite in a graphite-moderated gas-cooled reactor would be very serious at high temperature. Corrosion can also be a serious problem in helium-cooled reactors if the H_2O impurity level is sufficiently high.

The Magnox reactor (see Table 8.2) has natural uranium fuel, clad in magnesium. It is cooled by CO_2 and moderated with graphite. If natural uranium is used in a graphite-moderated reactor, it is essential that the element uranium not be diluted in the fuel by the addition of oxygen or carbon atoms. The Magnox reactor was designed for a maximum fluid temperature of about 400°C (850°F). Although this is 100°C above the top temperature of water-cooled

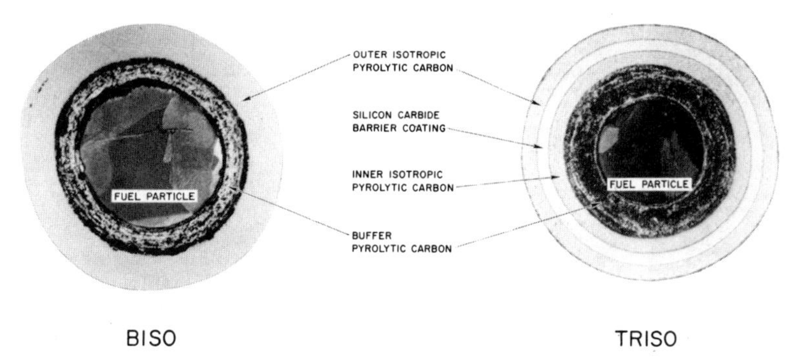

Figure 8.6b
Coated fuel particles for the HTGR. Courtesy of General Atomic Company.

reactors, the net station efficiency is scarcely any higher, largely due to the power consumed in the CO_2-circulating compressors.

The latest version of the CO_2-moderated reactor, the AGR, has enriched fuel in the form of UO_2. The fuel elements are clad in stainless steel to withstand high temperature. Use of an enriched, high-temperature fuel element means that the reactor power density and hence capital cost may be improved considerably. The AGR has been designed for fluid temperatures up to 566°C (1050°F) and thermal efficiency approaching 40 percent.

The HTGR, which is now operating in prototype form, could extend the performance range of gas-cooled reactors to temperatures of 1500°F (815°C) and possibly higher. With graphite used as both moderator and fuel container, the maximum temperature can be greatly elevated. Graphite increases in strength up to 4500°F (2500°C). With purified helium coolant, corrosion should be negli-

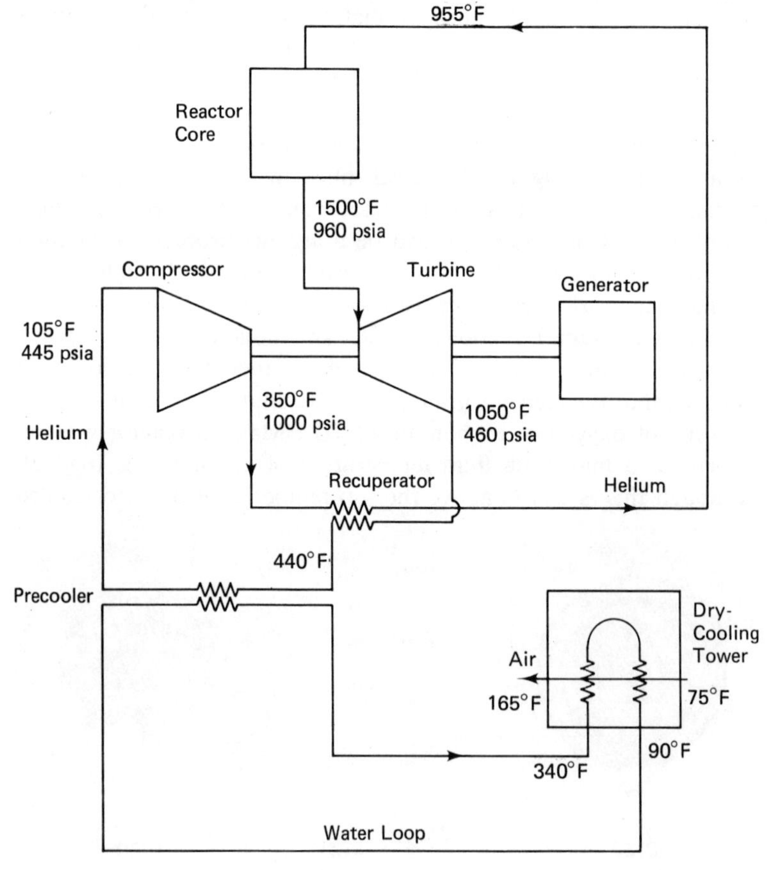

Figure 8.7
Closed-cycle gas turbine with HTGR (13). Overall thermal efficiency is 36.5 percent.

gible. A typical helium gas pressure level for the HTGR is 1000 psia. At such a high pressure level, heat transfer coefficients of up to 300 Btu/hr-ft^2-°F are feasible. With the pressure drop in the reactor core held to under 10 psia, pumping power is a small fraction of turbine power.

Given the potential for higher maximum fluid temperature, the HTGR can be adapted to a helium closed-cycle gas turbine. The gas turbine can make use of high maximum temperatures. It also exhausts heat at fairly high temperature, at least 200°C (350°F), so that cooling tower size can be much lower than for a steam plant of the same power. Even a dry-cooling tower might be reasonably low in cost.

The use of a helium gas turbine, as in Figure 8.7, would not necessarily provide a significant efficiency advantage. With 1500°F top temperature (a limit that could be raised with further reactor development), turbine and compressor efficiencies of 90 percent, and 90 percent recuperator effectiveness, the calculated cycle efficiency is still less than 40 percent. It is too early to draw firm conclusions on plant cost relative to a steam cycle, but the gas turbine plant would have an advantage in heat rejection for sites where cooling water is not available.

A considerable efficiency advantage could be obtained with a binary cycle, using a helium gas turbine with heat rejection to a steam turbine, as indicated in Figure 8.8. With this system overall efficiencies approaching or exceeding 50 percent are quite possible.

The HTGR shown in Figure 8.6 uses carbon- and carbide-coated fuel particles immersed in hexagonal graphite blocks in the reactor core. Helium coolant passes through holes in the graphite block. The coated fuel particles are small and spherical. The fuel particles, uranium carbide UC_2, are in the range of $200 \pm 50\ \mu$; thorium carbide ThC_2 fertile particles have diameters of $400 \pm 100\ \mu$. The purpose of the coatings is to contain the gases released due to fission within the fuel; this requires that the coating be able to accommodate the fuel expansion for long periods at high temperatures. One coating that is used in a number of the HTGR fuel cores consists of two layers: an inner, low-density buffer layer that allows the fuel matrix to expand and reduces the pressure buildup by absorbing some of the fission gases, and an outer, higher-strength layer whose function is to seal the particle so that radioactive products do not mix with the helium coolant.

8.6 FAST BREEDER REACTORS

As indicated in Chapter 1, low-cost reserves of uranium in the noncommunist world are potentially only about 1 million tons of U_3O_8. With continued rapid expansion of nonbreeder nuclear plants,

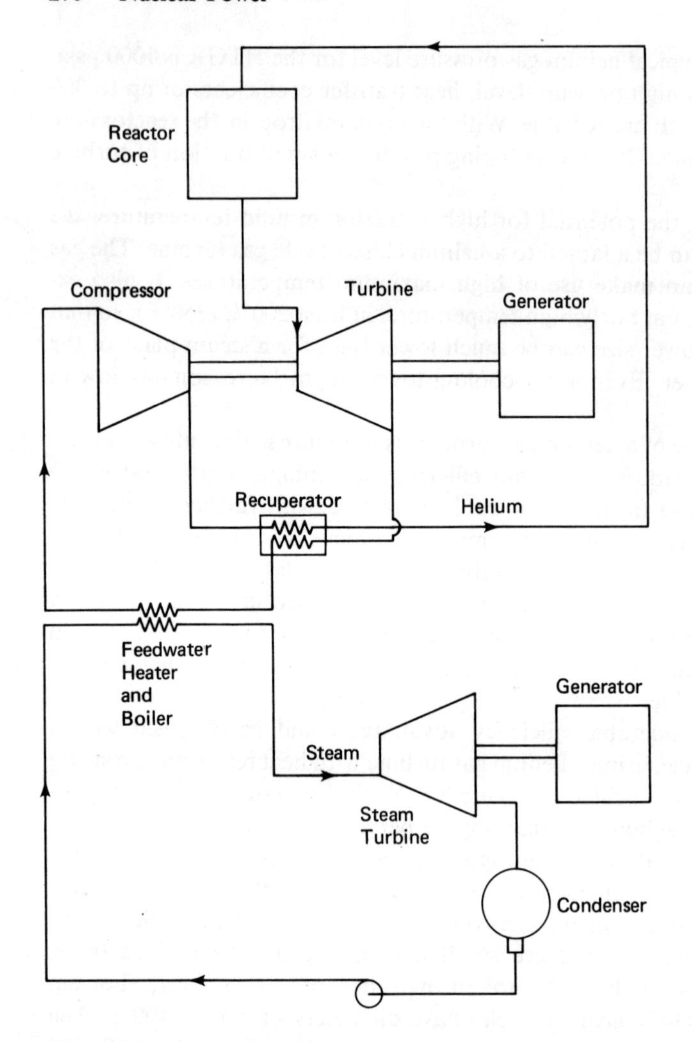

Figure 8.8
Helium–steam binary cycle with HTGR.

these reserves could conceivably be exhausted early in the next century, depending on the rate of discovery of new resources. Furthermore, this estimate refers to relatively low-price ores; if U_3O_8 prices rise from $20/kg to $200/kg, then perhaps 20 million tons will become available worldwide. However, these potential reserves may not be discovered. The breeder reactor, according to G. T. Seaborg and J. L. Bloom (14), could reduce by 1.2 million tons the amount of uranium ore that would be required in the next 50 years.

The term "breeding" indicates the conversion of atoms that are not fissionable to fissile ones. As pointed out earlier, U-235 atoms are the fissile material in thermal reactors. Their probability of fission following capture of low-energy neutrons is high. In contrast, U-238 atoms are called "fertile." Neutron capture does not cause fission of U-238, but initiates the transformation of this atom to Pu-239, which is fissile. Thus in the thermal uranium reactor three processes are simultaneously important:

1. fission of U-235:

U-235 + neutron → fission + energy + 2.42 neutrons

2. conversion of U-238 to Pu-239 fuel:

U-238 + neutron → Pu-239 + 2e

3. fission of plutonium:

Pu-239 + neutron → fission + energy + 2.87 neutrons.

The neutron production figures quoted above are the average numbers of neutrons produced in fissions due to incoming neutrons of thermal energies (about 0.025 eV). At higher energies the numbers released can be quite different. Not all of these released neutrons are available to cause new fissions. Some will be captured without causing fission; this is called "radiative capture." An important quantity, therefore, is the number η of neutrons potentially available to initiate further fission, which is equal to the number of neutrons released per fission minus the number that will be absorbed in radiative capture. For fissions caused by thermal neutrons, η is approximately 2.07 for U-235 and 2.11 for Pu-239. For a mixture of fuels, η is found by taking an appropriate average (15). The dependence of η on neutron energy is shown in Figure 8.9 for U-233 (produced from thorium), U-235, and Pu-239.

The conversion ratio (also called the breeding ratio) is defined as the number of atoms of fissile material produced per fissile atom destroyed. (If only U-238 and Pu-239 were present, Pu-239 would be the fissile material produced and also the fissile material consumed.) If the conversion ratio is greater than one, more fissile material is being produced than consumed; this is referred to as "breeding."

For the rate of production of fissile material to be just equal to its rate of consumption, the minimum value of η is 2. One of the

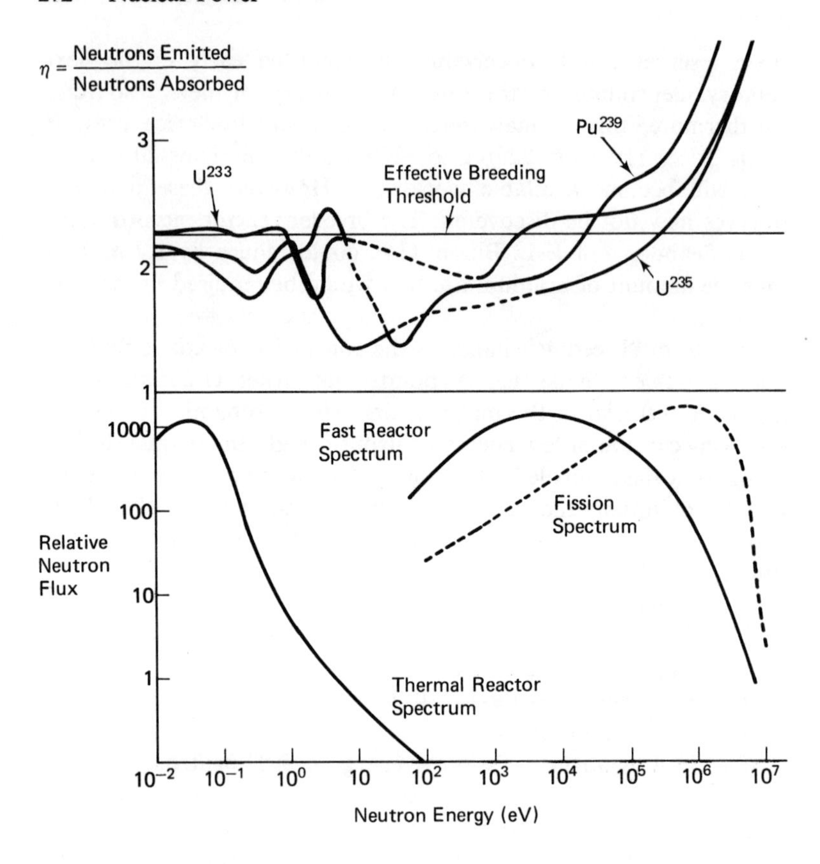

Figure 8.9
Breeding possibilities. After Mooradian (16).

neutrons produced by fission would be needed to convert a fertile atom to a fissile atom; the second would be required to cause a subsequent fission. In practice, η must be greater than about 2.2 for successful breeding, since some neutrons will escape from the reactor core or be absorbed in nonfertile substances in the reactor.

Possibilities for breeding are suggested by Figure 8.9. At thermal energies, breeding with Pu-239 is not realistic since η is appreciably less than 2.2. At high energies, characteristic of the fission neutron energy distribution shown in Figure 8.9, η is well above 2.2, so breeding of Pu-239 is quite feasible in fast reactors. The LMFBR is one practical example of a fast-neutron reactor in which breeding of Pu-239 is feasible.

Another possibility suggested by Figure 8.9 is breeding fissile U-233 atoms from fertile thorium in a thermal reactor, for which the following processes are important:

1. conversion of thorium to U-233 fuel:

Th-232 + neutron → U-233 + 2e

2. fission of U-233:

U-233 + neutron → fission + energy + 2.49 neutrons.

Even if true breeding were not quite possible in a thermal thorium reactor (i.e., if the breeding ratio were not quite equal to 1), conversion of Th-232 to U-233 could still supply large quantities of valuable nuclear fuel. The reactor fuel supply would of course need to be augmented by other fissile material such as U-235 or Pu-239, but the thorium "near-breeder" concept could still be an important means of extending nuclear fuel resources, should uranium reserves show signs of running low.

Table 8.5 shows relative fuel utilizations corresponding to various reactor types. For the thermal reactors, fuel utilization depends on whether enrichment is required and on the percentage of Pu-239 or U-233 left in the fuel when it is removed from the reactor, as well as on the thermal efficiency of the cycle. The negative numbers in the plutonium column indicate production of Pu-239 from U-238 simultaneously with the U-235 fission process. If the plutonium thus created can be exploited by fuel recycling, fuel utilization is augmented by nearly a factor of two for the CANDU reactor. Shown also in Table 8.5 are the relative fuel utilizations from the fast breeder reactor and three types of thermal reactor designed for Th–U-233 breeding. By far the lowest fuel consumption is associated with the fast breeder reactor. However, the Th–U-233 thermal breeder effectively amplifies uranium energy resources by an order of magnitude compared with present-day thermal reactors.

Although there are other types of fast breeder reactor, with coolants such as molten salt or inert gas, the sodium-cooled fast breeder is under most active current development in the United States and the United Kingdom.

The significant hazards of the sodium-cooled reactor include the

Table 8.5
Resource Utilization by Thermal and Fast Reactors

Date of Technology Availability	Fuel Cycle	Reactor Type	Running Net Consumptions (kg/MWe-yr)		
			U	Pu (or U-233)	Th
Present		Magnox	270	−0.54	
Present		AGR	147	−0.17	
Present	Pu export	LWR	171	−0.29	
Late 1970s		CANDU	110	−0.41	
Late 1970s		LWR	114		
Late 1970s	Pu recycle	CANDU	60		
1980s	fast breeder	FB	1	−0.23	
1970s	Th–U-233	LWR	65		2
Present		HWR	8		2
Late 1970s	Th–U-233	HTGR	115	−0.20	5.5

Source: Ref. (16).

reactivity of sodium with air and water and the large inventory of plutonium, which has a long half-life and could cause great harm if released in the atmosphere. On the other hand, the vapor pressure of sodium is low even at temperatures approaching 650°C (1200°F), so the problem of pressure vessel design is straightforward. The pressures required will be those associated with pumping the sodium through the compact fuel matrix. Sodium does not react with any of the materials in the primary circuit.

The plutonium hazard arises because the LMFBR requires a fuel recovery process for treating this highly toxic substance. However, the plutonium hazard is not peculiar to fast breeder reactors. Present-generation water-cooled reactors are constantly producing plutonium, which is stored with spent fuel wastes. The time may come when it is economical to recover this plutonium; the plutonium hazard will then also be associated with thermal reactors.

The power density of the LMFBR is about 600 MW/m^3, and the fast-neutron flux is about 100 times higher than in thermal reactors. This subjects fuel and structural materials to creep and swelling. The core typically consists of 30 percent fuel (PuO_2 and UO_2) by volume, 50 percent coolant, and 20 percent canning and structural material.

Figure 8.10 shows a cross section of the Clinch River fast breeder reactor now under construction; this is the first demonstration plant in the United States of the LMFBR. It is designed for 975 MW thermal and a net output of 350 MWe. The reactor vessel contains the fuel coolant, control rods, and structural assemblies.

The reactor is to consist of arrays of stainless steel fuel rods spaced in triangular pitch of 4.76 in. (120.9 mm). In the center zone of the reactor the fuel rods will contain sintered powder mixed PuO_2–UO_2 fuel pellets. Surrounding this inner zone is another containing the same fuel, but with higher enrichment. Surrounding these two zones, which comprise the core of the reactor, is the "radial blanket." This consists of an annular array of fuel rods containing depleted UO_2 pellets. Other depleted UO_2 fuel rods are located above and below the core. The purpose of the radial blanket is to shield the outer structures by absorbing and reflecting neutrons, as well as breeding fissile plutonium and generating thermal energy.

Sodium enters the reactor vessel near the bottom at 730°F (390°C) and leaves the outlet nozzle at 995°F (535°C). Flow through the reactor is distributed about 80 percent through the core, 12 percent through the radial blanket, and 2.5 percent through the control assembly and the balance in bypass.

Control rods are made of boron carbide cooled by sodium. During emergency shutdown these fall by gravity into the core. Afterward, reactor decay heat is transmitted via the sodium loops to the steam system; this is done by natural circulation in the event of loss of all electrical power.

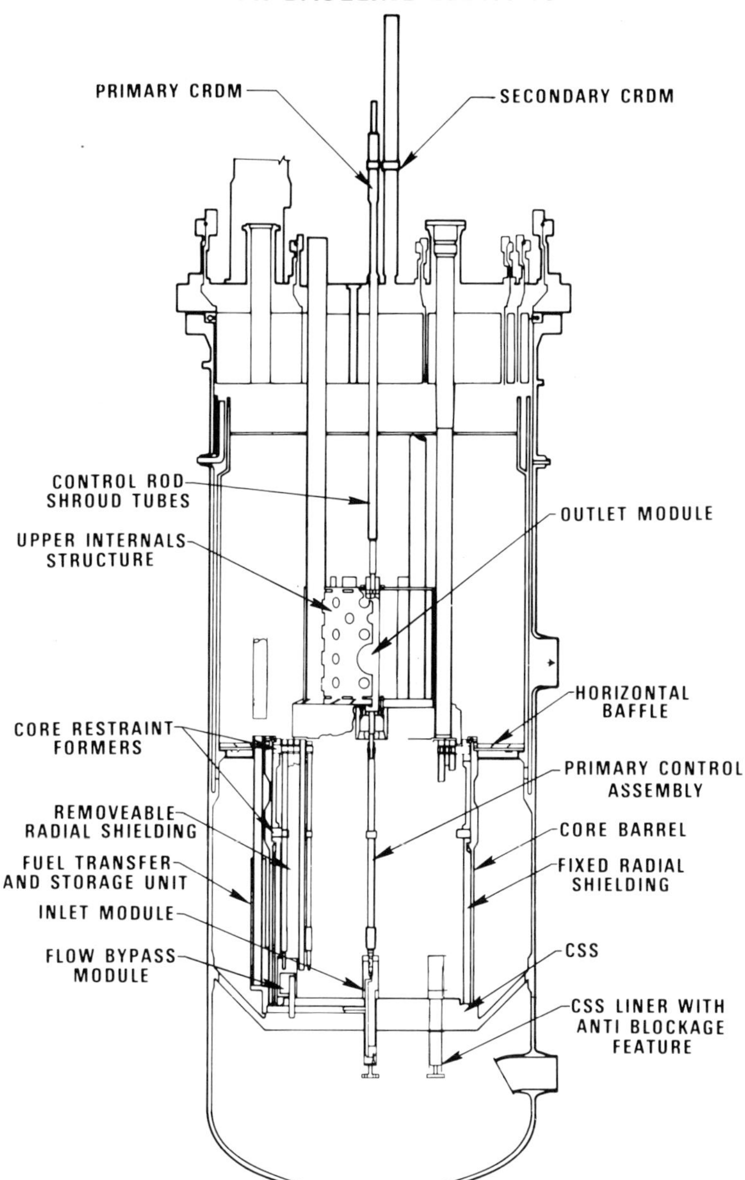

Figure 8.10
Liquid-metal-cooled fast breeder reactor (LMFBR) at Clinch River demonstration plant. Courtesy of Westinghouse Electric Corp.

The reactor vessel is made of stainless steel and contains an argon gas above the sodium level within the vessel. The sodium pumps in both the primary and intermediate loops are centrifugal pumps.

Cost estimates for the breeder reactor are necessarily very approximate at the present time. It is hoped that capital costs will not exceed those of present-generation thermal reactors by more than 25 percent.

The LMFBR cannot be put into utilization quickly. At least a decade will be required to provide assurances of safety of operation before breeding cycles can begin in earnest, doubling their output of fuel every 8 or 10 years. It appears that there will have to be a continuing program of thermal reactors for at least 20 years.

Figure 8.11 shows a fast breeder reactor coupled with a conventional steam cycle. Since the sodium coolant becomes highly radioactive in passing through the reactor, a second or intermediate sodium loop is used to transfer heat to the steam cycle. With the temperatures shown in Figure 8.11, overall efficiencies can approach 40 percent. The helium-cooled fast breeder reactor also has the advantage of potentially high thermal efficiency.

8.7 SUMMARY

Rising fossil fuel prices and the increasing costs of pollution control appear to make nuclear power cheaper than power from coal-fired plants. Capital costs of both coal and nuclear plants have escalated rapidly. Nuclear plants in the early years have suffered

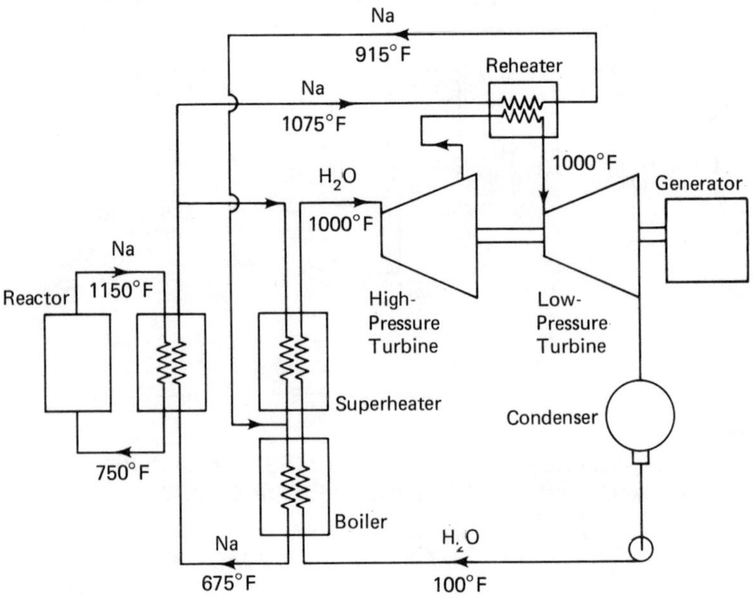

Figure 8.11
LMFBR with steam cycle.

from relatively long construction periods (seven to nine years) and relatively low availability during their first years of operation; these factors should improve considerably as time goes on. Light-water reactors are now being built in large numbers, so that, with increasing experience, equipment manufacturing costs may well decrease in future years. The rapid increase in the cost of construction is nevertheless a major concern.

The possibility of nuclear fuel depletion must be seriously studied. Present-generation thermal reactors, which extract about 1 percent of the fission energy available from uranium fuels, could exhaust presently known resources of high-grade ores within a few decades. Use of the much more abundant low-grade ores would greatly augment these resources and would not be prohibitively expensive. Also, it is possible that further exploration will lead to substantial revisions of uranium resource estimates. Quite apart from these possibilities, however, the development of fast or thermal breeders (or near-breeders) could raise overall fuel utilization by one or two orders of magnitude, and make it feasible to use relatively abundant thorium resources. Postponement of nuclear fuel depletion for a matter of centuries seems technically and economically feasible for any credible growth of world population and nuclear power capacity.

The major unresolved factor that might limit the growth of nuclear power is the safety of the giant-sized plants that are required for reasons of economy. So far, it appears that governments are generally willing to accept the risks of nuclear power in return for lowest-cost electricity and avoidance of increased air pollution due to fossil fuel combustion. Assessment of reactor failure probability is still a developing art, but its present results are reassuring. They indicate that with high standards of design and operation, water-cooled reactor plants pose negligible risk to populations. A continuing worry, however, is the hazard due to plutonium fuel processing, which could become widespread with the introduction of the breeder reactor or with the recycling of plutonium fuel for thermal reactors.

References

1. Robert Gilette. "Nuclear safety." *Science*, vol. 177, September 1, 1972, p. 771; September 8, 1972, p. 867; September 15, 1972, p. 970; September 22, 1972, p. 1080.

2. A. J. Surrey. "The future growth of nuclear power. Part 1. Demand and supply. Part 2. Choices and obstacles." *Energy Policy*, vol. 1, nos. 2, 3 (September and December 1973), pp. 208–224.

3. A. J. Surrey. "Power reactors '74." *Nuclear Engineering International*, April 1974.

4. Central Electricity Generating Board (U.K.). *Modern Power Station*

Practice. Vol. 8: *Nuclear Power Generation*. Oxford: Pergamon Press, 1971.

5. *Steam: Its Generation and Use*, 38th edition. New York: Babcock and Wilcox Co., 1972.

6. W. K. Davis, R. O. Sandberg, P. H. Reinker, A. H. Lazar, H. M. Winterton, J. M. West, A. E. Schubert, J. J. McNelly, J. C. Rengel, and J. T. Steifel. "United States light-water reactors: present status and future prospects." Paper A/Conf 49/P/034, *Proceedings of the Fourth United Nations International Conference on the Peaceful Uses of Atomic Energy*, Geneva, September 6–16, 1971.

7. Norman C. Rasmussen, Project Director. *Reactor Safety Study: An Assessment of Accident Risks in U.S. Commercial Power Plants*. U.S. Atomic Energy Commission Report WASH-1400 (draft version), August 1974; WASH-1400, NUREG-75/104 (final version), October 1975.

8. U.S. Atomic Energy Commission, Advisory Committee on Reactor Safeguards. *The Integrity of Reactor Vessels: Light Water Reactors*. Report WASH-1285, January 1974.

9. National Academy of Sciences, National Research Council, Division of Medical Sciences. *The Effects on Population of Exposure to Low Levels of Ionizing Radiation*. November 1972.

10. A. J. Mooradian and O. J. C. Runnalls. "CANDU: Economic alternative to fast breeders." Paper 4.1–1, *Transactions of the 9th World Energy Conference*, Detroit, Michigan, September 22–27, 1974.

11. W. G. Morrison, C. E. Bemon, E. K. Keane, and R. D. Wardell. "Pickering Generating Station." Paper 4.1–2, *Transactions of the 9th World Energy Conference*, Detroit, Michigan, September 22–27, 1974.

12. G. L. Brooks. *The Standardized CANDU 600 MWe Nuclear Reactor*. Atomic Energy of Canada Ltd., September 1974.

13. A. J. Goodjohn and R. D. Kenyon. "The high temperature gas cooled reactor: An advanced nuclear power station for the 1980's." Paper 73-Pwr-8 presented at the ASME–IEEE Joint Power Generation Conference, New Orleans, Louisiana, September 16–19, 1973.

14. Glenn T. Seaborg and Justin L. Bloom. "Fast breeder reactors." *Scientific American*, November 1970.

15. John R. Lamarsh. *Introduction to Nuclear Engineering*. Reading, Mass.: Addison-Wesley, 1975.

16. A. J. Mooradian. "Fission, fusion and fuel economy." *Proceedings of the Symposium on Energy Resources*, Royal Society of Canada, Ottawa, Ontario, October 15–17, 1973.

17. M. M. El-Wakhil. *Nuclear Engineering Conversion*. Scranton, Penn.: Intext Educational Publishers, 1971.

Supplementary References

R. G. Adams, F. R. Bell, C. F. McDonald, and D. C. Morse. "HTGR gas turbine power plant configuration studies." Paper 73-WA/Pwr-7 presented at the Annual Winter Meeting of the American Society of Mechanical Engineers, Detroit, Michigan, November 11–15, 1973.

R. Balent and R. J. Beeley. "The fast breeder reactor—Energy source for the future." *Energy Sources*, vol. 1, no. 2 (1974), pp. 189–200.

K. Bammert, J. Rurik, and H. Griepentrog. "Highlights and future developments of closed cycle gas turbines." Paper 74-GT-7 presented at the American Society of Mechanical Engineers Gas Turbine Conference, Zurich, March 31–April 4, 1974.

D. J. Bennett. *The Elements of Nuclear Power*. London. Longmans, 1972.

W. B. Lewis. "Large scale nuclear energy from the thorium cycle." Paper A/Conf 49/P/157, *Proceedings of the Fourth United Nations International Conference on the Peaceful Uses of Atomic Energy*, Geneva, September 6–16, 1971.

W. B. Lewis. "The nuclear energy industrial complex and synthetic fuels." *Proceedings of the Symposium on Energy Resources*, Royal Society of Canada, Ottawa, Ontario, October 15–17, 1973.

David J. Rose. "Nuclear electric power." *Science*, vol. 184 (April 1974), pp. 351–359.

George Schlueter. "Gas cooled reactors—A way of fulfilling the national objectives of nuclear reactor development." Paper 74-WA/NE-13 presented at the Annual Winter Meeting of the American Society of Mechanical Engineers, New York, New York, November 17–22, 1974.

B. I. Spinrad. *The Role of Nuclear Power in Meeting World Energy Needs*. Report IAEA-SM-16412, International Atomic Energy Agency, 1971.

Problems

8.1 Approximate figures for burnup (thermal energy released per ton of UO_2 fuel) are provided in Table 8.A for various enrichments in light- and heavy-water reactors.

Table 8.A
Burnup in Light- and Heavy-Water Reactors

	Initial Enrichment (weight % of U-235)	Final Enrichment (weight % of U-235)	Burnup (MW thermal-days/ton)
LWR	2.8	2.2	10,000
	3.3	2.3	20,000
	3.8	2.4	30,000
	4.3	2.5	40,000
HWR	0.71	0.18	10,000

Source: Ref. (17).

Given that complete fission of 1 atom of U-235 produces 200 MeV of energy, show that 1 g of U-235 produces 5×10^{23} MeV (2.3×10^4 kWh). Compare this with the burnup figures quoted in Table 8.A, recognizing that not all of the fission energy in thermal reactors comes from U-235: some energy results from fission of Pu-239 formed by absorption of neutrons by U-238 (1 MeV = 1.602×10^{-13} J).

Using the thermal efficiencies of Table 8.2, calculate the consumption of natural uranium (kg/MWe-yr) and compare this with the numbers given in Table 8.5. As a first approximation, ignore the loss of uranium in the enrichment process, and assume that the consumption of natural uranium is proportional to the enrichment. For comparison, estimate the number of kilograms of coal consumed per MWe-yr in a large coal-fired plant whose efficiency is 39 percent.

8.2 Using the data of Table 8.2, calculate the ratio of heat rejected to electrical energy produced (kW(thermal)/kWe) for the reactor types cited. Compare these results with the heat rejection to cooling water of a fossil fuel plant whose thermal efficiency is 40 percent, assuming that 10 percent of the thermal input is rejected through the stack with the exhaust gases and in other losses to the atmosphere.

8.3 Using the approximate numbers cited in Table 8.2, estimate the average working stresses in the cylindrical walls of the pressure vessels of the PWR, BWR, CANDU, and LMFBR liquid-cooled reactor described in Table 8.2. Assume for each a simple cylinder of the given internal diameter and wall thickness.

Calculate the power densities (MWe/m^3) of H_2O-, He-, and Na-cooled reactors based on the size of the pressure vessel, and compare the results with calculations based on the reactor core volume.

8.4 Show that

$$\eta = \frac{\nu}{1 + \alpha},$$

where ν is the average number of neutrons produced in a single fission reaction, η is the average number of neutrons produced per neutron absorbed in fuel, and α is the ratio of the rate of nonfissile neutron capture to the rate of fission captures ($\eta < \nu$ due to nonfissile capture of some neutrons).

For breeding it is found that η must be greater than 2.2 (to allow for a sustained chain reaction, replacement of fissile atoms consumed, and leakage or absorption by nonfuel atoms). Given the approximate constants in Table 8.B for thermal reactors, determine whether breeding of U-233 (from Th) or Pu-239 (from U-238) is a practical possibility.

Table 8.B
Approximate Constants for Thermal Reactors

	ν	α
U-233	2.492	0.0899
Pu-239	2.871	0.362

Source: Ref. (15).

8.5 With the approximate fuel constants for fast reactors given in Table 8.C and the definitions in Problem 8.4, determine the maximum possible conversion ratio c (defined as the number of fissile atoms produced per fissile atom consumed) and the corresponding gain $G = c - 1$. Show that the maximum fraction of neutrons produced that are available for breeding, f_b, is

$$f_b = 1 - \frac{1}{\eta},$$

and show how f_b varies with fuel for fast and thermal reactors.

Table 8.C
Approximate Fuel Constants for Fast Reactors

	ν	α
U-233	2.59	0.068
U-235	2.50	0.15
Pu-239	3.00	0.15

Source: Ref. (12).

8.6 a. A breeder reactor has a gain rate G, thermal power level P_t (in MW thermal), and fissionable fuel consumption rate W (in g/MW thermal-day). Show that if the original quantity of fissionable fuel in the reactor is M_0, and the newly created fuel is retained within the reactor, the time required for fuel quantity to double will be

$$t_d = \frac{M_0}{GWP_t} \text{ (in days).}$$

This is called the linear doubling time.

b. Alternatively, suppose the newly created fuel is continuously removed from the reactor and fed to other breeder reactors of gain rate G and fuel consumption rate W. In this case the total power level of the system is proportional to the total mass of fuel in existence. Show that the time required to double the mass of fuel in this reactor system is

$$t_d = \frac{M_0 \ln 2}{GWP_{t0}},$$

where P_{t0} is the thermal power level when the total fuel quantity is M_0. This is called the exponential doubling time.

c. Suppose that a fast breeder reactor is fueled by a mixture of U-238 and Pu-239, with an initial inventory of Pu-239 of 1000 kg. Suppose also that the fuel consumption rate is 2 kg/day and the gain rate is 0.18. Calculate:

i. the linear and exponential doubling times,

ii. the rate at which Pu-239 accumulates if retained within the reactor.

8.7 A thermal near-breeder reactor is fueled mainly with thorium, which is converted to U-233. The conversion ratio c (see Problem 8.5) is 0.92. The reactor operates continuously with U-235 supplied to maintain a constant stock of fissile material. What is the required ratio of U-235 feed to U-233 production rate (atoms per atom) if the energy of fission of the two isotopes is the same?

Calculate the rate of consumption of natural uranium (0.71 mass percent U-235) in kg/MWe-yr. Neglect the U-235 loss in fuel processing and assume a thermal efficiency of 30 percent. Compare this with the consumption of natural uranium in a natural uranium reactor with a burnup of 10,000 MW thermal-days/tonne. Check results with Table 8.5.

8.8 Suppose (as a rough approximation) that the thermal efficiency of a BWR is proportional to the Carnot efficiency $1 - (T_{min}/T_{max})$, where T_{min} is the condenser temperature and T_{max} is the temperature of the saturated steam which the reactor produces for the turbine.

Suppose the maximum temperature is allowed to rise from 300°C to 325°C. By how much may the cycle efficiency and maximum pressure rise? For a 1100-MWe reactor power unit (Table 8.2), by how much would pressure vessel thickness need to rise to keep wall stresses constant?

The equilibrium vapor pressure for H_2O is 8.67 MPa at 300°C and 12.2 MPa at 325°C. Assume a condenser temperature of 35°C and a plant efficiency with 300°C coolant temperature of 30 percent.

8.9 Compare the plant efficiencies cited in Table 8.2 with the respective maximum (or Carnot) heat engine efficiencies $1 - (T_{min}/T_{max})$, using for T_{max} the reactor coolant temperature and for T_{min} a nominal value of the condenser temperature (35°C).

Calculate the volume flow rate of coolant (m³/sec per 1000-MWe capacity) required for the PWR, HTGR, and LMFBR reactors. Use the data in Table 8.D.

Table 8.D
Reactor Coolant Parameters

Coolant	Density (kg/m³)	Specific Heat (kJ/kg/°C)
H_2O	800	4.6
He	2.24	5.1
Na	880	5.5

8.10 Shown in Figure 8.A in simplified form is the Fort St. Vrain HTGR nuclear–electric plant design.

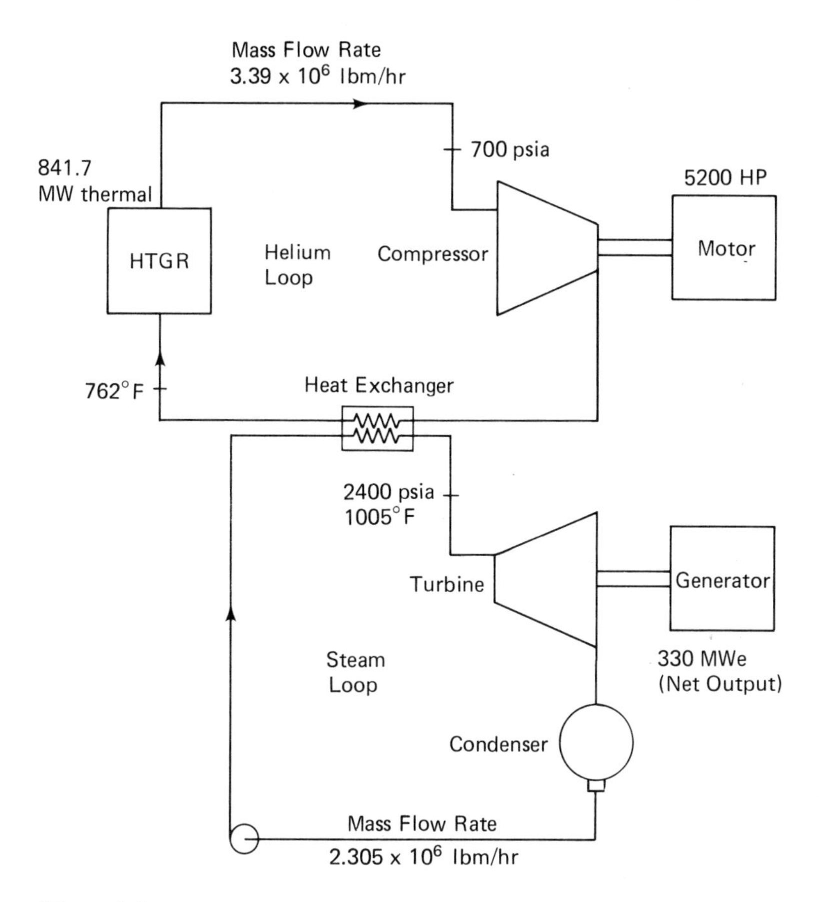

Figure 8.A
Schematic of the Fort St. Vrain HTGR nuclear–electric plant.

a. Calculate the overall plant efficiency.

b. Calculate the ratio of compressor power to electrical output power.

c. The fuel is a combination of thorium with 93 percent enriched uranium, and the burnup is 100,000 MW thermal-days/tonne. Estimate the annual fuel consumption.

d. The reactor core diameter and height are 5.9 m and 4.7 m, respectively. Compare the power density with the HTGR figure quoted in Table 8.2. The reactor pressure vessel is prestressed concrete 21 m in diameter and 32 m high.

9
Power from Fossil Fuels

9.1 INTRODUCTION

In 1973, over 80 percent of the total energy consumed for electricity generation in the United States was in the form of fossil fuels (see Table 9.1). In 1990, according to the U.S. Federal Power Commission, only 46 percent of the energy needed to make electricity will come from fossil fuels. Over the intervening years nuclear energy is expected to rise from 4 percent to 49 percent of the total; nuclear plant capacity may rise by a factor of 35.

The main reasons for rapid expansion of nuclear power include:

1. Shortage of oil and gas.

2. Shortage of low-sulfur coal, coupled with regulations prohibiting use of high-sulfur coal in many areas. Apparatus installed to remove sulfur from stack gases has proven costly and unreliable, and typically produces waste disposal problems.

3. Increases in coal prices. These are now so high that coal-fired power appears to be more expensive in many areas than nuclear power, despite the relatively high capital costs of nuclear plants.

4. A general expectation that the nuclear power plant system will be satisfactorily safe.

It may be seen from Table 9.1 that the annual use of coal is still expected to double over the period 1975–1990. Oil use could grow even faster, despite its cost and scarcity. This of course assumes that

Table 9.1
Contribution of Primary Fuels to Electric Power Generation in the United States

Fuel	1973	1980	1990
Petroleum (10^6 bbl)	557 (17%)	1000 (20%)	1300 (13%)
Natural Gas (10^{12} ft^3)	3.6 (18%)	2.9 (10%)	2.0 (3%)
Hydro (10^9 kWh)	271 (15%)	292 (9%)	319 (5%)
Coal (10^6 tons)	388 (46%)	550 (42%)	775 (30%)
Nuclear (10^9 kWh)	83 (4%)	600 (19%)	2913 (49%)
Total Electrical Energy (10^9 kWh)	1849 (100%)	3113 (100%)	5922 (100%)

Source: Ref. (1).

environmental regulations on coal and oil use can be satisfied. But it does not mean that oil and coal will be reasonably cheap sources of electrical energy. Rather, it reflects the conviction that nuclear power cannot grow very much more rapidly than is suggested in the table, and that total demand for electricity will grow more rapidly than nuclear capacity.

Thus, despite the attractions of nuclear power, fossil fuel plants will be with us for a long time, and there are many reasons for being concerned about technological improvement to provide clean power, especially from coal, which is known to be abundantly available.

There are many options for converting coal to clean fuel, as well as several possibilities for removal of pollutants during combustion or from the combustion products. All of these are costly, however, and need time for development to the point of guaranteed reliability.

Over the last century there has been intensive development of many prime movers whose energy is obtained by combustion of fossil fuels. Figure 9.1 shows a number of these and indicates the power range to which each is best suited. Electricity is generated almost entirely by those prime movers adapted to the largest power range (100 to 1000 MW/unit). The exceptions to this would include smaller units for peak load, emergency standby, or service in remote areas. Increasing size generally reduces unit capital cost and also labor and maintenance costs. It is for the larger units (with their low unit capital costs) that design for maximum efficiency has been most worthwhile. The operating efficiencies of steam turbine plants, for example, vary with size as shown in Table 9.2.

All costs—capital, fuel, labor, and maintenance—are lower with large-size units. Largeness also brings the advantage of relative ease

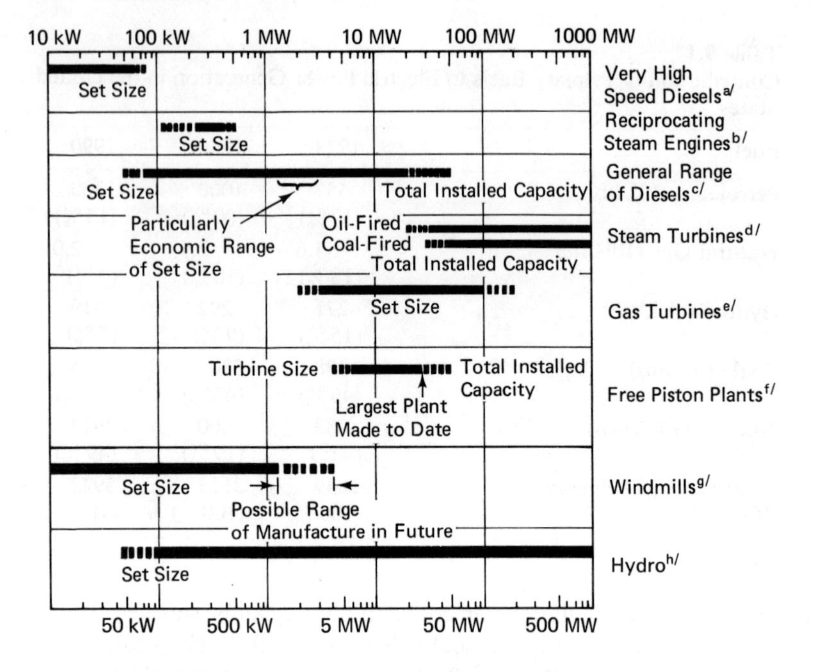

a/ Suitable for intermittent duty for a few hours daily in small rural centers where the cost of continuous day and night service cannot be justified.

b/ Suitable only where "free" waste fuel is available, where oil fuel is costly, and where load factor is fairly high.

c/ Range of suitability extends above about 20 MW if price difference between diesel fuel and boiler fuel is not great and if load factor is low. Above 40 or 50 MW installed capacity, steam turbines will usually be the cheaper.

d/ Suitable for large power plants generally. High load factors and large price differences between boiler fuel and diesel fuel will tend to lower the level of competition with diesels.

e/ Particularly suitable for low load factors where they can burn cheap fuel – e.g., for carrying short time peak loads. Can also compete with steam turbines for continuous operation, where load factor is low and fuel cheap, especially where water is scarce. Seldom competes with diesels below 2 MW sets.

f/ Below 5 MW set size, these seldom compete with the diesel. Particularly suitable for short term peak load duty. They can often theoretically compete with the steam turbine for continuous duty, especially where the load factor is low. They can compete with the straight gas turbine where fuel is costly and the plant factor high.

g/ Can occasionally be competitive with diesels if attendance costs can be kept low and if storage batteries (where installed) are well maintained; but less flexible in scope. Larger sizes more suited to parallel operation with other plant or to automatic load selection.

h/ Universal application over all plant sizes down to about 50 kW, provided it can compete with thermal alternative. This depends on capital cost per kilowatt and on plant factor. Each project will have its own solution. No generalizations possible.

Figure 9.1
Power plant applications. From Ref. (2).

Table 9.2
Typical Steam Plant Efficiencies

Power per Unit (MW)	Efficiency Range (percent)
0–0.1	7.5–11.5
0.1–1.0	10–19
1–10	17–24.5
10–50	22–31
50–100	27–34
100–1000	35–40

Source: Ref. (2).

of emissions control. Design of appropriate pollution control equipment, monitoring and maintenance by skilled personnel, and provision of tall stacks or power generation at sites remote from population, are all very much less costly per kilowatt of power for large units.

Figure 9.2 illustrates the way in which typical unit capital costs for prime movers have varied with size in the recent past. In the 100- to 1000-MW range, the major contenders have been steam and gas turbine plants. The gas turbine plants have a tremendous advantage in capital cost, but their overall efficiency is typically much lower.

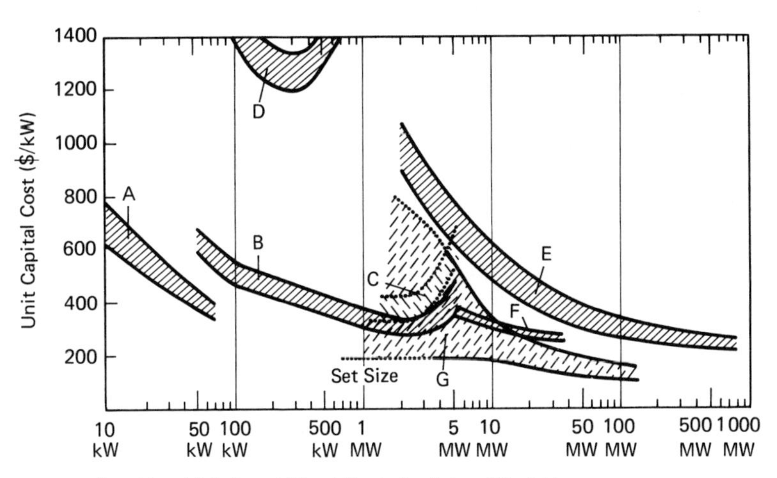

Figure 9.2
Approximate 1975 costs for thermal power plants. Adapted from Ref. (2).

For this reason the gas turbine is generally used only for peak load duty; operating for a small fraction of the year, the value of the extra fuel burned per kilowatt-hour due to low efficiency is less than the cost of the capital saved by buying a gas turbine rather than a steam turbine. Gas turbines also have the great advantage for peak load duty of being able to be started in minutes rather than hours, as is typical for large steam turbines. Steam turbine plants have been universally preferred for base load service with fossil fuels, owing to their high efficiencies and ability to burn low-grade fuels.

Major improvement in the utilization of fossil fuels can be made if electricity plants are used as sources of heat as well as of electricity. This is being done in many European countries; it can yield significant reductions in total fuel consumption as well as in waste heat rejection. If heating and electrical loads are balanced, overall fuel energy utilization can easily be doubled.

9.2 STEAM POWER PLANTS

9.2.1 Fuels and Combustion

The decision whether to build a fossil fuel or nuclear plant must follow a thorough analysis of

1. plant capital cost
2. fuel cost
3. operating and maintenance costs
4. reliability and safety
5. environmental effects.

If such an analysis shows that the fossil fuel plant is cheapest, the security of fuel over the life of the plant must be considered. The future availability and price of low-sulfur oil and coal are uncertain. Natural gas is virtually unobtainable for new electrical plants, and synthetic gas may not be freely available for a long time.

Gas is the most convenient fuel to burn since fuel storage, dust collectors, and ash-handling equipment are unnecessary. Corrosion and deposition on heating surfaces are generally negligible since natural gas can be cleaned of sulfur before entry to the pipeline delivery system. Given the growing shortage of natural gas, it would be very satisfactory if large supplies of clean synthetic gas could be produced cheaply from coal. Unfortunately, synthetic gas is expensive, and commercial processes suitable for large electrical power plants are still in the development stage.

The main causes of corrosion and deposition with fuel oil are sulfur and vanadium. These can cause deposits on boiler tubes and may lead to severe corrosion and interfere with heat transfer. Removal of these components from the liquid fuel has not proven commercially feasible, and oil-fired boilers must be washed during regular shutdowns for maintenance.

Figure 9.3
Coal pulverizers, which grind the coal to a fine powder. Coal particles are blown in to the furnace and instantly ignited. Reproduced courtesy of Ontario Hydro.

The use of coal brings many problems. Long-distance transport is costly, huge areas are needed for fuel storage, and windblown coal dust can be a serious problem in the neighborhood of the plant. Also, there is the need to collect and dispose of the ash, which may be 5 to 20 percent of the fuel weight. In some areas, progress has been made in utilizing coal ash in highway and building materials. Coal of almost any type can be burned successfully after being pulverized to very small particles (approximately 80 percent of the particles will pass a 200-mesh screen, the aperture width of which is less than 0.1 mm), which are blown into the furnace and ignite almost immediately (Figure 9.3). The pulverized coal method has been successful in many applications; it is efficient and flexible over a wide range of fuel firing rates. However, the costs of pulverizers and ash and dust collection equipment are considerable. Pulverized coal firing is not adaptable to removal of sulfur during combustion. For certain coals (with low ash-fusion temperatures), ash removal and dust collection as well as preparation are more economical with the cyclone furnace.

The great success of pulverized coal and cyclone combustion has probably inhibited development of a promising alternative—fluidized-bed combustion. In this scheme, a fairly closely packed volume of coal particles is suspended in space by the frictional forces of an air stream that passes vertically upward through the bed. Boiler tubes are immersed directly in the bed. The major advantages of fluidized-bed combustion are reduction in furnace size and control of pollutants. Sulfur removal by limestone injection in the bed appears feasible, and the formation of NO_x can be very much less than in pulverized coal combustors, since the temperature of the combustion zone can be greatly reduced by immersion of the boilers in the combustion zone. Fluidized-bed combustion is still in the development stage.

9.2.2 Plant Arrangements

Boilers for steam power plants are built in sizes up to 10^7 lb/hr of steam. A unit of this size would supply a single turbine–generator set producing about 1300 MWe.

In simplified form, Figure 9.4 shows the flow diagram of a modern fossil fuel generating plant, comprised of boiler and superheater combined, high-, intermediate-, and low-pressure steam turbines, condenser, feed pump, and feedwater heaters. Leaving the boilers at maximum pressure and temperature, superheated steam expands in the high-pressure turbine, then returns to the boiler for reheating to about the same temperature, though at much lower pressure. Following expansion in the intermediate-pressure turbine, the steam is again reheated before expansion in the low-pressure turbine. In many steam plants, depending on maximum pressure and temperature, only one reheat may be necessary. Reheating is generally necessary to prevent formation (in the expanding steam) of excessive moisture, which would damage turbine blades. It greatly increases the work per unit mass of steam passing around the cycle, and may effect a significant improvement in efficiency.

The condenser is an extremely important element in the steam cycle; by enabling the turbine to exhaust into a near-vacuum rather than against atmospheric pressure, it will allow the fuel consumption

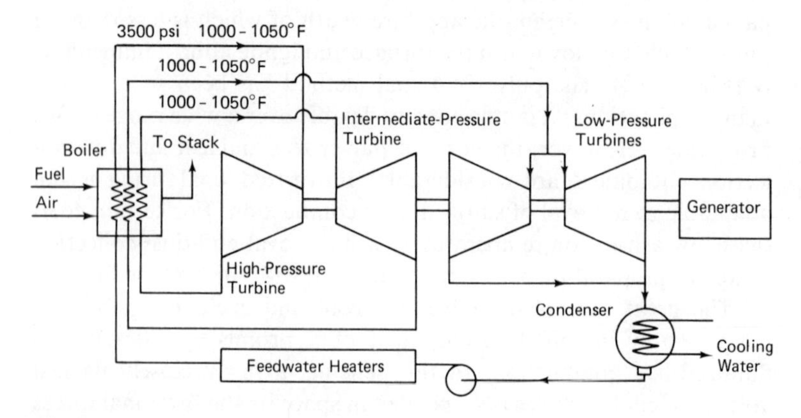

Figure 9.4
Conventional steam power plant.

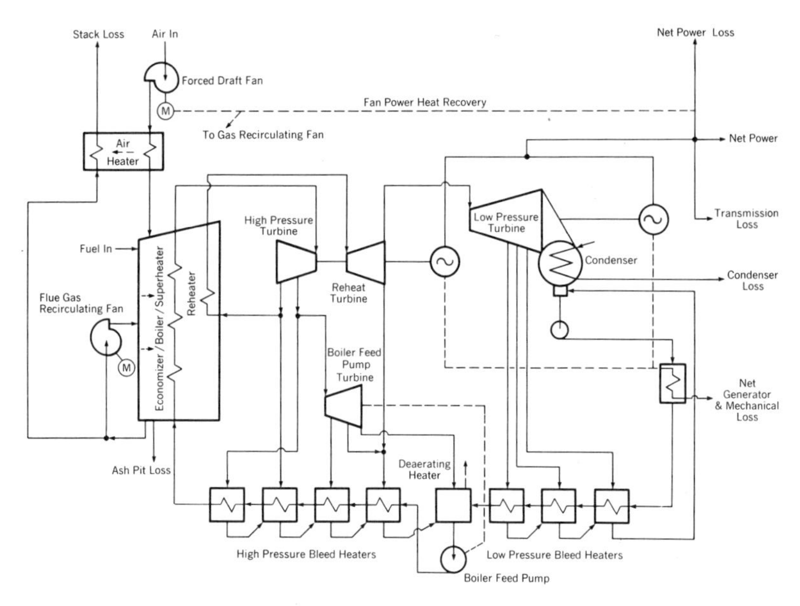

Figure 9.5
Power cycle diagram: fossil fuel, single reheat, 8-stage regenerative feed heating; 3515 psia, 1000°F/1000°F (24.1 MPa, 538°C/538°C) steam. From Ref. (3), courtesy of Babcock and Wilcox.

to be reduced by up to 40 percent. Cooling water requirements are of the order of 1000 ft^3/sec for a 1000-MWe plant. With cooling towers, this water is recirculated, 1.5 to 2.5 percent of the flow being lost to evaporation on each pass.

The feedwater heaters arc responsible for a substantial gain in efficiency. These are shown more clearly in Figure 9.5, which (though still very much simplified) indicates something of the complexity of the modern steam plant piping network. Here only one reheat is indicated, since steam passes directly from the reheat turbine to the low-pressure turbine. Much of Figure 9.5 is occupied with the feedwater heating units, to which steam (bled at various points from the turbines) is taken to heat the feedwater before it enters the boiler. Taking steam from the turbines decreases shaft power, but the compensating reduction in boiler heat input yields a significant gain in efficiency. With an infinite number of heaters, the nonsuperheat Rankine cycle could theoretically approach Carnot cycle efficiency. The eight feedwater heaters shown in Figure 9.5 will be responsible for an increment of 10 or 11 percentage points in the total efficiency. A ninth heater would contribute little extra improvement in efficiency.

9.2.3 Efficiencies and Steam Conditions

Figure 9.6 shows improvements over the years in average and best

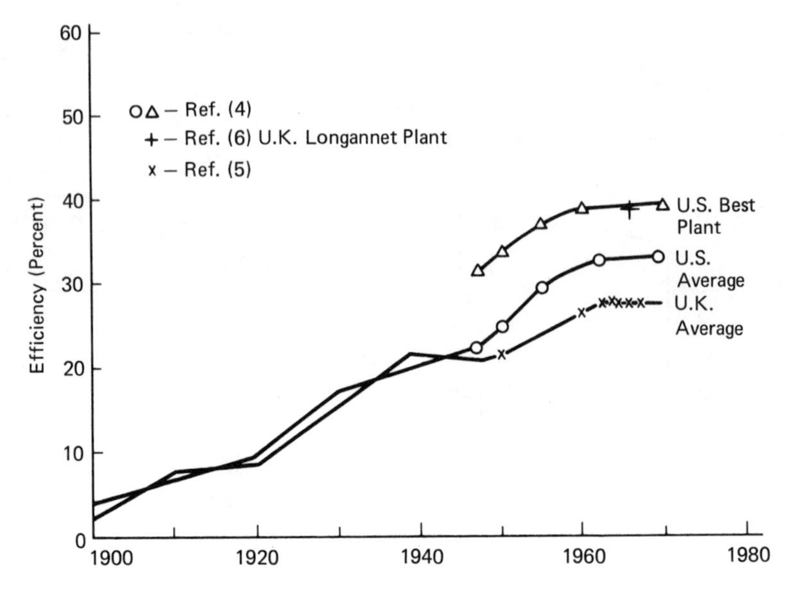

Figure 9.6
Thermal–electric power plant efficiency.

Figure 9.7
Steam boiler drum. Reproduced courtesy of Ontario Hydro.

plant efficiencies of steam power plants. Average efficiencies are substantially less than best plant efficiencies for several reasons. The system generally has a mixture of older, less-efficient plants, and newer ones. Perhaps more important, the system is typically designed for minimum power cost, not maximum efficiency. The substantially lower average plant efficiency in the United Kingdom does not mean that U.K. designers are less competent; it results primarily from the fact that electricity demand in the United Kingdom is very low in summer compared with winter, whereas in many parts of the United States summer and winter electricity loads are nearly balanced so that the plant is running steadily year-round. This makes it worthwhile to spend money to raise plant efficiency and cut fuel bills. Still

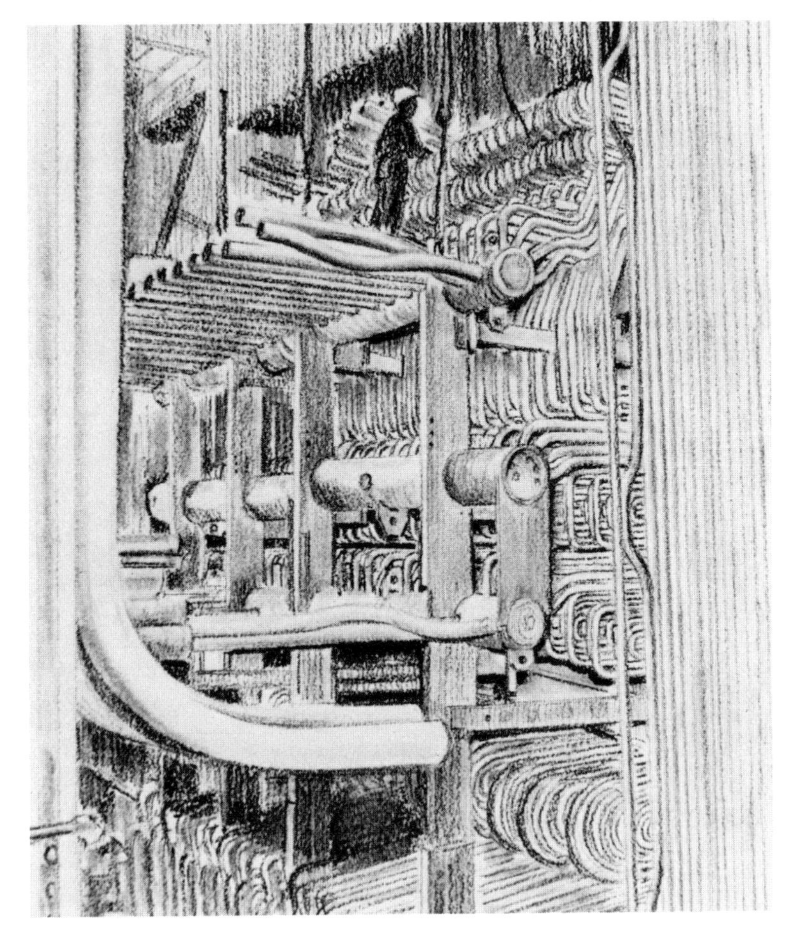

Figure 9.8
Superheater section of a large coal-fired boiler. Reproduced courtesy of Ontario Hydro.

Figure 9.9
High- and low-speed turbo alternators. Reproduced courtesy of Ontario Hydro.

Figure 9.10
Steam turbine rotor, with blades 2 to 34 inches in length. Reproduced courtesy of Ontario Hydro.

Figure 9.11
Generator rotor. Reproduced courtesy of Ontario Hydro.

another reason why average is less than peak efficiency is that when an individual plant runs at less than, say, 50 percent of full load, its efficiency will drop. It is not possible to run all plants at full load year-round.

The plateau in efficiency to which steam plants in both the United States and the United Kingdom have tended is due primarily to metallurgical limitations on maximum steam temperature. Raising steam temperature would require more expense in high-temperature boiler tubing and piping alloys than would appear to be justified by the attendant improvements in efficiency. (Figures 9.7–9.11 show the elements of a modern steam–electric plant.) At the same time, experience suggests that plant reliability will suffer if strenuous temperature increases are resorted to. This leaves little room for further improvement in system efficiency with conventional plants.

Table 9.3 shows how steam plant efficiency has improved in recent years in the United Kingdom as a result of increasing pressures, temperatures, and unit size. The steam conditions in Table 9.3 refer to the boiler pressure, followed by the temperatures of the steam at entry to the high-pressure turbine, and then the temperature at entry to the reheat turbine. Table 9.4 shows steam conditions and net plant heat rates (Btu/kWh) for high-performance steam–electric plants in the United States in 1972. It will be noted that almost all of these have steam temperatures around 1000°F (538°C).

A number of these plants have two stages of reheating with steam conditions in the range 3500 psi/1050°F/1050°F/1050°F (steam inlet pressure/throttle temperature/first reheat temperature/second reheat

Table 9.3
Steam Power Plants in the United Kingdom

Size of Unit (MW)	Steam Conditions (lb/in.^{2}/°F)	Year First Unit Commissioned	Design Thermal Efficiency (percent)	Capital Cost of Complete Station ($/kW)[c]
30	600/ 850	1948	27.6	167
60	900/ 900	1950	30.5	142
100	1500/1050	1956	33.7	145
100[a]	1500/ 975/ 950	1957	34.5	145
120[a]	1500/1000/1000	1958	35.6	127
200[a]	2350/1050/1000	1959	37.5	122
300[a]	2300/1050/1050	1963	37.8	103
375[b]	3500/1100/1050	1965	39.8	93
500[a]	2300/1050/1050	1965	39.2	83

[a]Reheat cycle.
[b]Supercritical temperature.
[c]With £1 = $2.50.
Source: Ref. (7). Reproduced courtesy of the Council of the Institution of Mechanical Engineers.

temperature). About half the new units have been designed for 3500 psia, and one-third for around 2400 psia. The Eddystone coal-fired plant is exceptional, with steam conditions of 5000 psi/1200°F/1050°F/1050°F. In 1962, this plant achieved a record heat rate of 8534 Btu/kWh, an annual average efficiency of 40 percent (overall thermal efficiency = 3413/heat rate). Its performance since 1962 has not been quite as good; in 1967, the annual average heat rate was 8874 Btu/kWh (efficiency 38.5 percent). The high temperature and pressure at the turbine inlet in this plant required extensive use of relatively expensive austenitic steel; overall, the Eddystone plant capital cost was unusually high (see Figure 7.4).

The economics of increasing steam temperature are not very favorable. None of the units scheduled for service in the United States between 1968 and 1974 has throttle temperature above 1000°F (538°C) (8), presently the upper limit for ferritic piping material. Russian designers have also decreased maximum steam temperature in recent years, roughly from 1050°F (565°C) to 1000°F (538°C). Now that fuel prices are so high, economically optimum steam temperatures may have risen a little, but this will not raise plant efficiency very much. Conventional steam plants have been developed to the point of diminishing returns. Even with steam conditions of 10,000 psi/1400°F/1300°F/1250°F, the maximum efficiency would only be about 45 percent.

The metal used in superheater and reheater tubing is subject to failure for several reasons, including corrosion by steam, corrosion by flue gas, and creep and creep-rupture.

Carbon steels are quite resistant to oxidation in steam and flue gases in temperatures up to 1000°F as long as stress levels are low. Accelerated corrosion may occur at lower temperatures when the materials are highly stressed (3).

For temperatures up to 1050°F (565°C), alloy steels with up to 1.25 percent chromium are usually adequate. Ferritic steels have sufficient strength at these temperatures. However, above 900°F migration of carbon to form graphite in heat-affected zones at weld boundaries can cause very serious problems after long service. Still higher temperatures require chromium–molybdenum steels (9 percent chromium) or stainless steels to provide sufficient oxidation resistance or higher mechanical strength.

Typical allowable temperatures and stresses for tube steels are shown in Table 9.5.

Particular problems with corrosion develop as the result of ash deposits on tubes. Coal ash corrosion has been found to be serious even with stainless steel alloys at a steam temperature as low as 1050°F (560°C), particularly with high-sulfur coal. The corrosion seems to depend mainly on the temperature of the ash deposit, and it is accelerated by the presence of alkali metals in the ash. Oil ash is also dangerously corrosive, even at temperatures as low as 1000°F (538°C), depending on the sulfate and vanadium content of the ash; sodium content is also significant.

The recent trend back toward 1000°F (538°C) as the maximum steam temperature has provided a reasonable degree of safety against corrosion problems for most oil and coal fuels and so has reduced overall plant maintenance costs. Plant outage due to the time required to replace boiler or superheater tubes is very costly, considering the value of the plant that is necessarily idle during that period.

The selection of steam pressure and temperature is a result of detailed studies of the dependence of generation cost on plant efficiency, fuel cost, capital cost, and outage and replacement cost. The cost of the superheaters, reheaters, and piping to the turbine is an important part of plant cost; this depends strongly on materials used, and therefore on allowable operating temperatures.

Figure 9.12 shows the average annual efficiencies of new plants entering service in the United States in the years 1970–1971. Little, if any, trend of improvement in efficiency with increasing size is apparent. The trend in recent years toward increasing unit size is shown by Tables 9.3 and 9.4. By 1970, 660-MW single units were in operation in the United Kingdom, and 1150-MW units in the United States. Increasing size leads to savings in capital cost per

Table 9.4
Fossil-Fuel Fired Steam–Electric Generating Units with Best Annual Heat Rates, United States, 1972[a]

Plant Name	Unit Number	Year in Service	Size (MW)	Btu per Net kWh	Net Generation (10^6 kWh)	Steam Conditions Pressure (psia)	Initial Temp.(°F)	Reheat Temp. (°F)
Coal-Fired under 9100 Btu								
Marshall	4	1970	650[c]	8546	4119	3500	1007	1000–1000
Marshall	3	1969	650[c]	8630	4339	3500	1007	1000–1000
New Genoa	1	1969	346[c]	8771	1672	3500	1010	1010
Marshall	2	1966	350	8774	2955	2400	1050	1000
Marshall	1	1965	350	8809	2755	2400	1050	1000
St. Clair	6	1961	353	8820	2374	2400	1050	1000
Sporn, Philip	5	1960	496	8903	2723	3500	1050	1050–1050
Amos	1	1971	816	8961	5773	3500	1000	1025–1050
Ft. Martin	2	1968	576	8974	3580	3675	1000	1000
Trenton Channel	9	1968	536[c]	8980	3850	2400	1000	1000
Gas-Fired under 9700 Btu								
Victoria	6	1968	261	9553	1269	2000	1000	1000
San Angelo	1[b]	1966	133[c]	9554	830	1450	1000	1000
Robinson, P. H.	1,2	1966,1967	485(ea.)	9597	5435(total)	3500	1000	1000
Cedar Bayou	1,2	1970,1972	765(ea.)	9659	7930(total)	3334	1010	1005
Nine Mile Point	4	1971	783	9684	5321	3690	1000	1000

Oil-Fired under 9600 Btu								
Brayton Point	3	1969	643[c]	8788	4076	3675	1000	1030–1055
Canal	1	1968	543	9031	3696	3500	1000	1000–1000
Possum Point	4	1962	239[c]	9044	1240	2400	1000	1000
Brayton Point	2	1964	241	9273	1821	2520	1000	1000
Brayton Point	1	1963	241	9288	1585	2520	1000	1000
Middletown	3	1964	239[c]	9350	1447	2675	1000	1000
New Boston	1	1965	359	9422	1928	2400	1000	1000
New Boston	2	1967	359	9422	2575	2400	1000	1000
Portsmouth	4	1962	239	9454	1156	2400	1000	1000
Hudson	1	1964	455	9518	2501	3500	1000	1025–1050

[a]Based on full year's operation at 50 percent and better annual capacity factor.
[b]Combined cycle unit.
[c]Tandem-compound units; all others are cross-compound.
Source: Ref.(9).

Table 9.5
Allowable Stresses in Tubes

Material	Maximum Allowable Temperature[a] (°F)	Allowable Stress (psi)							
		800°F	900°F	1000°F	1100°F	1200°F	1300°F	1400°F	1500°F
Carbon Steel	950	9,000	5,000	1,500					
Chromium Alloy (2.25% Cr, 1% Mo)	1,125	15,000	13,000	7,800	4,200	2,000			
Stainless Steel (18% Cr, 8% Ni)	1,400			13,750	9,750	6,050	3,700	2,300	1,400

[a]Based on maximum oxidation resistance of metal.
Source: Ref. (3).

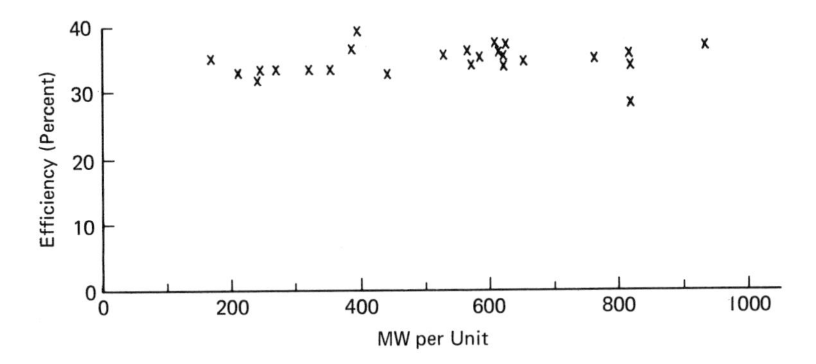

Figure 9.12
Annual average efficiencies of fossil fuel plants entering service in the United States, 1970–1971. Data from Ref. (9).

kilowatt as well as operating expense; the first is about 10 percent for an increase from 500 to 1000 MW (6). The saving in operating cost has little to do with efficiency; it results mainly from reduction in required manpower—about the same number of people are needed whether the unit is 300 or 1000 MW. There is some concern that larger units are less reliable than smaller ones, although this is difficult to decide, since, as units have been growing larger, steam conditions have been raised significantly. Despite increasing complexity with two-shaft or possibly three-shaft machines, still larger steam turbine units might eventually be built.

9.2.4 Exhaust Gas Losses

The hot combustion gases leaving the boiler give up much of their energy to heat incoming combustion air (in the air preheater) and feedwater (in the economizer). Yet overall combustion efficiency is considerably less than 100 percent, owing largely to residual sensible energy carried up the stack. With coal, gas, and oil fuels, and identical furnace conditions, the overall heat rate is said to decrease by about 2.5 percent for every 100°F decrease in exit temperature. Another way to express this is to say that the heat rate will decrease by about 2 percent for every 100°F increase in air preheating.

Two limitations to the extraction of energy from stack gases are first, that exhaust gases must be warm enough to provide sufficient buoyancy for the exhaust plume, and second, that the temperature of the exhaust gases must be above the acid dew point (otherwise SO_2 will combine with liquid H_2O to produce sulfuric acid, which will corrode the gas-side surfaces of the economizer). Of these two limitations, the latter generally controls the best attainable combustion efficiency.

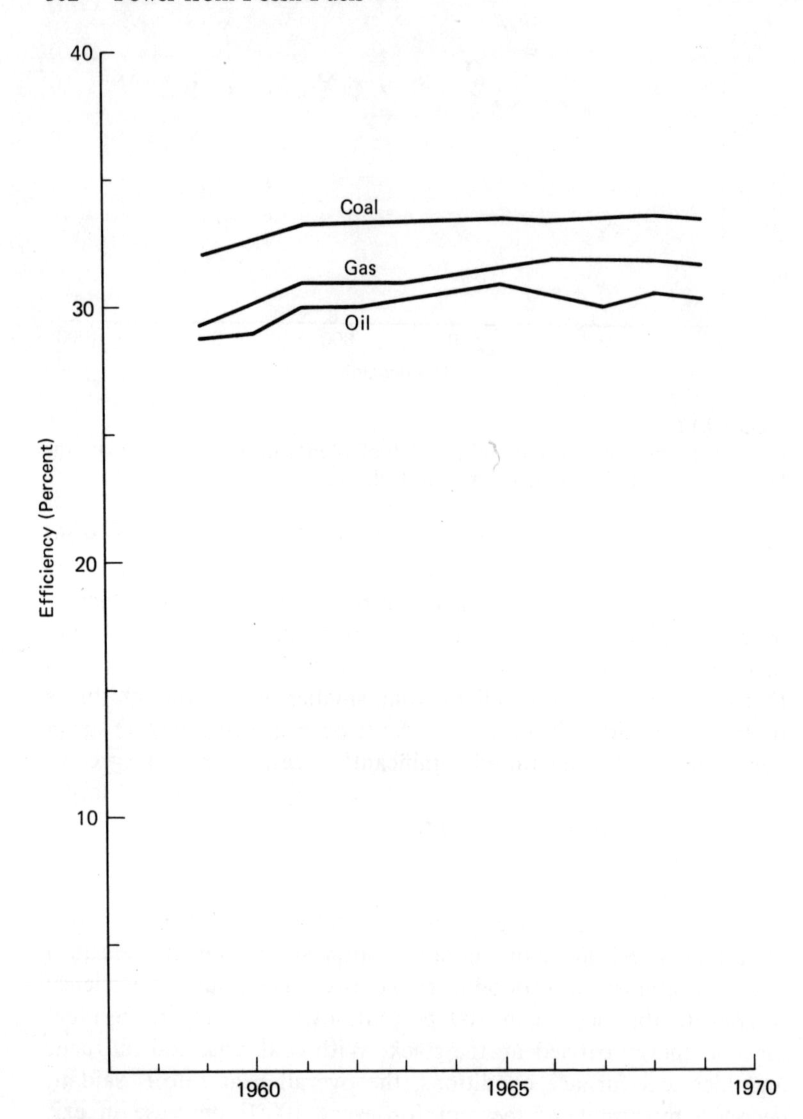

Figure 9.13
Average steam power plant efficiency in the United States. The efficiency is the ratio of electrical output to the higher heating value of the fuel. Data from Ref. (10).

Natural gas is usually "scrubbed" to remove sulfur prior to combustion, but even so, trace fractions will remain. Many oil and coal fuels contain 2 percent sulfur or more, which is difficult to remove; until the advent of strict SO_2 regulations in the United States, it was simply emitted with the other products of combustion.

Minimum metal temperatures are limited typically to:
240°F (116°C) for oil with 2 percent S,
173°F (79°C) for coal with 2 percent S (pulverized coal combustion or cyclone furnace),
177°F (81°C) for gas with 0.01 percent S.

Oil combustion products are more susceptible to low-temperature corrosion (on the gas side of the air preheaters). They usually contain vanadium, which accelerates conversion of SO_2 to SO_3, from which the sulfuric acid forms. Also, the oil products lack the alkali ash content of coal, which tends to neutralize acidity.

Some reduction in exhaust temperature can be allowed with improved gas temperature distribution and reduced excess air supplied to combustion (to inhibit conversion of SO_2 to SO_3), but there is little room for improvement in plant efficiency by reducing stack gas losses.

Figure 9.13 shows that average efficiencies of steam plants have varied significantly with the type of fuel. The main reason that the efficiency for gas tends to be lower than that for coal is not the difference in sensible energy corresponding to the difference in stack gas temperature. Rather it is the greater formation of H_2O in gas, which has a much higher H/C ratio than oil. In North America (unlike Europe) power plant efficiency is computed on the assumption that the latent heat of the H_2O in the exhaust products must be regarded as a loss.

9.2.5 Heat Rejection

In the United States and elsewhere, water quality standards severely restrict the use of rivers and lakes for cooling of large power plants. Very few rivers have sufficient flow capacity to absorb the thermal discharge from a 2000-MW plant. Sites on the shores of large lakes or on the seacoast are becoming progressively restricted. For example, hot water discharge from large new power plants will not be allowed into Lake Michigan. Large thermal discharges along coastlines may also be restricted where marine life would be adversely affected.

Table 9.6 shows relative costs of various methods of cooling fossil fuel steam plants, as evaluated by L. R. Glicksman (11). "Once-through cooling" in Table 9.6 refers to the use of water from lakes, rivers, or oceans to pass through the power plant. This method has minimum capital cost and is preferred where suitable water supplies are available. Cooling ponds are man-made lakes into which con-

Table 9.6
Fossil Fuel Steam Plant Cooling Costs, 1972

Method	Capital Investment ($/kW)	Typical Total Cost for Cooling (mills/kWh)
Once-Through Cooling	2–6	0.14
Cooling Ponds	3–8	0.20–0.24
Evaporative Cooling Towers		
Natural Draft	6–12	0.28–0.35
Forced Draft	5–9	0.2–0.3
Dry Cooling Towers	20–40	1.0–1.2
Forced Draft (direct)	15–20	0.4–0.7
Forced Draft (indirect)	15–30	0.8–1.3

Source: Ref. (11).

denser cooling water is discharged, and from which it is withdrawn. In the pond the water cools by radiation and conduction, but evaporation accounts for perhaps 40 percent of the heat load, so that makeup water is required. Cooling pond areas are typically large; Glicksman cites figures ranging from 1000 to 5000 acres for a 1000-MW plant. For smaller plants, sprays may be used to increase the cooling rate per unit area, but at the cost of additional evaporation. In many regions cooling ponds will be ruled out because of the unavailability or the cost of land.

Evaporative-cooling towers require a significant increase in capital cost relative to once-through cooling; dry-cooling towers are even more costly. The economically optimum power plant with a cooling tower will generally have a somewhat higher condenser temperature than one with once-through cooling, so that its overall cycle efficiency will be somewhat less. This factor must be taken into account when assessing cooling tower costs.

Cooling towers take the cooling water from the condenser and allow it to flow vertically downward through the tower over baffles and other solid surfaces to create liquid surface area. The air flow is upward, driven either by buoyancy forces alone (natural draft) or propelled by an induced draft fan located in the top of the tower. Air velocities and cooling rates are increased by the fan, so tower size and capital costs can be reduced, though at the expense of electrical power to drive the fan. The relative costs of forced and natural draft towers will depend on plant location and configuration. In an evaporative tower almost all of the heat loss is by evaporation. This means that water must be supplied to the plant at a rate of 2 to 3 kg/kWh, which is about 1.5 or 2 percent of the circulating water flow. The convection of unevaporated water droplets (called "drift") out of the towers is typically only 0.1 percent of the circulating water flow.

With the "direct" system the turbine exhaust steam is condensed directly in the cooling tower. This appears to be feasible only for relatively small plants, owing to the size of pipes necessary to transport the exhaust steam to the cooling tower surfaces. For larger units the indirect system is used. A large flow of circulating water condenses the steam by mixing the two together and transports the condensation energy to the cooling tower in a liquid stream. Relative costs are dependent on many factors of plant design.

Dry-cooling towers can eliminate the need for evaporative-cooling water and any undesirable local effects of the very large vapor plumes emitted from wet towers. However, they are three to four times as expensive as wet towers for conventional Rankine cycle plants, and are seldom used. For closed-cycle gas turbines, however, their unit capital costs may be considerably less than shown in Table 9.6, since gas turbine exhaust gas will typically be much higher in temperature than steam exhaust; much less heat exchanger surface area will be required per kilowatt of power. As noted earlier, gas turbine cycles do not compete very well in efficiency with steam cycles for fossil fuel plants, but they do appear to be suitable for gas-cooled nuclear plants. They may also be well adapted to the gas–steam binary cycle, where the efficiency is high; although the exhaust temperature is low, the steam cycle may produce one-third of the total power of the plant, so that the cooling tower costs per kilowatt of plant output are only about one-third those shown in Table 9.6.

Table 9.6 refers specifically to fossil fuel plants. For water-cooled reactors the cooling water thermal discharge per unit power output is about 1.7 times as high (per unit of power output) as that of the fossil fuel plant. The ratio of capital costs of cooling equipment is a like factor.

In summary, the additional cost of evaporative-cooling towers (relative to once-through cooling) for conventional fossil fuel steam plants is perhaps 3 to 5 percent of overall electricity generation costs. Dry-cooling towers would cost three to four times as much. Thus if evaporative-cooling towers are unacceptable in certain areas owing to fog and other effects on the local climate, alternative, though costly, solutions are available.

9.2.6 Other Working Fluids

Since the conventional steam power plant is in a mature state of development, its limitations can be seen quite clearly. The most important constraint is the peak temperature, which is limited to prevent oxidation and corrosion of the steel superheater tubes. The temperature could be raised by using more expensive steels, but the resulting increase in plant efficiency would be small and not likely worth the extra cost. Since this temperature limit—1000 to 1050°F (540 to 565°C)—is so much lower than the typical stoichiometric combustion temperature for fossil fuels, steam is not an ideal working fluid. Considerable gain in efficiency could be made if the peak temperature could be raised.

One solution to this problem would be to use a different fluid, one less corrosive to steel at elevated temperatures and with a higher critical temperature. Mercury has been used successfully in fairly large power plants, but the cost and toxicity of mercury, as well as its interaction with steel at less than 500°C, appear to rule it out for the future. Potassium has been suggested, but its cost and safety problems may be formidable.

Over the years a great many other substances have been proposed as working fluids. Carbon dioxide is one example; it can be compressed at its triple point of 88°F (30°C), where its work of compression would be very low, and then used in a supercritical cycle with maximum temperature 1150°F (620°C). The main problems with this cycle are corrosion of steel by CO_2 at high temperature and the need to reject heat below 100°F (38°C). Sulfur dioxide could also be a useful working fluid if it were not for the corrosion problem.

Dissociating gases such as N_2O_4 or Al_2Cl_6 have also been advocated as working fluids. The basic idea is that with pressures chosen so that during compression the gas will have high average molecular weight, the work of compression will be relatively low. During expansion, the gas is almost entirely dissociated, so that its work of expansion is relatively high. For the expansion, the Al_2Cl_6 would be almost entirely dissociated to $AlCl_6$ at peak temperatures above 1500°F (815°C).

The dissociation of N_2O_4 proceeds to NO_2, and to NO and O_2 above 1100°F (600°C). Work in the USSR indicates the possibility of an overall cycle efficiency of over 40 percent at temperatures of 1000 to 1050°F (540 to 565°C) (12).

A major difficulty with these dissociating gas cycles (in addition to the restriction to a narrow range of temperature and pressure) is that free oxygen or chlorine is a dissociation product. With even trace amounts of moisture present, severe corrosion could result.

Despite the awkwardness of its pressure–temperature–volume relationships, and its inherent limitation on cycle efficiency, water is not likely to be displaced as the principal working fluid for two-phase power plants. A combination of gas and steam cycles, to take advantage of the high-temperature operation feasible in future combustion turbines, is much more likely.

9.3 EMISSIONS CONTROLS FOR FOSSIL FUELS

9.3.1 The Polluting Emissions

The main pollutants from fossil fuel combustion are particulate matter, carbon monoxide, and the oxides of sulfur and nitrogen.

With normal control of the ratio of fuel to air in combustion, emissions of smoke and CO are negligible. Particulate matter in the form of ash or dust is effectively removed by electrostatic precipitators (except for extremely small particle sizes).

Most of the sulfur oxides are in the form of SO_2, which, as indicated in Chapter 6, can injure health. Official standards for maximum ground-level concentrations of SO_2 have been established (see Table 6.5). In many localities ambient concentrations have exceeded these levels and restrictive controls have been considered necessary. Sulfates are also of concern, though much less is known about their effects on health.

The oxides of nitrogen are mainly important in relation to the formation of photochemical smog. Ambient air standards have been stated for NO_x (see Table 6.5); these levels have also been exceeded in various areas.

9.3.2 Control Strategies

Constant emissions control Constant emissions control is used in the United States to limit power plant emissions, regardless of local atmospheric conditions. Emissions standards announced by the EPA (Table 9.7) are absolute. Enforcement of these standards means that fuels of more than 1 percent sulfur cannot be used at any time or place.

The problem in the United States with SO_2 emissions is that reliable methods are not generally available for meeting the absolute emissions limit shown in Table 9.7. There is insufficient low-sulfur coal available to meet electricity demand. Low-sulfur oil for new power generation plants is costly. Gas is in such short supply that it is generally unobtainable by electrical utilities.

Sulfur removal by scrubbing stack gases is costly and can be unreliable, and the requisite technology is not yet widely available. Coal

Table 9.7
EPA Emissions Standards for Fossil-Fuel Powered Steam Generation Units[a]

Pollutant	Fuel	Maximum Emissions per 10^6 Btu Heat Input (lb)	Ppm at Stoichiometric Fuel-to-Air Ratio
Sulfur Oxides	liquid	0.8	
	solid	1.2	
Particulates	all	0.1	
Nitrogen Oxides	gaseous	0.2	147
	liquid	0.3	270
	solid	0.7	560

[a]With heat input greater than 2.5×10^8 Btu/hr.
Source: Ref. (13).

refining to produce clean gaseous or liquid fuels could be done at a price, but many of the processes are still under development. Much time and money will be needed before the required capacity can be developed.

Whether stack gas scrubbing or coal refining will ultimately be the best solution has been much debated. One of the problems of stack gas scrubbing is that equipment must be developed for each kind of coal. It is typically expensive to operate and may be unreliable enough to reduce the load factor of the entire plant; this could be very costly. It may also cause major problems of solid waste disposal. In contrast, coal refining could supply clean fuels that can be burned efficiently and reliably with no extra station capital costs. Waste disposal problems could be reduced if coal refining were done near mining sites.

Emissions of nitrogen oxides from large new power plants in the United States are to be limited by proposed legislation, as indicated in Table 9.7. These limitations are severe in relation to typical emissions indicated in Table 9.8 for coal, oil, and gas combustors. As mentioned earlier, fluidized-bed combustion and low-Btu gas combustion both produce much lower quantities of NO_x than combustion of pulverized coal, owing to low combustion temperature. Alternative methods of lowering combustion temperature include two-stage combustion, flue gas recirculation, and control of the air–fuel ratio to a value as close as possible to stoichiometric (low excess air). Much development work will be required on plants to meet the stringent standards of Table 9.7, but costs of NO_x control are expected to be much less than those associated with control of SO_2 emissions.

Variable emissions control Until coal-refining facilities are widely available, variable emissions controls have been recommended by the U.S. Federal Power Commission (1). The idea is to allocate low-sulfur fuels to locations and times of poor atmospheric ventilation, and to use higher-sulfur fuels where they do not violate the air quality standards. Maximum use would be made of tall stacks, and a strategy would be developed for switching fuels and loads within large electricity generation systems. Subject to the environmental constraint, the system would still be used to produce power in the most economical way.

In Japan, continuous monitoring of air pollution is used along with knowledge of atmospheric conditions to project periods when air pollution could become excessive (15). At these times local controls are imposed on pollution sources. The controls are variable with locality and with atmospheric conditions.

Tall stack dispersion of pollutants (see Section 6.3) Evidence

Table 9.8
Nitrogen Oxide Emissions from a High-Intensity Combustor and from Various Types of Conventional Combustion Plants

Approximate Temperature of Flue Gas at the Hottest Point (°C)	Fuel	Amount of N Chemically Combined in Fuel (percent by weight)	Type of Plant	Composition of Combustion Air	NO_x Reported in Dry Gas (by volume ppm)
2300	coal	1.4	high-intensity combustor	oxygen-enriched air	10,000–13,000
1500–1700	coal	probably 1–2	a range of coal-fired power station boilers	air	200–1400
1500–1700	oil	not disclosed	a range of oil-fired power station boilers	air	110–800
1500–1700	natural gas	negligible	a range of natural gas-fired power station boilers	air	50–1500
1500–1700	cracked, residual fuel oil	1.0	an oil-fired power station boiler	air	425
	paraffinic fuel oil	0.2		air	215

Source: Ref. (14). Reproduced courtesy of the *Journal of the Institute of Fuel.*

has been provided from the United Kingdom that SO_2 can be easily dispersed by emission from tall stacks, so that it will be a minor contributor to ground-level emissions of SO_2 (16, 17). The long-range transport of SO_2 and sulfates has not yet been fully clarified; the extent to which sulfur compounds from the United Kingdom and other countries are transported and deposited in Scandinavia, for example, is under serious study.

9.3.3 Sulfur Removal from Stack Gases

Several methods of sulfur removal from stack gases are now being developed. In April 1973, a report by the Sulfur Oxide Central Technology Assessment Panel (SOCTAP) concluded that four processes were feasible for widespread application within the next five years (18, 19). These processes are wet limestone or lime scrubbing, magnesium oxide scrubbing with regeneration, catalytic oxidation, and wet sodium base scrubbing with regeneration.

With wet limestone scrubbing, pulverized limestone or lime is added to a scrubbing liquid which is sprayed into the exhaust gases. Most of the SO_2 is absorbed into the liquid, producing $CaSO_4$ and $CaSO_3$. The fly ash is also trapped by the scrubbing liquid, which is 5 to 15 percent solids in water. The quantity of waste products to be disposed of with wet limestone scrubbing is reported to be about 2.5 times the amount of fly ash from a typical coal-fired boiler. In waste disposal ponds, calcium sulfate settles out readily in solid crystalline form and can be recovered as a dry solid. Calcium sulfite, however, forms a sludge that does not settle out without special treatment. With wet limestone scrubbing about 1.5 times the stoichiometric quantity of limestone, or less, may be needed for adequate sulfur removal. A disadvantage of the wet scrubbing process is that spraying of the exhaust gases cools them to as low as 120°F, so that a reheating process will generally be necessary to restore buoyancy to the exhaust plume and provide adequate atmospheric dispersion. Figure 9.14 shows schematically the wet limestone scrubbing process in a pilot plant, but not the means for reheating the stack gases.

In the scrubber, SO_2 is absorbed and reacts chemically with water and limestone to form products that are transferred from the scrubber to the reaction tank. Here the chemical reactions go nearly to completion, resulting in disposable precipitates. Makeup slurry is added to the tank, and scrubbing liquid is sent back to the scrubber.

The thickener receives a mixture of 5 to 15 percent suspended solids in water, which are concentrated by sedimentation and removed to a pond or landfill. The calcium sulfite crystals may retain as much as 50 percent H_2O. Large areas are needed for removal of water and stabilization of these solids.

Limestone scrubbers are capable of removing up to 90 percent

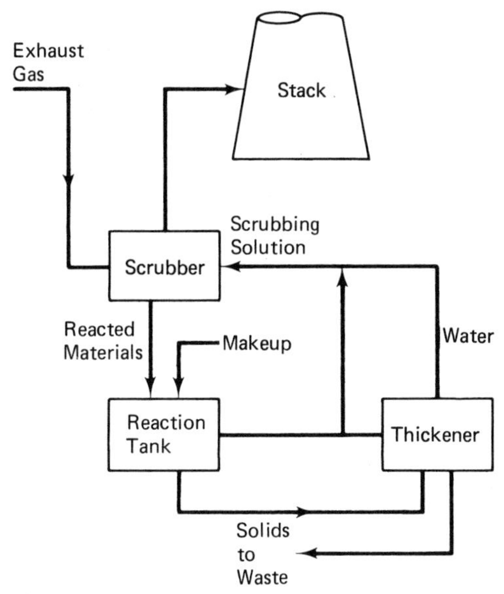

Figure 9.14
Wet limestone scrubbing. After Ref. (20).

of the SO_2 from flue gases, which may have 0.2 or 0.3 percent SO_2 (18).

Reheating the exhaust gases beyond the scrubber could be done by direct firing with natural gas or low-sulfur oil. Alternatively, air, from the preheater or steam-bled from the turbine may be mixed directly with the exhaust gases. Still another method would be heat exchange from stack gas entering the scrubber to the exit stack gas.

Estimates of capital costs for limestone and lime scrubbing systems have been reported in the range $27 to $46/kW (18). The stack gas sulfur-removal cost is equivalent to 1.1 to 1.2 mills/kWh. Major problems with the process are reliability and disposal of solid wastes.

Magnesium oxide scrubbing is similar to the lime scrubbing process except that the magnesium sulfate and sulfite salts are regenerated, producing a concentrated stream of SO_2, and MgO for reuse in the scrubber loop. Regeneration takes place at around 1000°C. Since the MgO is recycled, the fly ash must be efficiently removed from the flue gas before scrubbing. The SO_2 can be fed to a sulfuric acid plant. Magnesium oxide scrubbing has not been used as much as limestone scrubbing and is relatively expensive.

Catalytic oxidation is used to produce sulfuric acid from dilute SO_2 in the flue gas. One version of the process uses fly ash removal and electrostatic precipitation at high temperature, followed by conversion of SO_2 to SO_3 in a vanadium pentoxide catalyst bed. The gases are then cooled, and the SO_3 reacts with water in the flue gas to form a liquid which is 70 to 80 percent sulfuric acid and can be separated from the flue gases. This process produces large quantities of sulfuric acid, for which there may or may not be a market. It is reported to have at least twice the capital cost of other sulfur removal processes that are in advanced stages of development.

One of the wet sodium-base scrubbing processes uses a sodium sulfite–bisulfite solution to absorb SO_2 and convert sulfite to bisulfite. The scrubbing liquid is exposed to steam after absorption of SO_2 from thoroughly cleaned flue gas; steam stripping produces a strong stream of SO_2, and sodium sulfite crystals are recovered from the remainder of the solution by evaporation. These are returned to the scrubber. Again, the SO_2 can be used to make sulfuric acid or elemental sulfur.

Several other processes of flue gas sulfur removal are being worked on; some of these are recognized as being, at best, short-term methods of SO_2 emissions control from existing coal- or oil-fired plants. Flue gas desulfurization is inherently difficult because the sulfur is in dilute gaseous form, and because the total volume of flue gases emitted is so large. Though costly, it appears possible to remove 95 percent of the SO_2 at costs of perhaps \$40 to \$80/kW.

9.3.4 Fluidized-Bed Combustion

The most promising method suggested to date for removing sulfur during the combustion process is fluidized-bed combustion, which has been shown to be adaptable to coal firing. The "fluidized" combustion zone consists of a bed of coal particles suspended by a high-pressure airstream and cooled by steam pipes immersed in the bed. Limestone—perhaps 1.8 times the stoichiometric quantity—is added directly to the bed to react with sulfur. The sulfur removal is fairly efficient, and the combustion zone can be kept cool enough to reduce the formation of NO_X. Fluidized-bed reactors provide a very effective method of gas–solid particle contact. The relative gas velocity is high, and particles move freely so that heat transfer rates per unit volume are high. This process might have been developed commercially years ago, had not pulverized coal combustion proven so successful in the days before SO_2 emissions were subject to strict controls.

Figure 9.15 shows a typical fluidized-bed reactor for coal combustion. The pressure of the combustor may be in the range 1 to 25 atm. Air enters the combustion zone through a perforated plate and passes through at velocities of 1 to 5 m/sec. The solid particles have maximum sizes of 5 mm and are composed mostly of ash, lime, or limestone, with less than 1 percent in the form of unburned coal that is continuously fed in. Most of the heat of reaction is transferred to water or steam pipes immersed in the bed. The bed depth may be 0.6 to 1 m and may have a pressure drop somewhat higher than a

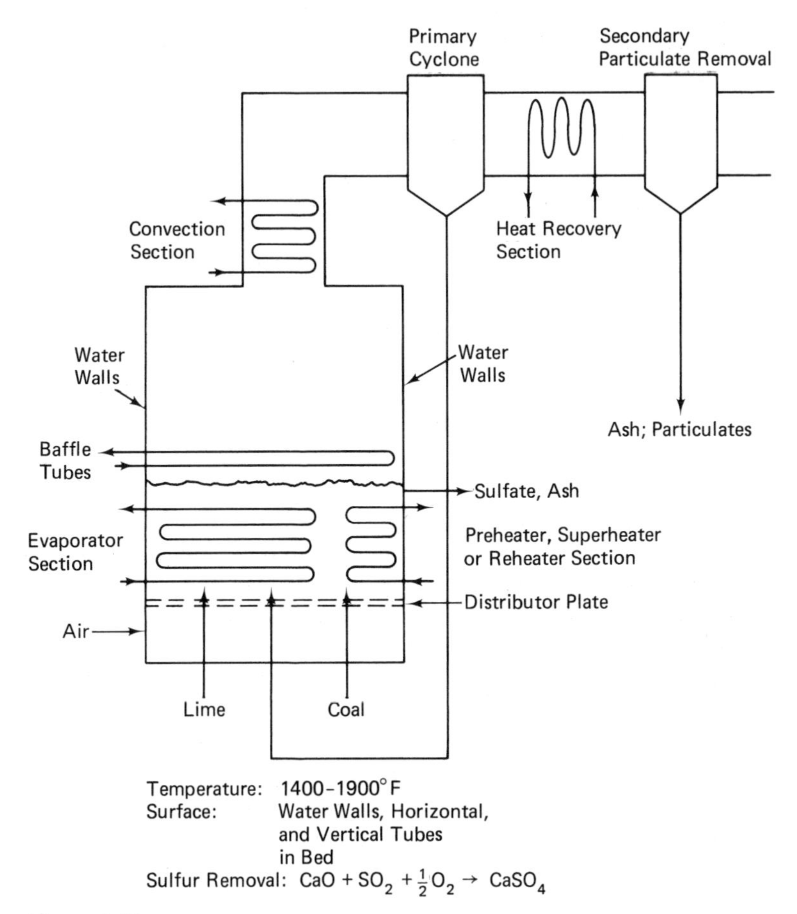

Figure 9.15
Fluidized-bed combustion boiler. From Ref. (20), courtesy of the World Energy Conference.

conventional boiler. The bed can be made much more compact by operating at elevated pressure. The required air compressor could be driven by a turbine supplied with combustion gas; the hot gas could be sufficiently clean and cool to assure satisfactory turbine life. It is estimated that pressurized fluidized-bed combustors could cost 10 percent less than conventional equipment (21).

9.3.5 Gasification of Liquid and Solid Fuels

The removal of sulfur from gasified fuel is much easier than from flue gas, since the volume of the gas to be cleaned is one order of magnitude less. Also, the sulfur may be largely in the form of H_2S, which is easier to scrub than SO_2. Further, the combustion temperatures of "low-Btu" gasified coal and oil are low enough that formation of NO_x is not a serious problem. Removal of ash and alkali metals from gasified fuel is also relatively straightforward.

The capital costs of the gasified fuel process are relatively high, however, and the energy required for gasification is a serious penalty unless gasification and power plants are closely integrated.

The gasification of fuels generally involves consumption of part of the heating value of the fuel prior to the purification process, which is often performed at low temperature. If the gases are cooled for cleaning, this could mean loss of 25 to 35 percent of the heat of combustion. If gasification and power generation are done at the same site, however, this energy can be utilized, as in a steam boiler.

Gasified fuels differ characteristically in energy content. Synthetic natural gas (SNG) is composed mainly of methane and has a high energy content, approaching the 1000 Btu/scf of natural gas. It therefore has the advantage that it can be transported economically by long-distance pipeline. However, it generally requires large supplies of oxygen and an extensive plant to convert the fuel into methane. For these reasons it is very expensive. Power gas is so-called because it is much less expensive than SNG and is adaptable for application to large thermal power plants. It has an energy content in the range 100 to 200 Btu/scf and is often called low-Btu gas. Gasification can be done with air and steam or air alone, and no attempt is made to produce methane; the proportions of CO and H_2 vary over a wide range, as is suggested by Table 9.9, which shows the composition of five fuels derived from gasification of coal by air.

The widely used Lurgi process for gasification of coal dates back to 1934. Versions of this process have been used to produce "town gas" with a heating value of about 500 Btu/scf. The Winkler process for coal gasification is even older, the patent dating to 1922. Thus

Table 9.9
Analysis of Low-Btu Gaseous Fuels (% by volume)

Constituent	Natural Gas (reference)	Lurgi	Fluidized-Bed Coal Gas	Winkler No. 1	Winkler No. 1
N_2	0.5	30.2	50.4	55.3	1
CO_2	1.8	10.7	0.5	10	19
CO		10.7	31.8	22	38
H_2		15.7	15.6	12	40
CH_4	93.3	4.4	0.5	0.7	2
C_2H_6	3.5				
H_2O		27.8	0.5		
H_2S		0.5	0.7		
Other	0.9				
Total	100	100	100	100	100
Higher Heating Value (Btu/scf)	1030	150	163	117	272

Source: Ref. (22).

the idea of destructive distillation of fuel in order to convert it to a clean gaseous product is not new, but the stringency of present-day pollution regulations is creating heightened interest in these processes.

Since the product gas is clean, it can be used at high temperatures without metal corrosion. This strongly suggests its use in gas turbines coupled with a low-temperature steam cycle. The relatively low energy of the gas means that flame temperatures will be much lower than with solid or liquid fuels; for this reason NO_x formation and emission may be negligible. The combined gas and steam cycle with power gas will be discussed later, but it may be noted here that with present technology it has at least as high an overall efficiency as a conventional steam plant. It shows considerable promise of future improvement.

Much recent work in fuel gasification has been concerned with the use of residual oils, which are too high in sulfur for burning in conventional plants under present U.S. SO_2 regulations and in the absence of adequate SO_2 removal technology. The basic ideas for coal gasification are very similar, but additional problems arise with coal due to its high ash content, possible caking tendency, and the variability of coal composition, which might necessitate considerable process adjustment to accommodate each type of coal.

One process developed for the production of power gas from residual oils is the Texaco Partial Oxidation Process (23). Air from a compressor is introduced with the oil to the gasification chamber and reacts with the fuel to form a mixture of CO and H_2. At the same time, the sulfur in the fuel is converted to H_2S or COS. The ash in the fuel is scrubbed from the product gas by water, after cooling. The gas then passes to an acid removal unit, where the H_2S and COS are removed and converted to either elemental sulfur or sulfuric acid. Essentially all ash, salts, particulate matter, and sulfur in the fuel can be removed. The process is said to be adaptable to any liquid feedstock.

Fluidized-bed gasification is also being developed, both for oil and coal gasification. D. H. Archer describes the scheme shown in Figure 9.16. For oil gasification, 20 to 25 percent of the oxygen needed to complete combustion is supplied to the bed. The heat released preheats the air and fuel feeds to the bed, and steam is not required.

Coal gasification in a fluidized bed requires that the coal first be dried and heated to prevent sticking of the coal particles. The sulfur in the coal is transformed to H_2S during the gasification process; the H_2S then reacts with limestone particles in the bed. Limited combustion with air releases thermal energy and some combustion products. Steam and CO_2 also react with the coal to form the fuel gases CO and H_2. Reaction rates are very dependent on pressure and

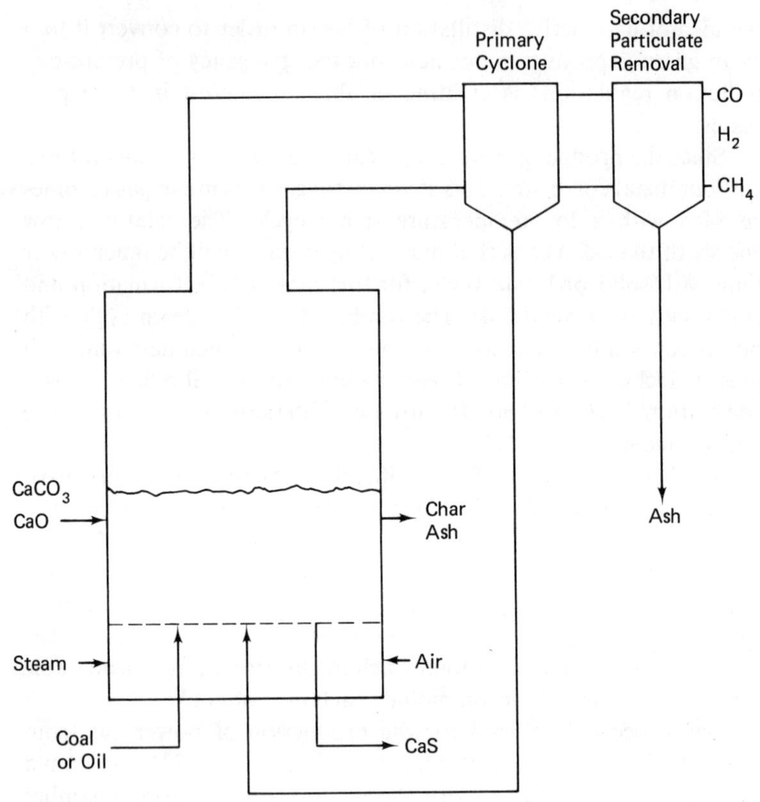

Figure 9.16
Fluidized-bed gasifier. From Ref. (21), courtesy of the World Energy Conference.

temperature, as well as on fuel composition; typical temperatures are 1500 to 1700°F (815 to 925°C). Numerous variants of the fluidized-bed gasification of coal have been studied or are under development (24), but much work remains to be done before widespread commercial application of the process becomes feasible. Archer indicates that the fluidized bed must be operated at high pressure before it will demonstrate substantial advantage over stack gas sulfur removal (21).

In a very different approach to coal gasification, Rockwell is developing a molten salt process, shown in Figure 9.17. The fuel is gasified while immersed in a sodium carbonate melt at a temperature of about 1800°F (580°C) and a pressure of 20 to 25 atm. Air is blown into the melt, which retains the ash and sulfur and does not require pretreatment of the coal or close sizing of particles. To remove the sulfur and ash from the melt, a stream of it is removed continuously and quenched with water. The ash is removed by filtering, and the sulfur compounds by conversion to H_2S and then into elemental sulfur. Sodium carbonate is recovered from the

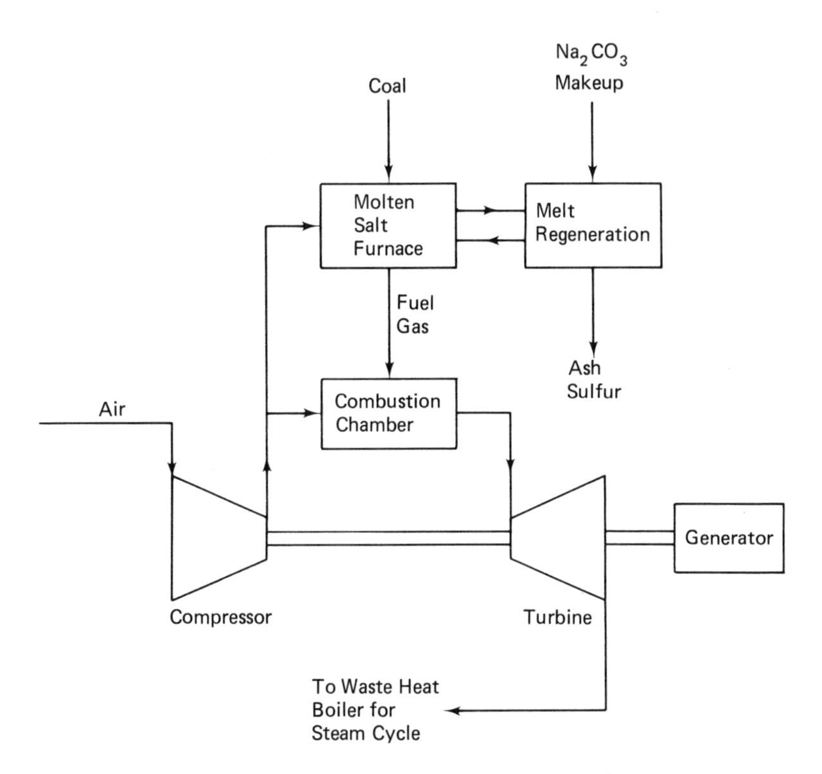

Figure 9.17
Schematic of Rockgas Na_2CO_3 molten salt gasification with gas–steam binary cycle. After Ref. (25).

solution by crystallization and evaporation, before return to the melt. The molten salt provides high heat transfer rates, a good medium for dispersion of coal particles and air bubbles, and a good sink for sulfur and ash. The principal problem is containment of the highly corrosive high-temperature melt.

9.4 GAS TURBINES

Gas turbines have been used increasingly in recent years for electricity generation during periods of peak demand. Single-shaft open-cycle machines are typical; these have efficiencies in the range 20 to 27 percent and a capital cost per kilowatt about half that for large, high-performance steam plants. Unit sizes range up to 100 MW. Some are a combination of gas turbines derived from designs for aircraft propulsion; others are specially designed for industrial use. These machines have another advantage for peak power generation—rapid startup time. For simple open-cycle gas turbines the startup time is only 2 to 8 min; for engines with heat exchangers the required startup time may be 10 to 20 min.

Although a great variety of cycle configurations may be used

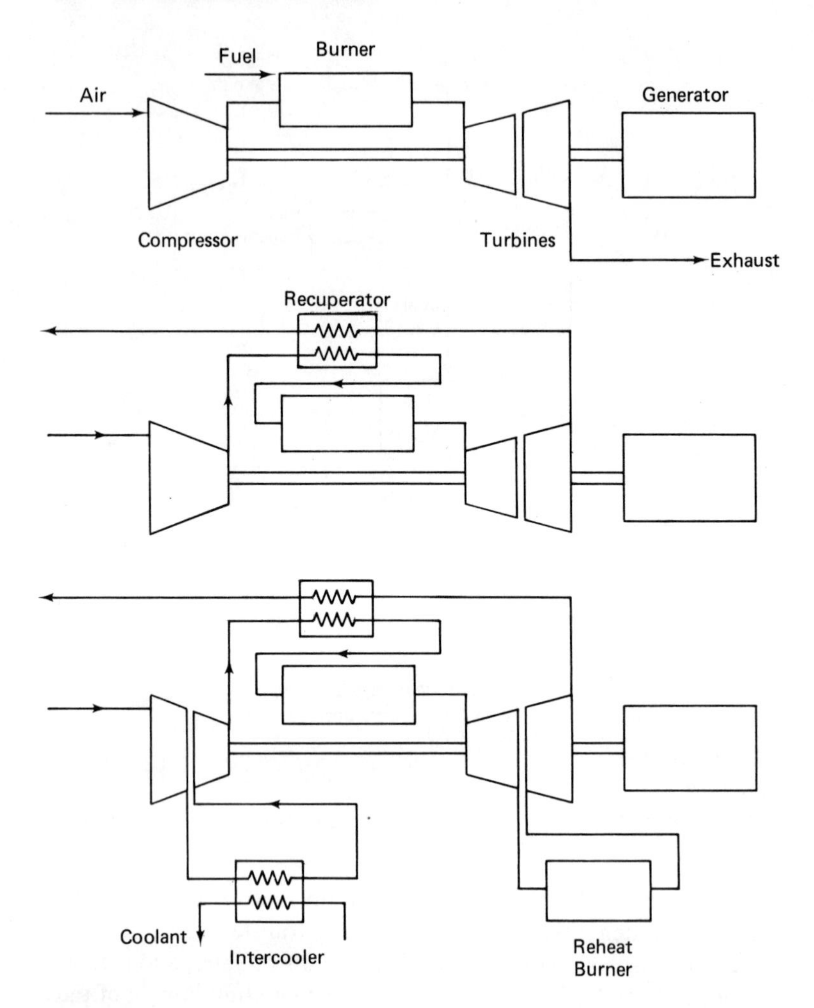

Figure 9.18
Open-cycle gas turbines.

(including recuperators, reheaters, intercoolers, and two- or even three-shaft layouts), the units are usually simple. A single-shaft engine is generally used, and possibly a stationary heat exchanger for exhaust heat recovery. Figure 9.18 shows three of the possibilities. The overall efficiencies of electrical power generation using gas turbines are in the range 15 to 25 percent for simple open-cycle engines, 21 to 29 percent for open-cycle recuperative engines, and up to 33 to 35 percent for two- and three-shaft engines with intercooling and reheat. Future increases in turbine inlet temperature could raise these efficiencies considerably. The single-shaft engine may not have nearly as good part-load efficiency as the two- or three-shaft engine, but it is mechanically much simpler.

Turbine inlet temperatures for continuous operation have

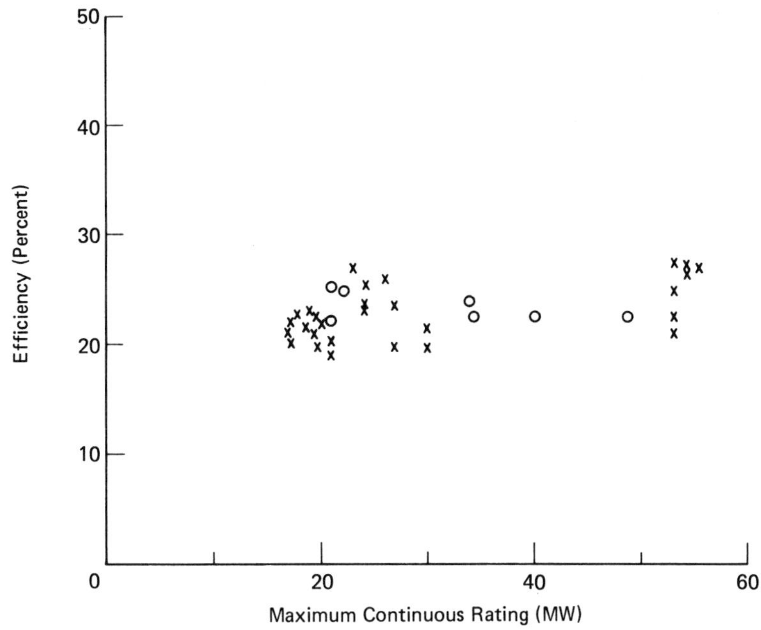

Figure 9.19
Efficiency of open-cycle gas turbines for electrical power entering service in 1970–1972. Data from Ref. (27).

covered a wide range over the past few years, from 1200 to 1650°F (650 to 900°C). With high-grade fuel and the best high-temperature materials, temperatures up to 1900°F (1050°C) are presently feasible. With the advanced blade-cooling techniques now being applied to aircraft gas turbines, much higher temperatures are possible.

The industrial gas turbine has, in principle, the advantage of burning a wide range of fuels, from residual and crude oils to methane and hydrogen gaseous fuels. However, most users have chosen to operate with light distillate oil and natural gas. With residual oils, solid ash may deposit on the blades and liquid ash may corrode the blades. Sodium, vanadium, and sulfur contaminants may force frequent turbine overhauls and necessitate low turbine inlet temperatures, unless elaborate fuel treatment is used (26).

Closed-cycle gas turbines would eliminate the problem of turbine durability with a wide range of fuels, but would bring in the extra capital costs of the combustion heat exchanger. Problems of corrosion and deposits on the combustion side of the exchanger may force the closed-cycle turbine to operate at fairly low peak temperature. Even with clean fuels, the maximum working fluid temperature may be limited ultimately to 1600 to 1700°F (870 to 925°C) due to heat exchanger and ducting temperature limits. Free choice of working fluid pressure level would enable the size and cost of the rotating machinery to be considerably reduced. So far, closed-cycle combustion-fired gas turbines have had little or no application to electrical power generation. However, the helium-cooled high-temperature reactor may be well suited to closed-cycle gas turbine operation (see Section 8.5).

Figures 9.19 and 9.20 show the efficiencies and capital costs of gas turbines installed in the United States in the years 1970–1972 for peak electricity generation. There is no obvious effect of unit size on cost of efficiency up to the single-unit maximum of about 55 MW. Larger gas turbines are reported in utility operation, but these are assemblies of smaller units. Gas turbines with capacities well above 100 MW are now being installed in combined gas and steam turbine plants; these will be discussed later.

The scatter in Figure 9.20 results partly from the nature of the market, which is relatively new, narrow, and rapidly growing, and partly from the nature of the product. The designer has considerable freedom in selecting engine features, turbine inlet temperature, stress levels, etc. Nevertheless, the efficiencies lie in a relatively narrow band, as shown in Figure 9.19.

The potential for future development of gas turbines is indicated

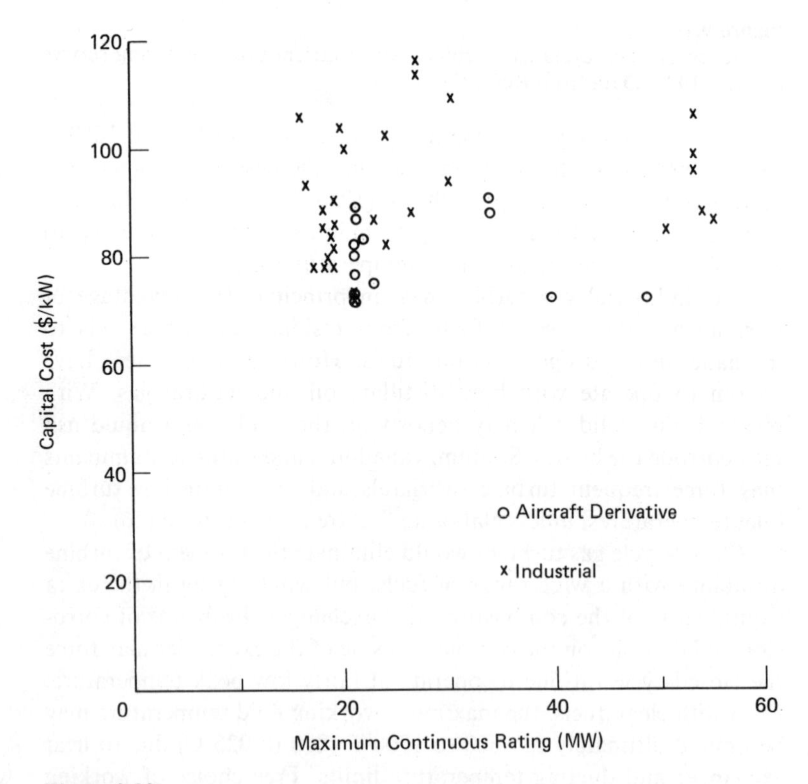

Figure 9.20
Capital costs of open-cycle gas turbines for electrical power entering service in 1970–1972. Data from Ref. (27).

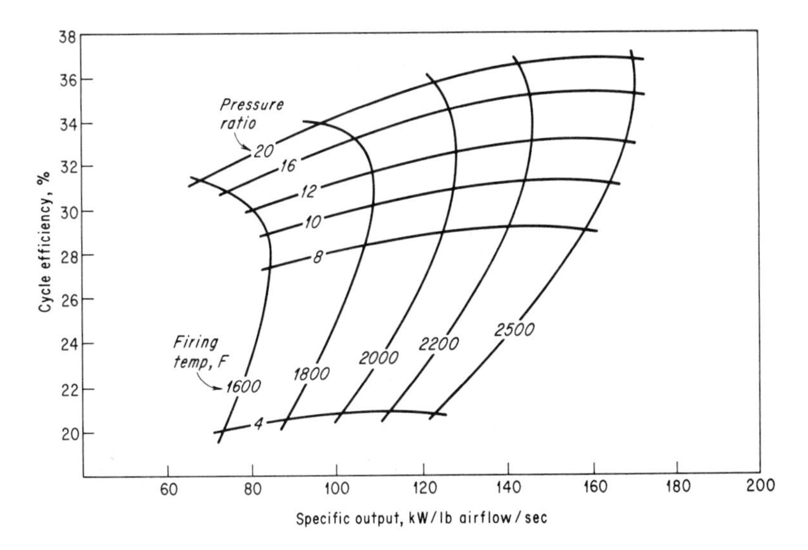

Figure 9.21
Efficiency and specific work for single-cycle gas turbines: polytropic compressor and turbine efficiencies 89 and 88 percent, respectively; combustion efficiency 100 percent; compressor cooling air bleed 2 percent at 2500°F (1370°C). Reprinted by permission from *Power* (January 1974).

in Figures 9.21 and 9.22: Figure 9.21 indicates the cycle efficiency and specific power output estimates of the Curtiss Wright Corp. for gas turbines, as a function of compressor pressure ratio and turbine inlet temperatures. The calculations on which Figure 9.21 was based incorporate the following assumed efficiencies:

polytropic compressor: 89 percent
polytropic turbine: 88 percent
combustion: 100 percent.

They also assume variable air extraction from the compressor for cooling the seal structure and turbines. At 1600°F (870°C) turbine inlet temperature, 2 percent of the compressor air is required for cooling. At 2500°F (1370°C) a total of 10 percent would be required. The maximum pressure ratio shown in Figure 9.21 is not extreme; certain aircraft engines have pressure ratios up to 30:1. It may be feasible to have even larger pressure ratios with the extremely large compressor that would be required for a 1000-MW gas turbine. As shown, however, the cycle efficiency does not increase rapidly with compressor pressure ratio.

Specific power (kilowatts per pound of air per second) is a good indicator of the machine capital cost per kilowatt, which would drop substantially with any increase in turbine inlet temperature.

A turbine inlet temperature of 2500°F (1370°C) represents a very large increase for current industrial turbines. Figure 9.22 shows the historical trend in turbine inlet temperature for various classes of

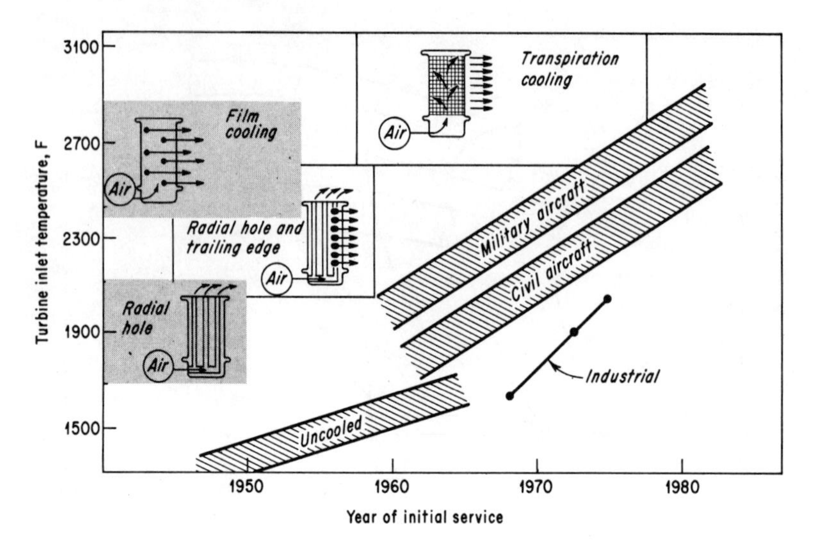

Figure 9.22
Historical trends of turbine inlet temperatures for military and civilian aircraft and industrial gas turbines. Reproduced by permission from *Power* (January 1974).

engines. For steady operation, 1600°F (900°C) seems about the maximum feasible inlet temperature for uncooled gas turbine engines. For continuous operation, present gas turbines are limited to inlet temperatures of approximately 1850°F (1015°C). According to A. J. Giramonti, maximum turbine blade temperatures have advanced 20°F (11°C) per year because of improvements in materials (28). In recent years, however, the use of turbine cooling has permitted temperature advances as high as 70 to 80°F (35 to 40°C) per year. Aircraft already operate at turbine inlet temperatures as high as 2200°F (1230°C) during cruise. Giramonti predicts that industrial gas turbines will operate at 2200°F (1230°C) in the near future, at 2600°F (1430°C) in the 1980s, and at 3000°F (1650°C) or more by the 1990s. Figure 9.22 suggests that the upward trend may easily continue over the next 15 years, and that industrial turbine inlet temperatures could almost catch up to aircraft engines.

Yet, as Figures 9.21 and 9.22 suggest, it will be a long time before the industrial gas turbine could be expected to compete in cycle efficiency with the large steam plant (cycle efficiency 40 to 41 percent, electricity generation efficiency 39 percent). For this reason the major development effort required to produce very large gas turbines of ultrahigh turbine inlet temperature is not under way. Significant progress is being made with cooling techniques on machines of up to 100 MW, so that in a few years' time turbine inlet temperatures of 2200 to 2400°F (1200 to 1300°C) will be feasible. However, the day when the gas turbine can compete in efficiency with the steam plant for base load is not now foreseen. This would require much more progress on high-temperature cooling to achieve long blade life while using heavy fuels.

In summary, the gas turbine is already playing a valuable role in peak load power generation. The open-cycle gas turbine has relatively low installation costs and should have a great potential for development. It should eventually be scaled to larger sizes and should show substantial gains in overall efficiency with future application of advanced blade-cooling techniques now being used for aircraft gas turbines. However, it appears to have serious disadvantages in competing with steam for base load power generation; its overall efficiency is significantly less unless an elaborate cycle is used, and this may destroy its capital cost advantage. Typical power and cycle efficiency values at the present time are 60 to 70 MW and 30 percent, though gains in both power and efficiency are likely. For example, the target heat rate for the 100-MWe FT50 engine now being developed by United Technologies Corp. is 10,000 Btu/kWh. In a combined gas and steam cycle plant the overall cycle efficiency may approach 45 percent with present technology.

The gas turbine would be well suited to operation with clean, low-energy synthetic gas, but since this fuel is very expensive, the low efficiency of the gas turbine would be a serious disavantage. For base load power generation with fossil fuels, the great hope of the gas turbine lies not in displacing the steam turbine, but in joining it.

9.5 BINARY CYCLE PLANTS

In large steam plants great care is taken to use as much as possible of the energy in the exhaust gases from the boiler. Some is used to heat the air before it enters this boiler; some is used in what is called the economizer to heat feedwater before it enters the boiler. The proportion is determined by minimizing the total cost of the boiler and superheater tubes, and of the economizer. Almost all of the stack gas energy above the acid dew-point temperature can be extracted at reasonable cost.

In a gas turbine, depending on the turbine inlet temperature and pressure ratio, the turbine exhaust temperature could be in the range 800 to 1200°F (400 to 600°C). Depending on the cycle pressure ratio, part of this energy may be used to heat air from the compressor before it enters the combustor. The required heat exchanger is usually expensive, and the exhaust energy that corresponds to temperatures less than the compressor exit temperature must still be wasted.

Alternatively, the gas turbine exhaust energy can be used to make steam. With steam temperatures limited to 1050°F (766°C), as mentioned earlier, the exhaust heat alone may be sufficient to supply a

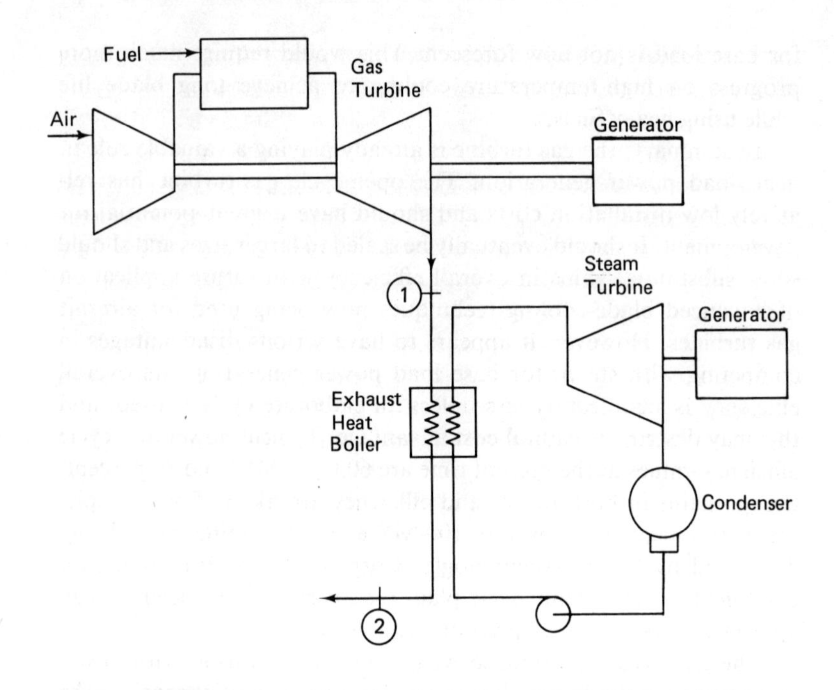

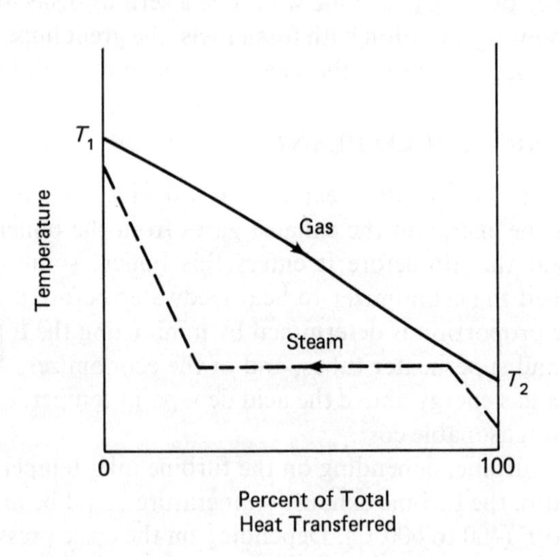

Figure 9.23
Gas–steam binary plant.

high-performance steam plant. Also, since the exhaust from gas turbines is oxygen-rich, supplementary firing could be used to increase the steam temperature to the desired level. Such supplementary firing is less efficient than simply using exhaust heat, and should be unnecessary with future turbine inlet and outlet temperatures. Since the addition of the steam cycle need have little effect on gas turbine power output, it is clear that adding a "bottoming" steam cycle can substantially augment the output and overall efficiency of a gas turbine power plant.

There appear to be practical limits to the pressure ratio which the gas turbine can develop; even for large gas turbines, the last-stage compressor blades are undesirably small for pressure ratios greater than 25 or 30. With increasing turbine inlet temperature (as more effective means of blade cooling are developed) and a fixed pressure ratio, the turbine exhaust temperature rises. For the gas turbine plant this means that a recuperator can be increasingly helpful in raising plant efficiency. For the gas–steam binary cycle it means that steam temperatures can be raised to the point where steam or gas corrosion of the boiler tube steel is significant. The potential for future improvement in performance of gas–steam cycles appears to lie in the possibility of a substantial rise in the gas turbine inlet temperature.

9.5.1 The Gas–Steam Cycle

A gas–steam combined cycle is shown in Figure 9.23. Here the conventional gas turbine operates at relatively high temperature, discharging its exhaust into a steam generator. The energy in the exhaust is used for feedwater heating, boiling, and superheating. In the scheme shown in Figure 9.23 the exhaust energy cannot be used in the most effective way. The larger the average temperature difference between exhaust gas and steam, the less work can be produced in the steam cycle. Further, the steam pressure may need to be kept low in order for the minimum difference between gas temperature and feedwater temperature to be positive. This pressure limitation also restricts the power output of the steam cycle. One way to avoid these difficulties is to use two steam circuits in parallel. The high-pressure loop uses the exhaust gas (while it is in the high-temperature range) for feedwater heating, evaporation, and superheating, as shown in Figure 9.24. When the exhaust gases are cooled, they are used for feedwater heating and evaporation for a low-pressure steam supply. Figure 9.25 shows the high- and low-pressure steam being fed to different parts of the same turbine and exhausting to a common condenser. This reduces the average difference between exhaust gas and steam temperature (Figure 9.24 vs. Figure 9.23) and allows much more flexibility in selecting pressures and temperatures

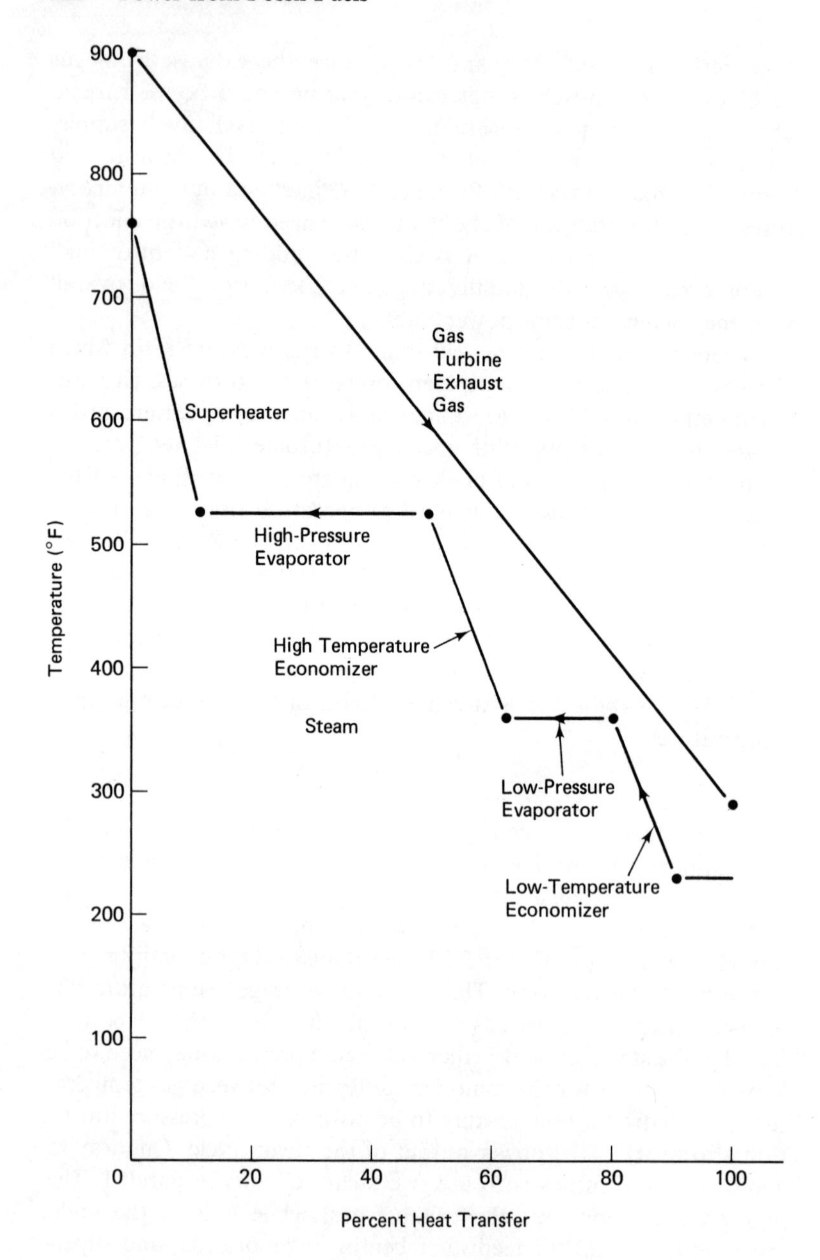

Figure 9.24
Steam generation with high- and low-pressure loops of a gas–steam binary cycle.

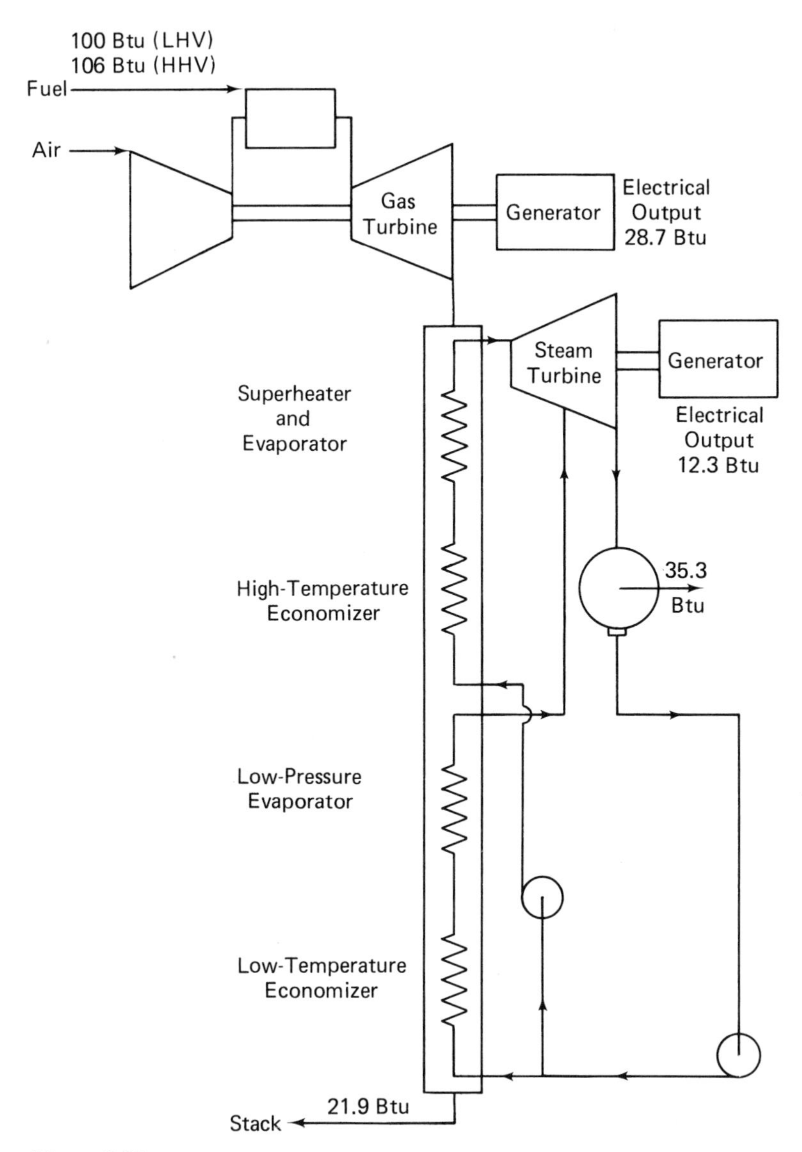

Figure 9.25
Energy flows in a gas–steam binary cycle. Overall efficiency is 41 percent, based on the lower heating value.

for the steam cycle. It leads to significantly higher overall plant efficiency.

Figure 9.25 also shows typical energy flows for an operating example of a combined cycle plant that uses gas turbine exhaust energy (without supplementary firing) to generate steam. The maximum steam temperature shown in Figure 9.24 is moderate. More than two-thirds of the power comes from the gas turbines, but the supplement from the steam turbine raises plant efficiency to 41 percent (based on the higher heating value of the fuel) or 38.7 percent (based on the lower heating value). This is equivalent to an overall heat rate of 8820 Btu/kWh, which is comparable with the best heat rates for conventional steam–electric plants (see Table 9.4). An operating 125-kW plant of this type is described in Ref. (29).

Adding the steam cycle to the simple gas turbine can raise overall cycle efficiency by 10 or 12 percentage points. In the example above the maximum temperatures for the steam and gas turbines were moderate. If the steam inlet temperature were raised to 1050°F (565°C) and the gas turbine inlet temperature were raised to 2300°F (1371°C) (as suggested by Figure 9.22 for the 1980s), combined gas–steam cycle plants should be able to reach overall efficiencies near 50 percent, or best rates as low as 6900 Btu/kWh.

The efficiency potential of the gas–steam cycle will not be a compelling advantage if gas turbine fuels are too expensive or if supply is unreliable. The distillate fuels typically used for gas turbines are unlikely to be available to supply a huge increase in combined cycle plant capacity, especially if used for base load. Attempts to fire gas turbines with coal have met with discouraging difficulties in blade deposits, corrosion, and erosion. Then, too, there are the problems of sulfur and NO_x emissions. As pointed out earlier, little low-sulfur coal and oil is available, and sulfur removal from stack gases is proving difficult and costly. It is possible that the best means of removing sulfur from coal and oil may be to first gasify these fuels and then remove the sulfur by a scrubbing process prior to combustion. The cleaning of gasified fuel, as opposed to exhaust gases, has the great advantage that the volume of gas to be cleaned is much less. If gasification is done at high pressure, the volume of gas to be cleaned could be smaller still.

9.5.2 The Gas–Steam Cycle with Fuel Gasification

Taken together, these factors suggest that the main hope for widespread use of a highly efficient gas–steam cycle lies in integration with a coal gasification plant. This would provide an abundant supply of clean gaseous fuel to the turbine, enabling durable operation with low emissions. With clean fuel the turbine could have a lengthy operating life even at relatively high temperatures. Gasification would mean that a wide range of coal and oil fuels would be available for power production. Crude oils can be burned directly in gas turbines, but only with the use of a fuel treatment system to remove or inhibit alkaline metals that corrode the hot surfaces of the turbine.

With crude oil gasification, lower-cost, high-sulfur crudes can be used. The alkaline metals can be almost completely removed. The fuel sulfur is converted to hydrogen sulfide, which is more easily removed from the fuel than is SO_2 from the exhaust gases. Also, the product low-Btu gas, having a stoichiometric temperature considerably less than oil or coal, is a negligible source of NO_x.

The capital cost of the gasification plant (perhaps $200/kW) is one obstacle to implementation, and the energy required for gasification is another. Production of power gas of low energy content (around 150 Btu/scf) means conversion of much of the fuel to CO and H_2, with conversion of a considerable part of the chemical energy of the fuel into sensible energy. Use of this sensible heat for power generation requires close integration of gasification and power generation; they should be on the same site.

Figure 9.26 shows a simplified view of the steam–gas binary cycle integrated with fuel gasification. Heat for steam generation is obtained from the hot exhaust gases of the simple open-cycle gas turbine. Waste heat from the fuel gasification process is used to heat

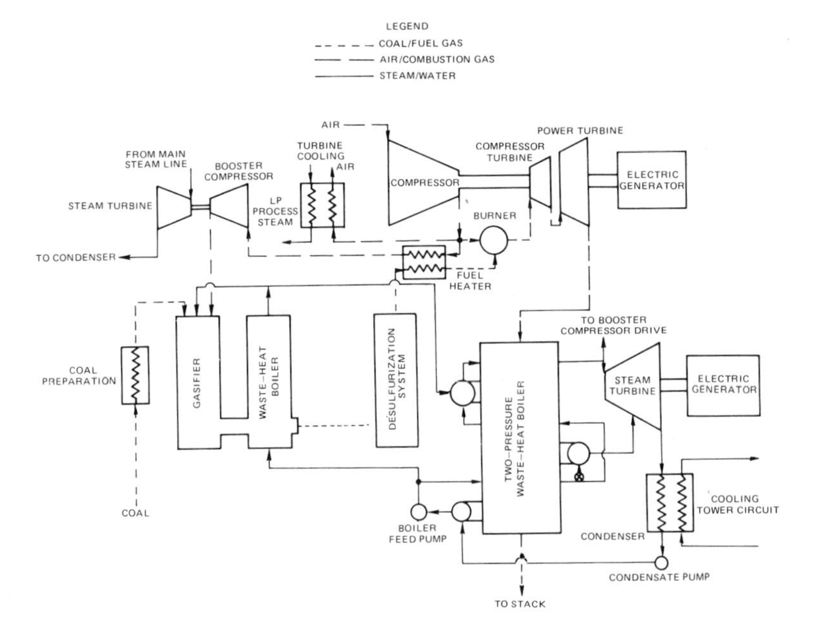

Figure 9.26
Integrated coal gasification: the COGAS Power Station. Reproduced courtesy of United Technologies Corp., United Technologies Research Center.

the boiler feedwater and also the steam used in the gasification process.

Figure 9.26 indicates the flows into each of the three main subsystems. For the gas turbine, part of the airflow at the compressor exit goes to the burner and part to the gasifier via a booster compressor. A small fraction is taken from the compressor to cool the gas turbine, after some of its energy has been used to heat process steam.

The second subsystem, the gasifier, receives a flow of coal, air, and possibly steam (not shown) to increase the hydrogen content of the fuel. The coal feed is converted into products that include carbon monoxide, hydrogen, methane, and water vapor, During subsequent scrubbing and desulfurization, the sulfur (in the form of H_2S and COS) is largely removed, as well as most of the CO_2. The ratio of CO to H_2 in the gaseous fuel depends on fuel composition and the ratios of air and steam to fuel, as well as on pressures and temperatures of operation.

The third subsystem shown in Figure 9.26 is the steam turbine system. If the gas turbine inlet temperatures are high enough, the exhaust gases can heat the steam to its limiting temperature. Follow-

Table 9.10
Parameters for Residual Oil Gasification with Combined Gas–Steam Power Generation

	Level of Technology	
	Mid-1970s	Early 1980s
Hot Gas Efficiency (percent)	74	76
Gas Turbine Power (MW)	132	96
Turbine Inlet Temperature (°F)	2200	2600
Compressor Pressure Ratio	16	20
Steam Turbine Power (MW)	110	136
Steam Turbine Inlet		
Temperator (°F)	870	980
Pressure (psia)	1250	1500
Net Station Power (MW)	228	309
Net Station Efficiency[a] (percent)	36	40
Net Station Heat Rate (Btu/kWh)	9480	8532
Estimated Capital Costs[b] ($/kWe)		
Fuel Processing (96% S removal)	69.7	54.4
Gas Turbines	41.7	38.8
Steam System	79.8	71.5
Miscellaneous Equipment	39.5	32.0
Total Capital Costs	230.7	196.7

[a]Using the higher heating value of the fuel.
[b]Based on mid-1970s dollar value, not including escalation or interest during construction.
Source: Ref. (28).

ing expansion in the steam turbine and condensation, part of the condensate goes to the waste heat boiler in the gasifier and part directly to the gas turbine exhaust boiler for feed heating, boiling, and superheating.

Table 9.10 shows systems described by Giramonti for gasification of high-sulfur residual fuel oil prior to combustion in a combined gas–steam cycle (28). Two levels of turbine operating conditions are shown, with corresponding heat rates. Also indicated are estimated direct capital costs, which show the gasification (fuel processing) system to be the most expensive component.

Table 9.11 shows corresponding performance parameters for coal gasification integrated in a gas–steam cycle plant. The comparison of performance for the three systems is based on current gas turbine performance. J. Agosta et al. estimate that increasing gas turbine inlet temperature could decrease plant heat rate to 9050 Btu/kWh (30). Utilization of coal would require equipment for crushing and dissolving the fuel. This would mean somewhat higher capital costs than suggested by Table 9.10. Compared with a conventional coal-fired plant, the gas–steam plant with coal gasification would have the advantages of low emissions, due to sulfur removal following gasification, and low temperature combustion (low NO_x formation rate). Also, with most of the power coming from gas turbines, its starting time could be relatively rapid. Further, plant construction times could be reduced since so much of the plant could be prefabricated and built in modular units. Gas–steam binary plants of 260 MW, fueled by natural gas, are already operational (31). The next step would be inclusion of gasification equipment so that a wide range of fossil fuels could be used.

Table 9.11
Performance of Combined Cycle Plant with Coal Gasification

Performance Factor	Conventional Steam Plant	Steam Plant with Gasification	Steam and Gas Turbine Plant with Gasification
Steam Pressure (psia)	2415	2415	1265
Steam Temperature (°F) throttle/reheat	1000/1000	1000/1000	900
Gas Turbine Power (MW)			400
Gross Station Power(MW)	508	526.4	541
Net Station Power (MW)	462	492.7	534.2
Heat Rate (Btu/kWh)	9700	12,480	10,720

Source: Ref. (30).

9.5.3 Other Binary Cycles

As has been shown, a steam "bottoming" cycle can substantially augment the output and the efficiency of a gas turbine power plant. But even in the ideal case it cannot utilize all of the theoretically available work in the gas turbine exhaust gases. This is because the enthalpy–temperature relation for constant-pressure heating of the H_2O stream is quite different than the enthalpy–temperature curve for constant-pressure cooling of the exhaust gases.

Figure 9.24 illustrates that even with the use of high- and low-pressure evaporators and economizers, substantial differences must exist between the temperatures of the hot gases and of the steam. In principle, other fluids could be found that would be superior to steam in this respect.

For low-temperature "bottoming" cycles, certain organic fluids do indeed appear to be better than steam (32, 33). However, few, if any, of these are suitable for long-term operation at temperatures above 750°F (400°C) owing to chemical deterioration of the fluid at elevated temperatures; they would be ideally applicable only to the recuperated gas turbine cycle. In this case the turbine exhaust gases would be cooled by heating the compressor exit airflow before being used as a heat source for the organic cycle. With this arrangement, mixed organic bottoming cycles have thermodynamic advantages. Much effort in recent years has been devoted to selection and identification of such fluids, recognizing that thermodynamic properties are not the sole criterion of suitability. Other very important characteristics are chemical stability (at high temperature), compatibility with the construction materials, safety, and cost. Dowtherm, one of the organic fluids capable of operation at highest temperatures, is limited to less than 700°F (370°C) because of thermal decomposition. Fluorinal 85 (85 mole percent trifluoroethanol CF_3CH_2OH and 15 percent water) is useable with peak temperatures up to 600°F (320°C). For the same peak temperature, the organic cycle could produce more work than the steam system (33). However, the top temperature limitation is serious; the organic cycle could only be compatible with an advanced gas turbine and the use of a large recuperator whose capital and maintenance costs would be substantial.

The mercury–steam binary cycle has been used successfully in the past in competition with earlier generations of steam plants. The cost, toxicity, and scarcity of mercury as a working fluid, the complexity of the binary Rankine plant, and subsequent improvements in steam plants generally have now made that option relatively unattractive.

Between 1917 and 1948 seven mercury vapor cycle plants were built in the United States. Experience showed that at temperatures above 900°F (480°C) mercury tends to take iron–chrome–nickel alloys into solution and then forms deposits on low-temperature surfaces. At this temperature limit, the mercury–steam binary cycle is not as efficient as the steam cycle operating up to 1000°F (538°C).

Another possibility which is currently being advocated is the potassium–steam binary cycle (Figure 9.27). Potassium top temperatures of 1540°F (838°C) are postulated, with constant-temperature heat rejection to steam at 1000°F (538°C) and overall plant efficiencies of 50 percent or so. The vapor pressure of potassium is relatively low, only around 2 atm at 1540°F; at around 1000°F potassium would have the pressures and densities common to many steam turbines operating at around 240°F (115°C). Early experimental work on development of potassium Rankine cycle power generation has been done for proposed space vehicles.

In tests conducted by NASA, small-scale potassium boilers, turbines, condensers, and pumps operated quite successfully. Whether large potassium–steam power plants can be built safely and economically is open to question. The availability, cost, toxicity, and flammability of potassium as a working fluid are problems to be considered.

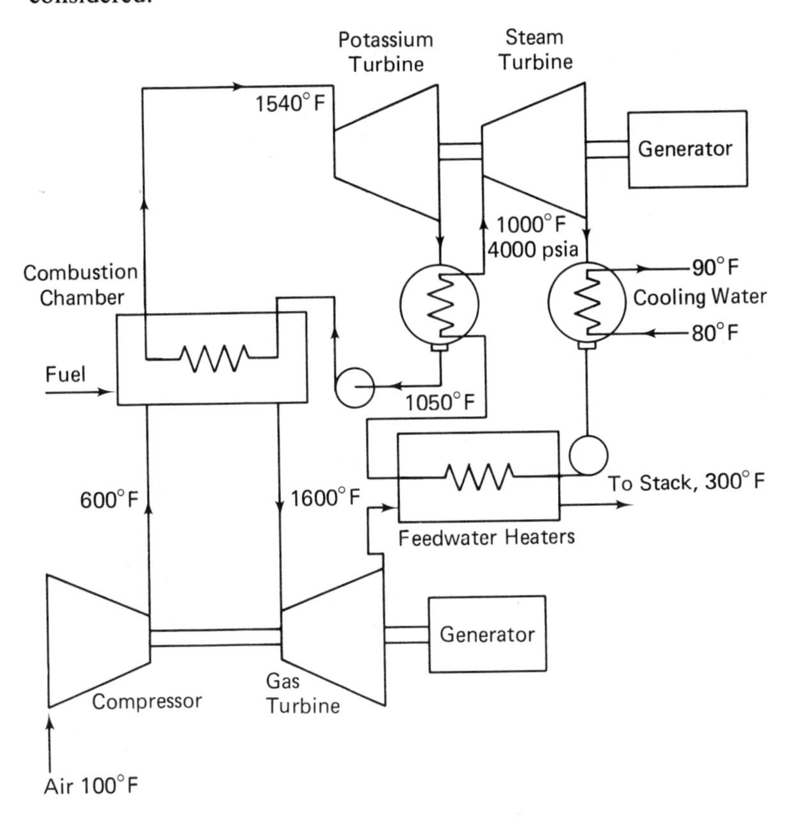

Figure 9.27
Potassium–steam binary cycle with combustion chamber supercharged by a gas turbine. After Ref. (34).

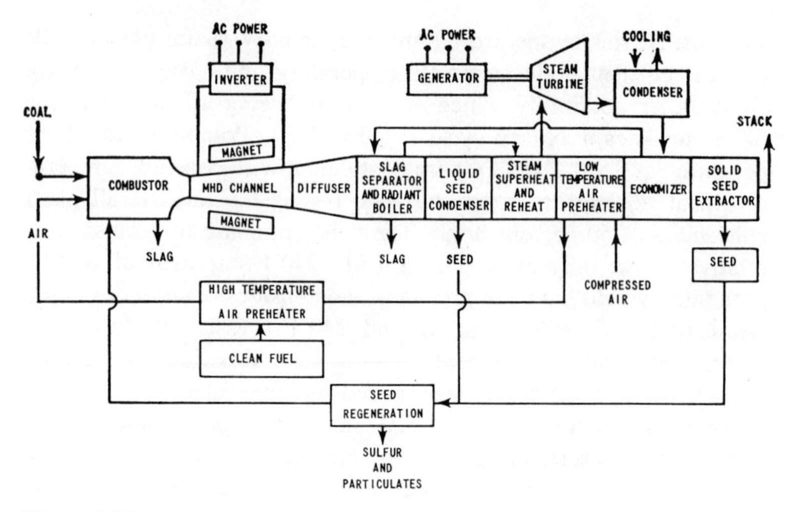

Figure 9.28
Coal-fired magnetohydrodynamic power generation with steam turbine cycle. From Ref. (40), courtesy of the American Power Conference.

Cesium is another potentially useful fluid for the high-temperature part of a steam binary cycle. Fraas (35) reports that the lowest practical condenser temperature for potassium is around 1020°F (550°C), whereas for cesium it is about 840°F (450°C).

Still another topping cycle that has been proposed is the magnetohydrodynamic cycle (MHD) (36–40). The MHD generator converts the kinetic energy of an electrically conducting gas stream directly into electrical energy with the aid of a magnetic field and electrodes placed in the walls of the channel. Figure 9.28 shows a schematic of a coal-fired MHD channel.

In principle, the MHD generator may operate at extremely high top temperatures since it is not subject to the temperature limitations associated with rotating parts in turbomachinery; the stationary walls are much easier to cool than are the blades of a turbine. However, it is extremely difficult to make the generator efficient. With fossil fuel combustion products it is in general necessary to add "seed" (an alkali metal such as potassium) to the stream in order to make the fluid conductive to electric current. It is difficult to control the current distribution in the fluid, especially in the wall region, where overheating or short-circuiting can occur. Open-cycle MHD generators raise problems with reclamation of seeding material and emissions of ash and other pollutants from fossil fuel combustion. With coal combustion as the MHD energy source, molten ash in the combustion products aids in recovery of the alkali metal seed.

Calculations of combined MHD–steam plant performance show overall efficiency of 50 percent if the MHD generator extracts 20 percent of the enthalpy from the fluid passing through it and has an isentropic efficiency exceeding 70 percent. Best performance so far appears to be about 8 percent enthalpy extraction, with isentropic efficiency of 45 percent, operating at a power level of 30 MW for 1 min, in a unit built by the Avco-Everett Research Lab (40). In principle, the efficiency of an MHD generator is very sensitive to channel size. The limitation on enthalpy extraction is due to the limiting axial electrostatic field in the generator, above which electric arcs form on the electrode wall, damaging the electrode and insulator surfaces. Meeting acceptable performance criteria will require close control of electrode and wall breakdown, as well as minimization of heat transfer and viscous losses in the channel and electrode losses or other voltage drops. For acceptable conductivity with coal, combustion temperatures must be 3600°F (2000°C) or above, even with alkali metal seeding. This necessitates a preheater to raise the fluid temperatures to perhaps 2160°F (1200°C) before combustion. Many problems remain to be solved before the fossil fuel MHD generator reaches the point of engineering feasibility.

The largest MHD power generator built to date is the U-25 plant in the USSR (36), for which the design power level is 25 MW. The fuel is natural gas; the oxidizer is air enriched by up to 40 percent with oxygen from an oxygen plant. The air is preheated to about 2160°F (1200°C) before entering the combustion chamber with the fuel and potassium carbonate (K_2CO_3) seed material. The generator channel has segmented electrodes that transmit power to a multielement DC–AC inverter. Products of combustion flow through the channel to a heat recovery steam generator supplying a closed-cycle steam turbine. The products then enter a hydromechanical cleaning system which is said to recover 99 percent of the seed material. A fan then sends the products to an exhaust stack. The emphasis in the USSR program is on developing a generator (to 5000-hr life), a superconducting magnet system, and a coal-fired combustion system.

In the United States the AVCO Mark VI MHD generator has been designed for 500 kW of power. The fuel is light fuel oil and the oxidizer is oxygen-enriched air. Simulated coal firing has been done with coal of 10 percent ash content injected with the fuel oil. The ash injection causes a thin, glassy layer to be formed on the channel walls; at power levels up to 400 kW this provided a good protective coating for the electrode, giving hope that one of the difficulties in extending the life of MHD generator channels—damage to electrodes by arcing—may be on the way to solution. Channel life heretofore has been in the range 10 to 30 hr.

Burning coal, even with maximum preheating and oxygen-enriched air for maximum combustion temperature, still requires substantial injection of seed material to provide satisfactory electrical conductivity of the working fluid. A typical seed rate is 1 g-mole

K_2CO_3/kg coal; this seed must be recovered if the cost of operation is to be reasonable. Tests conducted by the U.S. Bureau of Mines (37) show that 99.5 percent of the seed can be recovered by various techniques, including mechanical cyclones, venturi scrubbers, baghouse filters, and electrostatic precipitators. The particles appear to be in the range 1 to 5 μ.

The injected potassium reacts quickly with the SO_2 in the combustion gases to form K_2SO_4 particles, which are readily captured. In this way up to 99.8 percent of the SO_2 was removed from the combustion gases in the Bureau of Mines experiments, leaving only 5 ppm of SO_2 in the exhaust gases. Estimates of the economics of seed recovery and sulfur recovery for these very high temperature processes indicate capital costs of only about $10/kW, or about one-fifth the cost of sulfur removal from stack gases in conventional plants.

The NO_x problem is apparently less severe than one might expect. Although the combustion gas has very high equilibrium concentrations of NO_x, experiments show that with two-stage combustion, NO_x levels can be reduced well below the EPA requirements (Table 9.7). Although maximum temperature is high, since the rate of temperature drop in the channel is sufficiently low that the gases are nearly in equilibrium as they cool, the high-temperature concentrations of NO_x are not "frozen" as in the rapid cooling of automotive combustion. The staging of combustion is beneficial. With one stage of combustion, measured concentration of NO_x in the exhaust was 4086 ppm. With two stages of combustion (95 percent stoichiometric air-fuel ratio in the first stage, and secondary air admitted at 2000°F), the exhaust NO_x was reduced to 150 ppm, or 0.12 lb $NO_x/10^6$ Btu input. This is well below the EPA level of 0.7 lb $NO_x/10^6$ Btu.

It is too early to project capital costs for MHD plants; it is not expected that they will cost much more per kilowatt than conventional fossil fuel steam plants. The pace of MHD research and development is currently slow. A. E. Sheindlin and W. D. Jackson report that work on the major MHD projects begun on or shortly after 1960 in Japan, Poland, the United States, the USSR, and several Western European countries (Federal Republic of Germany, France, and the United Kingdom) "has been discontinued, or reduced to a modest level as a result of national energy policies regarding the use of fossil fuels for electric power generation" (36).

The efficiency of power plants using various cycles has been estimated by A. P. Fraas (see Figure 9.29) making allowances for losses in turbines, compressors, and heat exchangers. Conventional steam plants are limited (by maximum temperature) to perhaps 40 percent overall efficiency, but higher efficiencies are within reach for other cycles.

The cycles that appear to have the greatest chance of practical

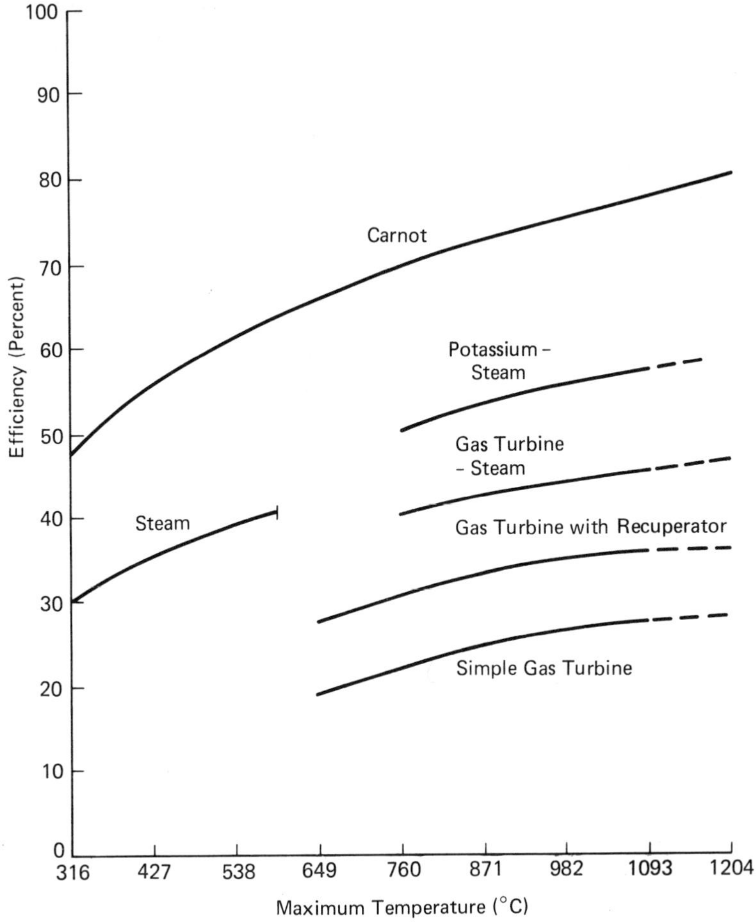

Figure 9.29
Cycle efficiencies. After Ref. (33).

success in raising steam plant efficiency above 50 percent are the gas–steam binary cycle and the potassium–steam cycle. Of these, the first has already been proven successful at plant efficiencies of 40 percent. Raising the efficiency 10 points will mean, however, raising the turbine inlet temperature from 1650–1750°F (900–950°C) to possibly as high as 3000°F (1650°C), a very formidable challenge. The potassium–steam cycle would require maximum temperatures around 1500°F (820°C), but poses major problems of safety and quality control, and no large-scale plant experience is available.

9.6 FUEL CELLS

A fuel cell is a steady-flow chemical reactor in which the reaction energy is transferred largely to a flow of electrons instead of to the thermal energy of the products. Since the fuel cell is not a heat engine, its efficiency is not limited by its maximum operating temperature. The maximum possible efficiency (for the hydrogen–oxygen fuel cell) is about 93 percent at room temperature. With natural gas, a fuel cell has a theoretical maximum efficiency close to 100 percent. The practical maximum efficiency will be much less than this figure (perhaps 40 to 50 percent) owing to energy losses in fuel processing and within the fuel cell, but still may exceed the efficiency of a heat engine using the same constituents and operating between practical upper and lower temperature limits.

The principal disadvantage of the fuel cell is that (in contrast to combustion) its chemical reactions must take place at solid surfaces, where they are limited by relatively slow processes of diffusion. Thus power densities of fuel cells are limited by the physics of diffusion, the availability of durable catalysts, and the closeness with which porous electrode surfaces can be packed within the volume of the cell.

An important advantage of the fuel cell is that it is efficient in small sizes and at part load, so that power generation within the home is an attractive possibility. This would permit energy supply for electricity via underground gas pipelines instead of relatively expensive overhead transmission lines. Other advantages are low noise and minimal pollutant emission. Another advantage is that, unlike other prime movers, its efficiency at part load is even higher than at full load. The actual efficiency depends greatly on the fuel used.

Although the idea has been long established, the first substantial demonstration model of a fuel cell, a 6-kW hydrogen–oxygen cell, was built by F. T. Bacon in 1959. Figure 9.30 shows the Bacon cell schematically. It contains two porous electrodes separated by an electrolyte of potassium hydroxide or sodium hydroxide. At the left-hand side of the cell, hydrogen moves through the porous anode to the interface, where it interacts with hydroxyl ions to form H_2O and release electrons. Passing to the right-hand side of the cell through an electrical load, the electrons interact with oxygen diffusing through the positive electrode and with water molecules to form the hydroxyl ions. Suitable catalysts must be embedded in both electrodes to enable the reactions to proceed at reasonable rates. Also, the porosity of the electrodes and their proximity are key factors in determining power density. Losses are due to resistance to current flow within the electrodes and the loss in energy necessary for ions to diffuse at finite rates toward the electrode surfaces. Both of these losses vary with current density, and the efficiency of the cell may thus be appreciably higher at part load than at full load.

The Bacon cell was developed by Pratt and Whitney Aircraft for use in the Apollo spacecraft. These units developed overall efficiencies of 72.5 percent; operating on hydrogen and oxygen, they

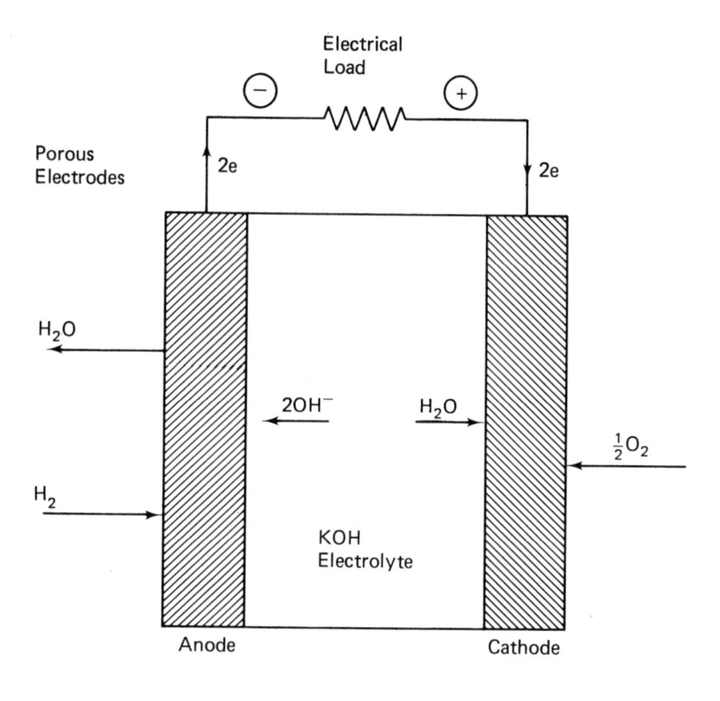

Figure 9.30
Hydrogen–oxygen fuel cell.

produced drinkable water as a useful by-product. They developed a maximum power of 3 kW and about 800 kWh of electrical energy. The total weight of fuel and cell was much less than the minimum total weight of an equivalent prime mover driven by hydrogen–oxygen combustion.

The Apollo fuel cell experience is not directly applicable to terrestrial power generation since these cells were specially developed for extreme lightness, with expensive fuels, high-cost structural materials and catalysts, and a limited life. Since the cell size was small, the particular difficulties of flow and temperature control in very large cells were not encountered.

Since abundant supplies of pure hydrogen fuel appear unlikely for quite some time, a more practical possibility is the use of synthesis gas (CO and H_2) in fuel cells. This can be made from coal or oil and would be directly acceptable to the fuel cell if sufficiently clean. Best efficiency with synthesis gas is expected to be lower (at 42 percent) than with pure hydrogen, but expected capital costs are still very modest (Figure 9.31).

Methyl alcohol is an example of a fuel that requires a separate fuel processor to produce a hydrogen-rich mixture for the fuel cell. This is expected to add about 25 percent to the system capital cost.

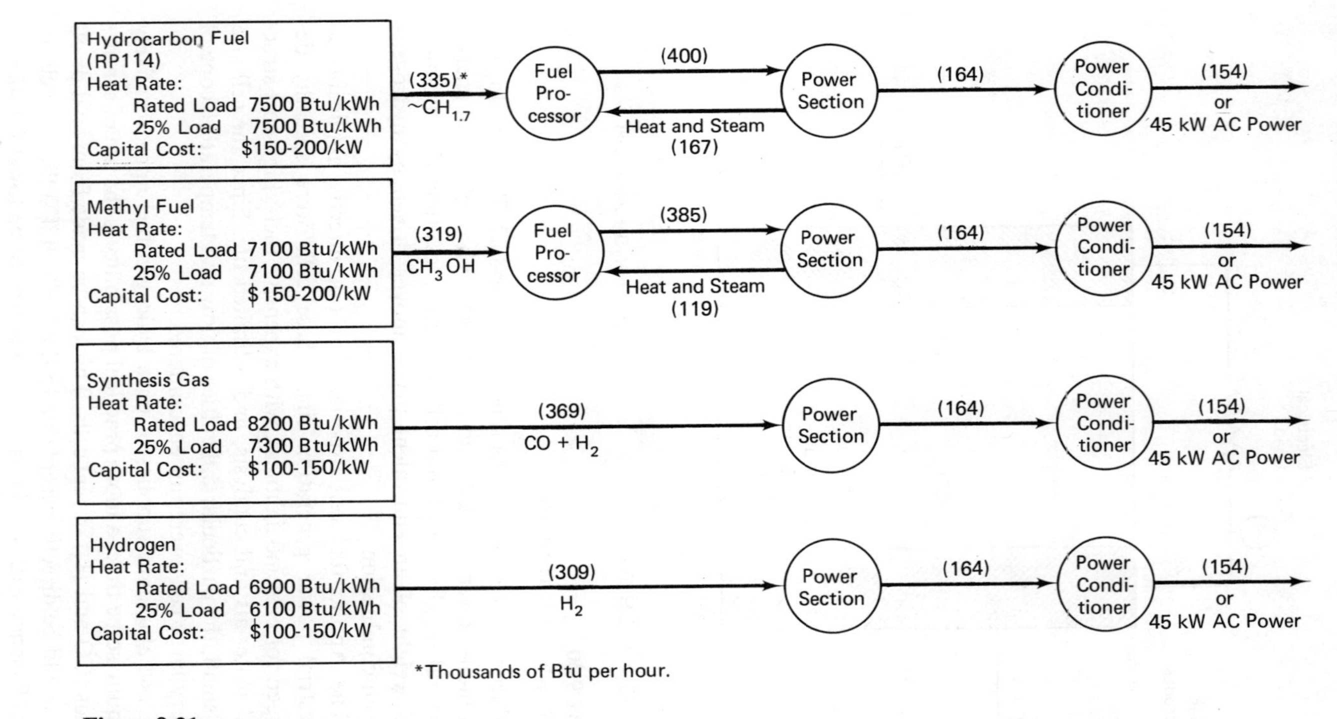

Figure 9.31
Future possibilities for the fuel cell. From Ref. (41), courtesy of the Electric Power Research Institute.

The RP114 system in Figure 9.31 refers to a fuel cell under development by United Technologies Corp. (42). It uses a molten carbonate electrolyte and operates at a cell temperature of 1200°F. The electrodes are porous nickel. At the anode the reaction is

$$H_2 + CO_3^{=} \rightarrow H_2O + CO_2 + 2e^-.$$

The cathode reaction is

$$\tfrac{1}{2}O_2 + CO_2 + 2e^- \rightarrow CO_3^{=},$$

and so the overall reaction is

$$H_2 + \tfrac{1}{2}O_2 \rightarrow H_2O.$$

An advantage claimed for this system is reduced cost, because noble metal catalysts are not required. It is hoped that by the 1980s this system will have been developed with a cell life of 40,000 hr (41) and capital costs under $250/kWe (42) for use with a variety of hydrocarbon fuels. Development work continues on methods of desulfurization and high-temperature fuel reformation to produce the hydrogen-rich fuel composition required by the cell. The heat rate of 7500 Btu/kWh shown in Figure 9.31 is equivalent to an efficiency of 46 percent.

Table 9.12 shows the best power densities obtained to date in various fuel cells. It should be remembered that these figures pertain to relatively small units and may not be appropriate to extremely large arrays of fuel cells. In general, these power densities are low relative to the volume of burner, turbomachinery, and condenser in a conventional steam plant. Yet, taking into account the vast building typically used, the overall steam plant power density is hardly more than 3 kW/m^3. Thus it is not clear that central power stations using fuel cells need be larger than conventional steam plants, though they may be much more tightly packed.

For future electricity generation by fuel cells, the main interest will likely be in fuels derived from coal. Synthetic natural gas, power gas, methanol, and other liquid fuels are all possibilities. A basic

Table 9.12
Fuel Cell Power Densities: Best Performances to Date

		Power Density		
	Temperature (°C)	(W/cm^2)	(W/kg)	(kW/m^3)
H_2–Air[a]	250	1	100	20
Hydrocarbon–Air[a]	200	0.1	25	10
Coal–Air[a]	1000	1	200	20
Hydrazine–Air[b]	100	0.1	120	40

[a]With high catalyst loading.
[b]This fuel has carcinogenic properties.
Source: Ref. (43).

choice must be made between dispersed power generation and a central power station in which the fuel cells and coal gasifier are closely integrated for maximum overall efficiency.

9.7 SUMMARY

Conventional steam plants for coal, oil, and gas fuels have been developed to high efficiency and reliability. The shortage of natural gas and low-sulfur oil and coal means that much work will be required to adapt conventional plants to increasingly severe emissions standards.

Control of SO_2 emissions has proven difficult and costly. In the absence of complete information on the effects of sulfates and ultrafine particulates on health, it is not clear that tall stacks provide adequate dispersion of pollutants. Flue gas sulfur removal is very expensive and has so far proven unreliable in service. Fluidized-bed combustion could be an adequate low-cost solution to the SO_2 problem, but will require major development of combustion systems. Coal and oil gasification is costly but permits clean burning of a wide range of fuels. The overall efficiency with gasification is reasonable if the power plants are closely integrated for heat recovery.

Restrictions on NO_x emissions will necessitate further modifications, which should be feasible at reasonable cost. Both fluidized-bed combustion and the use of low-Btu gas are associated with low NO_x emissions.

Drastically increased fuel costs make power plant efficiency more important than ever. Conventional steam plants have been developed to the point of diminishing returns in efficiency, but combined gas and steam cycles are presently competitive in cost and efficiency and show considerable potential for future development. An increase of best plant efficiency from 40 to 50 percent over a decade or so is a reasonable possibility. The combined gas–steam plant integrated with coal or oil gasification may be the best answer for clean, efficient utilization of fossil fuels over the next two or three decades. Several other binary cycles have been suggested for raising efficiency, but these are not of proven feasibility and would require major development. Fuel cells also have the potential advantage of high efficiency, but much work will be needed to show that fuel cells can operate economically and durably with available fuels.

References

1. John N. Nassikas (Chairman, U.S. Federal Power Commission). Address to the 57th Annual Convention of the National Coal Association, Washington, D.C., June 18, 1974.

2. United Nations, Department of Economic and Social Affairs. *Small Scale Power Generation*, New York, 1967.

3. *Steam: Its Generation and Use*, 38th edition. New York: Babcock and Wilcox, 1972.

4. *Fuels for the Electric Utility Industry 1971–1985,* A Report of the National Economic Research Associates, Inc., to the Edison Electric Institute, 1972. See also *54th Semi-Annual Electric Power Survey*, A Report of the Electric Power Survey Committee of the Edison Electric Institute, October 1973.

5. *CEGB Statistical Yearbook 1972.* London: Central Electricity Generating Board, 1972.

6. "How efficient are power stations?" *Engineering*, November 15, 1968.

7. A. R. Cooper. "Steam in central power stations: CEGB practice." *Proceedings of the Institution of Mechanical Engineers Conference on Steam Plant Engineering: Present Status and Future Trends*, May 2–4, 1963.

8. Federal Power Commission. *The 1970 National Power Survey*, 4 vols. Washington, D.C., December 1971.

9. Federal Power Commission. *Steam–Electric Plant Construction Cost and Annual Production Expenses, Twenty-Fifth Annual Supplement, 1972.* Washington, D.C., April 1974.

10. National Coal Association. *Steam-Electric Factors.* Washington, D.C.

11. L. R. Glicksman. "Thermal discharge from power plants." Paper 72-WA/Ener-2 presented at the Annual Winter Meeting of the American Society of Mechanical Engineers, New York, New York, November 26–30, 1972.

12. A. K. Krasin et al. "Nuclear power stations with gas-cooled fast reactors using dissociating nitrogen tetroxide as coolant. Paper A/Conf 49/P/431, *Proceedings of the Fourth United Nations International Conference on the Peaceful Uses of Atomic Energy*, Geneva, September 6–16, 1971.

13. *Federal Register*, vol. 36 (December 23, 1971), p. 24875.

14. J. T. Shaw. "Progress review no. 64: A commentary on the formation, incidence, measurement, and control of nitrogen oxides in flue gas." *Journal of the Institute of Fuel*, vol. 46 (April 1973).

15. Hitoshi Ishida. "Present state of air pollution and its regulation in Japan." Paper 2.1–4, *Transactions of the 9th World Energy Conference*, Detroit, Michigan, September 22–27, 1974.

16. A. J. Clarke, D. H. Lucas, and F. F. Ross. "Tall stacks: How effective are they?" Paper presented at the Second International Clean Air Conference, Washington, D.C., December 1970.

17. D. H. Lucas. "The effect of emission height with a multiplicity of pollution sources in very large areas." Paper 2.1–2, *Transactions of the 9th World Energy Conference*, Detroit, Michigan, September 22–27, 1974.

18. James T. Dunham. "High sulfur coal for generating electricity." *Science*, vol. 184 (April 1974), pp. 346–351.

19. *Final Report of the Sulfur Oxide Central Technology Assessment Panel on Projected Utilization of Stack Gas Cleaning Systems by Steam–Electric Plants.* Environmental Protection Agency Report No. APTD-1569, 1973.

20. "SO_2 removal systems." *Power*, September 1974, pp. S-2–S-24.

21. D. H. Archer, D. Berg, and E. V. Somers. "Fluidized-bed gasification

and combusion for power generation." Paper 4.1–17, *Transactions of the 9th World Energy Conference*, Detroit, Michigan, September 22–27, 1974.

22. R. B. Schieffer and D. A. Sullivan. "Low-Btu fuels for gas turbines." Paper 71-GT-21 presented at the American Society of Mechanical Engineers Gas Turbine Conference, Zurich, March 30–April 4, 1974.

23. W. B. Crouch, W. G. Schlinger, R. D. Klapatch, and G. E. Vitti. "Recent experimental results on gasification and combustion of low Btu gas for gas turbines." Paper 74-GT-11 presented at the American Society of Mechanical Engineers Gas Turbine Conference, Zurich, March 30–April 4, 1974.

24. Arthur M. Squires. "Clean fuels from coal gasification." *Science*, vol. 184 (April 1974), pp. 300–315.

25. C. A. Trilling. "Coal gasification: Atomic International's Rockgas process." Paper 74-WA/Pwr-11 presented at the Annual Winter Meeting of the American Society of Mechanical Engineers, New York, New York, November 17–22, 1974.

26. H. Hoch and H. N. Sharon. "Low emission power systems." Paper 4.1–14, *Transactions of the 9th World Energy Conference*, Detroit, Michigan, September 22–27, 1974.

27. Barry R. Korb and Joseph Kivel. "Air pollution control under the Clean Air Act and its energy implications." Paper 2.1–9, *Transactions of the 9th World Energy Conference*, Detroit, Michigan, September 22–27, 1974.

28. A. J. Giramonti. "Advanced COGAS power systems for low pollution emissions." Paper presented at the American Chemical Society Symposium on Novel Combined Power Cycles, Dallas, Texas, April 1973. See also "Advanced COGAS systems for low Btu power generation." Paper presented at the Connecticut Clean Power Symposium, West Hartford, Connecticut, May 13, 1972.

29. W. J. O'Donnell and P. J. Schwalie. "A 125-MW unfired combined cycle electric generating unit from concept to operation." *Combustion*, April 1974, pp. 23–27.

30. J. Agosta, H. F. Ilian, R. N. Lundberg, O. G. Tranby, D. J. Ahner, and R. C. Sheldon. "The future of low-Btu gas in power generation." *Proceedings of the American Power Conference*, vol. 35 (1973), p. 510.

31. P. A. Berman. "Construction and operation of a PACE combined cycle power plant." Paper 74-GT-109 presented at the American Society of Mechanical Engineers Gas Turbine Conference, Zurich, March 30–April 4, 1974.

32. G. Angelino and V. Moroni. "Perspectives for waste heat recovery by means of organic fluid cycles." *Journal of Engineering for Power*, April 1973, pp. 75–83.

33. D. T. Morgan and J. P. Davis. "High efficiency gas turbine/organic Rankine cycle combined power plant." Paper 74-GT-35 presented at the American Society of Mechanical Engineers Gas Turbine Conference, Zurich, March 30–April 4, 1974.

34. A. P. Fraas. "A potassium–steam binary vapor cycle for better fuel economy and reduced thermal pollution." *Journal of Engineering for Power*, January 1973, pp. 55–63.

35. A. P. Fraas. "Topping and bottoming cycles." Paper 4.1–12, *Transac-*

tions of the 9th World Energy Conference, Detroit, Michigan, September 22–27, 1974.

36. A. E. Sheindlin and W. D. Jackson. "MHD electrical power generation: An international status report," Paper 4.1–13, *Transactions of the 9th World Energy Conference*, Detroit, Michigan, September 22–27, 1974.

37. K. Rosa, S. Pelly, G. Enos, K. Kessler, and J. Kleplis. "Recent MHD generator testing at AVCO-Everett Research Laboratory, Inc." Paper presented at the Annual Winter Meeting of the American Society of Mechanical Engineers, New York, New York, November 17–22, 1974.

38. A. E. Sheindlin, editor-in-chief. *MHD Electrical Power Generation.* 1972 Status Report, Joint ENEA/IAEA International Liaison Group on MHD Electrical Power Generation. *Atomic Energy Review*, vol. 10 (1972).

39. D. Bienstock, P. O. Bergman, J. M. Henry, R. J. Demiski, J. J. Demeter, and K. D. Plants. "Magnetohydrodynamics: Low air pollution power generation." Paper 73-WA/Ener-3 presented at the Annual Winter Meeting of the American Society of Mechanical Engineers, Detroit, Michigan, November 11–15, 1973.

40. W. D. Jackson and P. S. Zygelbaum. "Open Cycle MHD Power Generator." *Proceedings of the American Power Conference*, vol. 37 (1975).

41. Arnold P. Fickett. "An electric utility fuel cell: Dream or reality?" *Proceedings of the American Power Conference*, vol. 37 (1975).

42. J. M. King, Jr. *Advanced Fuel Cell Technology for Fuel Cell Application*: South Windsor, Conn.: United Technologies Corp., Power Systems Division, August 1975.

43. Eugene G. Kovack, ed. *Technology of Efficient Energy Utilization.* Report of a NATO Science Committee Conference held at Les Arcs, France, October 8–12, 1973.

Supplementary References

Leonard G. Austin. "Fuel cells." *Scientific American*, October 1959.

"Increased gas turbine efficiency." *Power*, January 1974.

E. S. Miliaras. *Power Plants with Air-Cooled Condensing System.* Cambridge, Mass.: MIT Press, 1974.

A. I. S. Tantram. "Fuel cells past and present." *Energy Policy*, vol. 1, no. 4 (March 1974), pp. 55–56.

Problems

9.1 Using the relative capital costs shown in Figure 9.2 for gas turbine and steam plants, and total capital and operation and maintenance costs of 17 percent per year, find the annual capacity factor for which steam and gas turbine generation costs are equal. Assume that plant efficiencies are 29 and 39 percent for the gas turbine and steam plant, respectively. Consider the cost of distillate fuel for the gas turbine as $\$2.50/10^6$ Btu and the cost of coal for the steam plant as $\$1/10^6$ Btu.

9.2 Compare the efficiencies of simple open- and closed-cycle steam plants to show the effect of the condenser (Figure 9.A) on the efficiency of the steam cycle. The heat transfer q per unit mass of steam passing through the boiler is

$$q = h_2 - h_1,$$

where h_2 is the enthalpy of the steam leaving the boiler and h_1 is the enthalpy of steam entering the boiler. The work done by the steam in the turbine is given by

$$w_t = h_2 - h_3,$$

where h_3 is the enthalpy of the steam leaving the turbine. The work absorbed in the pumps is small, and may be taken to be 15 Btu/lb.

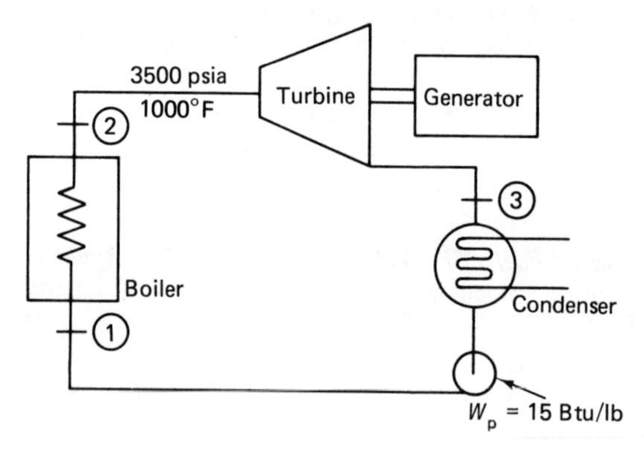

Figure 9.A
Steam cycle with condenser.

One function of the condenser is to lower the turbine exhaust pressure so that more work can be done by the turbine. (Another important function is to return the highly purified H_2O to the boiler and turbine, rather than treating a large new supply of feedwater for each pass through the turbine.) The steam property data in Table 9.A are associated with high efficiency of expansion in the turbine.

Table 9.A
High-Efficiency Parameters for Boiler Steam

Station	Pressure (psia)	Temperature (°F)	Enthalpy (Btu/lb)
1	3510		73.0
2	3500	1000	1420.9
3 (with condenser)	0.95	212	866.3
3 (with condenser removed)	14.7	100	994.7

Calculate the efficiency of power generation, allowing in each case for the pump work and for an electrical generator efficiency of 97 percent.

9.3 Determine the effect of one reheat on the efficiency of a simplified steam plant, as shown in Figure 9.B.

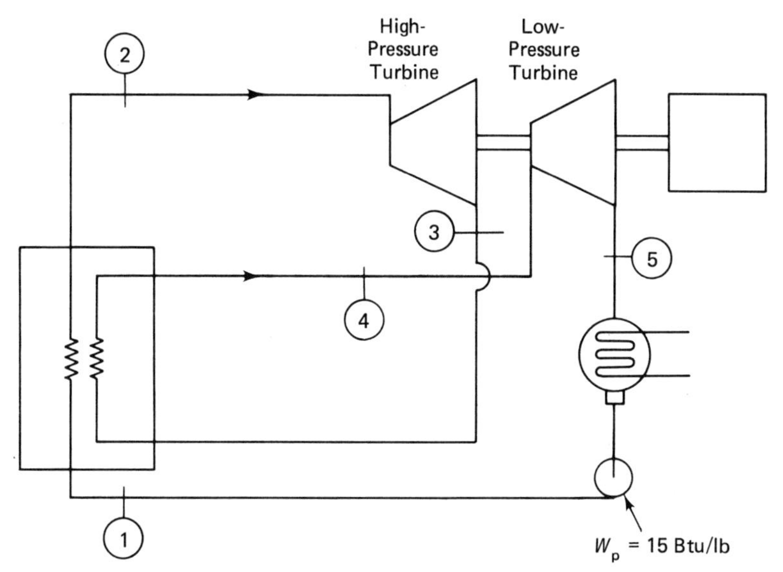

Figure 9.B
Simplified steam plant with reheat between stages.

Using the data in Table 9.B, determine the efficiency of the cycle, with and without reheat. Also compare the total work per pound in the two cases. As for Problem 9.2, calculate heat transfer and work

Table 9.B
Properties of Steam Plants with and without Reheat

Station	Pressure (psia)	Temperature (°F)	Enthalpy (Btu/lb)
With Reheat			
1	3510		73.0
2	3500	1000	1420.9
3	542		1233.1
4	540		1519.6
5	0.95	100	1007.3
With No Reheat			
1	3510		73.0
2	3500	1000	1420.9
5	0.95	100	866.3

from enthalpy differences. Assume that combustion and electricity generation efficiencies are 90 and 97 percent, respectively.

9.4 Show that a single feedwater heater using steam extracted from a turbine can raise the efficiency of a simple steam cycle. Let the ratio of flows at stations 2 and 3 in Figure 9.C be m in one case and zero in another. In the feedwater heater the bled steam mixes freely with the feedwater, and the mixture is then sent to the boiler. The cycle efficiency is

$$\eta = 1 - \frac{Q_{\text{out}}}{Q_{\text{in}}} = 1 - \frac{Q_{\text{out}}}{Q_{\text{out}} + W_{\text{out}}}.$$

(Show that this may be written

$$\eta = 1 - \frac{h_3 - h_4}{(h_1 - h_4) + m\,(h_1 - h_2)}$$

if the pump work is neglected.)

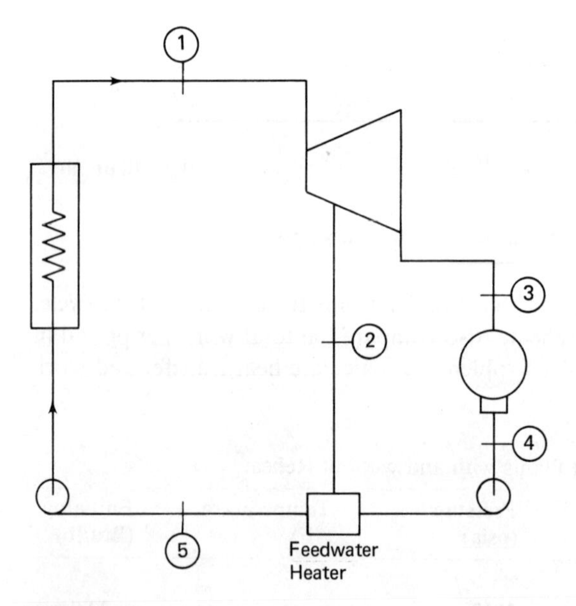

Figure 9.C
Steam cycle with feedwater heater.

9.5 Show how the efficiency of a simple gas turbine increases with turbine inlet temperature. In the past two decades, design improvements (turbine blade cooling and higher-temperature turbine blade materials) have permitted maximum temperature to increase from 1500 to 2000°F. Future development may allow peak temperatures of 2500°F or more.

Consider the simple open-cycle gas turbine shown in Figure 9.D.

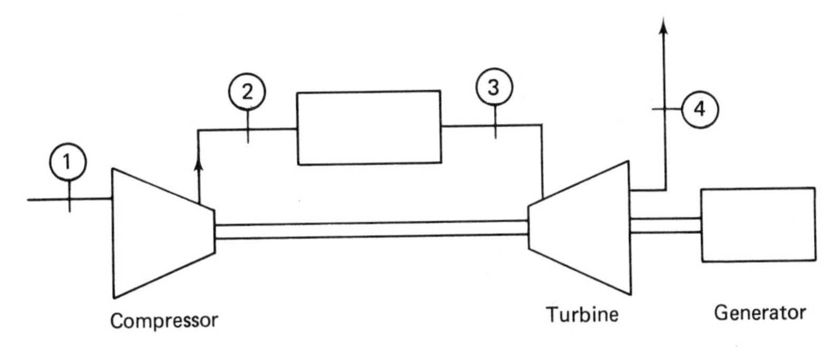

Figure 9.D
Simple open-cycle gas turbine.

It operates with temperatures and pressures as indicated in Table 9.C. Estimate the electricity generation efficiency using the following simplifying assumptions:

1. Mass flow rate through turbine is approximately equal to mass flow rate through compressor.
2. Compressor work per unit mass is $C_{pc}(T_2 - T_1)$, where $C_{pc} = 0.24$ Btu/lb-°R.
3. Turbine work per unit mass is $C_{pt}(T_3 - T_4)$, where $C_{pt} = 0.27$ Btu/lb-°R.
4. Fuel energy input is $C_{pb}(T_3 - T_2)$, where $C_{pb} = 0.27$ Btu/lb-°R.

Next, examine the effect of increasing the turbine inlet temperature to 2040°F. The compressor pressure ratio is unchanged, and so T_2/T_1 and T_3/T_4 (absolute temperature ratios) do not change. In both cases the electrical generator efficiency is $\eta_g = 0.97$.

Table 9.C
Gas Turbine Pressures and Temperatures

Station	Temperature (°F)	Pressure (atm)
1	70	1
2	663	10.1
3	1540	10
4	780	1

9.6 Shown in Figure 9.E is a simplified sketch of a gas–steam binary cycle, for operation with a clean gaseous fuel. Generator 1 produces 500 MWe.

Given the conditions in Table 9.D, estimate the mass flow rate of gas turbine exhaust, the steam flow rate, the power output of generator 2, and the overall plant efficiency. For the gas turbine, assume that:

1. Mass flow rates through compressor and turbine are identical.

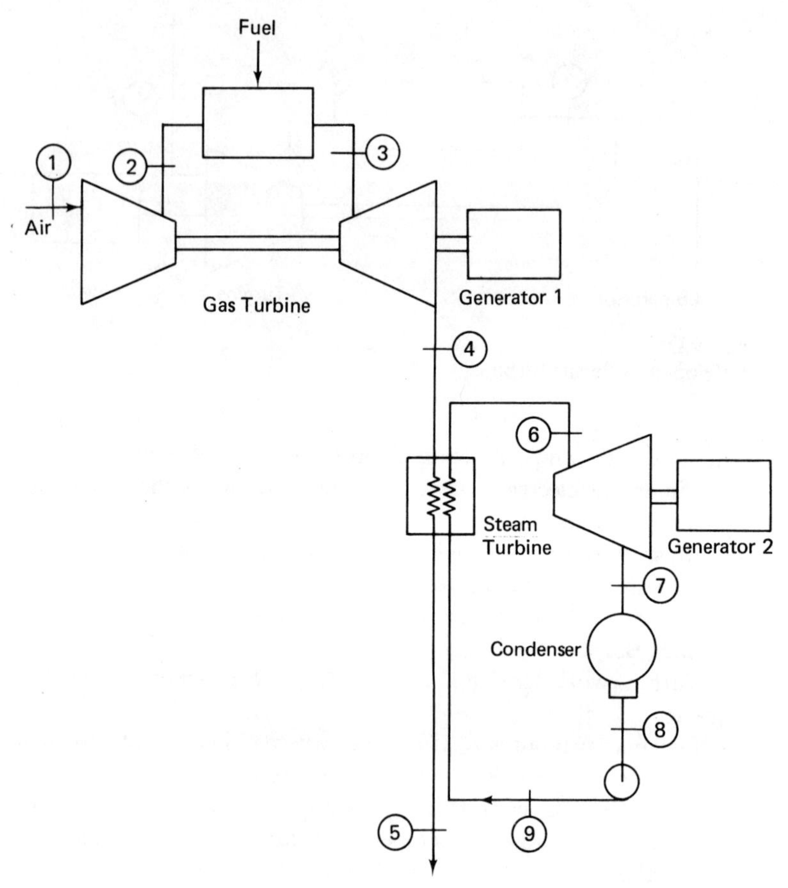

Figure 9.E
Simplified gas–steam binary cycle.

Table 9.D
Properties of a Gas–Steam Binary Cycle

Station	Pressure (atm)	Temperature (°F)	Enthalpy (Btu/lb)
1	1	70	
2	10	620	
3	9.8	1900	
4	1.03	900	
5	1	300	
6	20.4	840	1442
7	0.0647	100	1015
8	0.0647	100	68.05
9	20		69.3

2. The specific heat of the compressor flow is 0.24 Btu/lb-°R.
3. The specific heat of the turbine and heat exchanger gas is 0.27 Btu/lb-°R.
4. Combustion efficiency is 0.97.
Assume that the electrical generator efficiency is 97 percent.

9.7 The following data are provided by Fraas (34) for the potassium–steam–gas turbine cycle shown in Figure 9.21.

Air flow rate:	3.24×10^6 lb/hr
Total heat input to combustion chamber:	37.4×10^8 Btu/hr
Heat input to potassium:	27.4×10^8 Btu/hr
Compressor	
inlet temperature:	100°F
outlet temperature:	600°F
Gas turbine	
inlet temperature:	1600°F
outlet temperature:	875°F
Steam flow rate:	2.28×10^6 lb/hr
Heat input to steam	
from potassium:	2.34×10^9 Btu/hr
feedwater:	1.09×10^9 Btu/hr
Heat rejected to cooling water:	2.06×10^9 Btu/hr.

Determine the power output of the gas, potassium, and steam turbines and the overall plant efficiency. For the gas turbine make the same assumptions as in Problem 9.6. Again, assume an electrical generator efficiency of 97 percent, and neglect pumping power for the steam and potassium cycles.

9.8 Calculation of fossil fuel plant efficiency depends significantly on whether "higher" or "lower" heating values are used, that is, on whether the H_2O in the combustion products is considered to be in the liquid or the gaseous state.

The composition of three fuels is shown in Table 9.E.

Suppose each is burned in the correct proportion of air and that the products are CO_2, H_2O, N_2, and SO_2 (this constituent absent for gas fuel). Assume that the coal ash is 100 percent removed during and following combustion.

Determine in each case:

a. The number of moles of H_2O formed per lb or scf of fuel burned.

b. The difference between higher heating value (HHV) and lower heating value (LHV) in Btu/lb-mole of fuel (at 77°F the latent heat of H_2O is 1050 Btu/lb).

c. Given the following HHVs for coal, gas, and oil fuels, determine the corresponding LHVs:

Coal: 25×10^6 Btu/ton
Gas: 1030 Btu/scf
Bunker C Oil: 18,000 Btu/lb.

Table 9.E
Composition of Coal, Gas, and Bunker C Oil

Coal (% by weight)		Gas (% by volume)		Bunker C Oil (% by weight)	
C	72.8	CH_4	90	C	87.9
H_2	4.8	C_2H_6	5	H_2	10.3
O_2	6.2	N_2	5	S	1.2
N_2	1.5			O_2	0.5
S	2.2			N_2	0.1
H_2O	3.5				
Ash	9.0				

9.9 Consider the performance of the MHD–steam power cycle shown schematically in Figure 9.22. Estimate overall efficiency under the following conditions:

Combustion Gases

Combustion temperature:	2300°K
Combustion pressure:	20 atm
Average specific heat:	1.18 kJ/kg-°K
Temperature at exit of MHD unit:	2040°K
Temperature at exit of channel:	420°K.

Steam Cycle

Maximum pressure:	163 atm
Maximum temperature:	565°C
Electricity generation efficiency:	39 percent.

Estimate heat input from the enthalpy difference of combustion between combustion and atmospheric temperatures. Assume no power loss in the inverter; that is, all of the enthalpy drop in the MHD unit is converted to electrical power output. Also assume no heat loss during heat exchange from combustion gases to steam.

9.10 It can be shown that the maximum work obtainable from the hydrogen–oxygen fuel cell shown in Figure 9.30 is equal to the difference between the free energy of the reactants and the free energy of the products. For the reactants shown, this free energy difference is 57.7 kcal/g-mole of H_2 at 25°C. The heat of reaction, or the heat transfer that would be required if reactants at 25°C were to burn and then be cooled to 25°C, is -68.3 kcal/g-mole H_2 (assuming the H_2O is in liquid form at 25°C).

a. Show that the maximum possible efficiency in this case is 85 percent (defining efficiency as maximum work divided by the negative of the heat of reaction).

b. Estimate the maximum cell voltage, recalling that the charge per gram-mole of electrons is 96,500 coulombs (1 faraday).

10
Solar Power

10.1 INTRODUCTION

As indicated in Chapter 1, solar energy is sufficiently abundant to meet world energy needs for a long time to come. Though fossil fuels can be considered a form of stored solar energy, we distinguish here between those depletable resources and other forms of solar energy that can be used continuously. The term "solar power" will be used in this discussion to refer to electricity generation by direct use of solar radiation, use of energy supplied to the atmosphere and oceans, or continuous use of photosynthesis energy supplied to organisms or vegetation.

Solar power could eliminate most of the serious environmental problems associated with fossil fuel and nuclear power. Chemical air pollution and dispersion of radioactive particles could be minimized; possible long-term chemical and thermal effects on the atmosphere due to power production would not be a serious worry; land damage due to mining of fuel would be avoided. Power costs could be stabilized, with free fuel in abundant supply.

Several ideas for solar–electric power have been developed, including

1. photovoltaic cells, developed for spacecraft
2. solar heat engines, which generally require concentration of

solar radiation to heat the working fluid to temperatures required for reasonable efficiency
3. wind turbogenerators
4. ocean heat engines, which utilize vertical temperature differentials in the ocean
5. heat engines that operate with photosynthetic fuel such as algae, trees, or other vegetation.

Working prototypes of all of these methods have been constructed over the years, and much is known about their technical limitations (1). The main uncertainties concern the possibilities for technical innovations (e.g., in solar cell materials) and for cost reduction through mass production (e.g., of solar collectors).

Solar energy has been used quite widely in the world for hot water heating (2, 3). The economics of solar home heating in the United States have been studied by R. A. Tybout and G. O. G. Löf and others (4, 5); their results suggest that if gas and oil prices continue their recent rapid ascent, solar heating will soon be competitive with gas and oil over a substantial part of North America. This conclusion rests on the availability of flat-plate glass-covered collectors that retain 80 percent of incident solar energy and cost in the range of \$2 to \$4/ft^2 (1973). The capital cost of this and a warm water storage tank would constitute most of the cost of the system. More recent estimates of the costs of high-volume production of flat-plate collectors fall in the range \$4.70 to \$5.80/ft^2 (6). These figures imply such high initial capital costs, as compared with those of other heating systems, that a long period of successful operation (25 to 30 years at least) would be needed to justify the initial investment.

The use of solar radiation for electrical power generation is not as close to feasibility as solar heating. The difficulty is not in the total supply of solar energy nor in the lack of energy conversion devices. A demonstration of solar power was provided by a solar-powered steam engine in Egypt as early as 1913. Designed by Schuman, it produced 55 hp, using 13,000 ft^2 of cylindrical trough collector with 2 to 3 percent overall efficiency. The problem is that solar energy is relatively diffuse and the cost of collection is high; the cost of concentration to temperatures appropriate to heat engines is much higher still. Beyond that comes the cost of the energy storage required because the incidence of solar radiation is so variable. Several methods of energy storage are technically feasible, but the cost of such storage systems can be considerable. A storage system would not be needed if the system were used merely to supplement the output from, say, a nuclear plant. The difficulty is that solar power capacity will most likely be unavailable at periods of peak demand; this would mean that just as much nuclear capacity would be required if peak demand were to be met. Since capital costs of nuclear plants are so high and fuel costs relatively low, the value of the solar plant in this case would be doubtful at best.

The intensity of solar radiation intercepted by the earth is 1.36 kW/m^2. Allowing for atmospheric absorption of part of this energy, the average at the earth surface with the sun overhead is around 1.0 kW/m^2. Allowing for day-to-day variations, the average intensity of solar radiation incident at ground level is about 0.17 kW/m^2 in the United States and 0.14 kW/m^2 in Canada. (In Canada annual average values range from 0.103 kW/m^2 at Aklavic, Northwest Territory, to 0.168 kW/m^2 in Lethbridge, Alberta (7). Sites in Arizona have up to 0.27 kW/m^2.)

The smallness of these quantities can be appreciated by thinking of the value of the electricity that could be generated annually from the energy impinging on one square meter of land area. Suppose (as upper limits) a plant availability of 90 percent, collector efficiency of 85 percent, conversion efficiency of 40 percent, and an electricity cost of 10 mills/kWh. The total value of the electrical energy generated per year would then be $0.17 \times 0.9 \times 0.85 \times 0.4 \times 8760 \times 0.01 = \$4.55/m^2/yr$. Table 10.1 shows optimistic and pessimistic estimates of justified investment in solar power if electricity is worth 20 mills/kWh: less than $57/m^2 ($5.30/ft^2) would be justified in the most optimistic case. This figure is less than the current cost of stationary low-temperature flat-plate solar collectors.

Case 1 in Table 10.1 indicates high conversion efficiency, 0.40. Solar cell efficiencies are considerably lower than this, not much above 0.10, and have been costing $100,000/kW capacity. For solar–

Table 10.1
Solar Power Economics: Two Cases

	Case 1	Case 2
Assumptions		
Radiation Density[a] (kW/m^2)	0.17	0.10
Plant Availability	0.90	0.80
Collector Efficiency	0.85	0.50
Conversion Efficiency	0.40	0.10
Value of Electricity (mills/kWh)	20	20
Capital Charge Rate (percent)	15	15
Operation and Maintenance Cost (mills/kWh)	1	3
Electrical Power Density[a] (kWe/m^2)	0.052	0.004
Annual Value of Electricity Produced ($/m^2)	9.00	0.70
Justifiable Investment		
$/m^2	57.00	4.00
$/kWe	1100	1000

[a]Annual average.

thermal plants, temperatures well above 300°C would be required. These plants would need relatively expensive focusing collectors and probably tracking mechanisms as well. Adding the cost of energy conversion and storage equipment, realistic capital cost would certainly exceed the justifiable investment shown in Table 10.1. If concentrating solar collectors (built to track the sun and to withstand wind loads) could be made for $\$100/m^2$ and have an average power density of $0.03\ kW/m^2$, the plant cost would be \$3330/kW, not taking into account the capital costs of energy transfer, conversion machinery, and storage, and the energy loss due to storage.

The land requirement for solar power is formidable. For an average power output of 1000 MW, the total collector area required would be in the range 20 to 100 km^2, or perhaps 20 to 100 times the land area required by fossil fuel or nuclear plants.

The approximate cost figures cited above suggest that solar power is far from competitive with fossil fuel and nuclear power. There is little doubt that it is technically feasible, but its costs may be an order of magnitude higher.

If world population rises far beyond its present level, and if the environment seriously deteriorates, there may come a time when the cost of solar power will seem reasonable. There is also the possibility of an ingenious technical innovation that will change the economic picture. It is with these thoughts in mind that we review the essential features of present-day ideas for large-scale solar power generation.

10.2 PHOTOVOLTAIC CELLS

Photovoltaic cells are made of semiconductors that generate electricity when they absorb light. As photons are received, free electrical charges are generated that can be collected on contacts applied to the surfaces of the semiconductors. Because solar cells are not heat engines, and therefore do not need to operate at high temperatures, they are adapted to the weak energy flux of solar radiation. Operating at room temperature, these devices have theoretical efficiencies of the order of 25 percent (8). Actual operating efficiencies are less than half this value and decrease fairly rapidly with increasing temperature. The development status of various photovoltaic cells is discussed in Ref. (9).

The best known application of photovoltaic cells for electrical power generation has been in spacecraft, for which the silicon solar cell is the most highly developed type. The silicon cell consists of a single crystal of silicon into which a doping material is diffused to form a semiconductor. Since the early days of solar cell development, many improvements have been made in crystal growing and doping, electrical contact and cell assembly, and production methods. Large

Table 10.2
Parameters of Silicon Solar Cells Used on Two Spacecraft

	Rangers 6 and 7	Mariner (Mars)
Maximum Power (W)	226	680
Number of Cells	9792	28,224
Area of Array (m^2)	2.3	6.5
Power Density (kW/m^2)	0.100	0.104
Efficiency (percent)	7.1	7.4

Source: Ref. (11).

numbers of cells have been manufactured with areas 2 × 2 cm, efficiencies approaching 10 percent, and operating at 28°C (10, 11).

Table 10.2 shows the performance of two silicon solar cell power systems for space vehicles. The efficiency is the power developed per unit area of array divided by the solar energy flux in free space (1.4 kW/m^2). The efficiency would have been greater than 8 percent if it referred to cell area rather than array area. A silicon solar cell responds best to light of wavelengths near 0.8μ and will have somewhat higher efficiency in sunlight at ground level. The efficiencies in Table 10.2 refer only to the array and not to the electrical power conditioning system. These cells are all mounted in flat arrays. Concentrators could be used to increase the power output of individual cells (to decrease the number of cells required for a specified spacecraft power level), but this would raise the problem of radiating heat from the cell to keep it at an adequately low operating temperature. For spacecraft, reflectors that concentrate the impinging radiation flux by factors of two to five could be a means of decreasing cost. Mounting cells in the bottom of a simple trough with slanted sides could concentrate sunlight by a factor of two.

For terrestrial applications, silicon solar cells have shown operating efficiences of about 12 to 15 percent; perhaps 17 percent is possible (6). Though silicon is one of the earth's most abundant materials, it is expensive to extract (from sand, where it occurs mostly in the form SiO_2) and refine to the purity required for solar cells. The great barrier to solar cell application lies in the costs of the cells themselves. Solar power for spacecraft has cost in the range $100,000 to $1 million/kW. The costs of these cells would need to be reduced by three orders of magnitude to make them competitive with nuclear power. Naturally, there is doubt on this point, though the possibility cannot be ruled out. W. R. Cherry and F. H. Morse believe that silicon cells could be adapted to terrestrial power systems for as little as $10,000/kW, making modest use of concentrators (12). Manufacturers claim that arrays of interconnected cells could be produced now for $20,000 to $30,000/kW, and that this could be reduced to $2000 by 1979 (6). Reducing the price of silicon cells is difficult because of the cost of making single crystals. One very promising method is being developed to produce continuous thin ribbons of single-crystal silicon to reduce fabrication costs. Cells made from this ribbon have so far shown efficiencies of around 8 percent.

Several other kinds of photocells are in the laboratory stage of development. Cadmium sulfide and CdS/Cu_2S cells are other possibilities. So far, efficiencies have been in the range 3 to 8 percent, and these cells have been less durable than silicon cells owing to degradation with exposure to oxygen, water vapor, and sunlight, especially at elevated temperatures. The active part of the CdS cell is a thin polycrystalline layer of CdS, about 10μ thick, on which a layer of a Cu_2S compound perhaps 0.1μ thick is grown. These cells can be made by deposition on long sheets of substrates, a process that might be adaptable to inexpensive mass production. Controllable low-cost fabrication of reliable, relatively efficient CdS cells has yet to be demonstrated.

Photovoltaic cells could be applicable to either small or large power plants, since they function well on a small scale, and may be adaptable to local energy generation on building rooftops. The costs of energy storage and power conditioning equipment might, however, make generation in large stations the most economical method.

10.3 THERMAL CONVERSION

Use of a heat engine for solar power requires high temperature for reasonable efficiency. Figure 10.1 shows efficiencies of various heat engines as a function of maximum temperature; it suggests that maximum efficiency falls off rapidly with decreasing temperature, in much the same way as Carnot efficiency. Figure 10.1 pertains to a variety of heat engines, mostly of small size. Large-scale plants may well be capable of developing to higher efficiency; a temperature of at least 250°C would likely be needed for an overall plant efficiency greater than 20 percent.

The temperature to which a working fluid can be raised by solar radiation depends upon the relative thermal absorptivity and emissivity of the collector, as well as on its geometry and orientation relative to the sun. Flat-plate collectors appear to be restricted to temperatures less than 100°C above ambient temperature (14), so that a concentrator is required to increase the intensity of the energy flux. This is most unfortunate. A concentrator generally

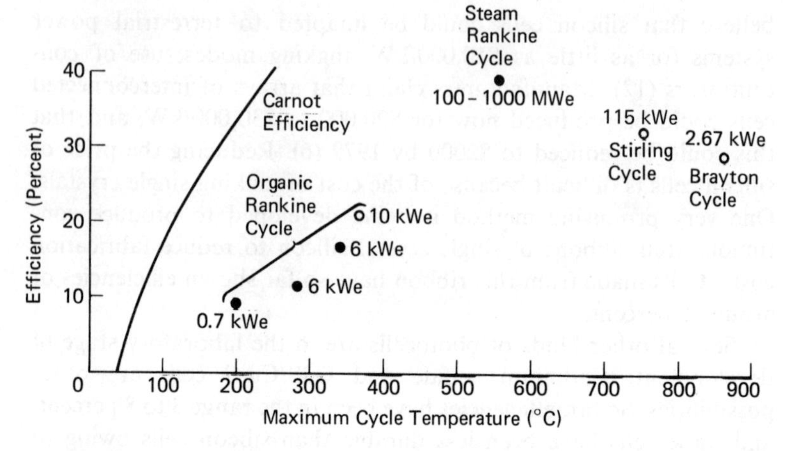

Figure 10.1
Heat engine efficiencies. Filled circles: built and tested. Open circles: calculation based on tests. Data from Ref. (13).

requires a mechanism for tracking the sun; this means that the structure must be nonrigid, creating a problem of providing sufficient structural strength (at low cost) to withstand wind loads. Then, too, the concentrator would be effective mainly in the "beam," or directed component of sunlight. A substantial fraction of solar radiation at the earth's surface is diffuse; this scattered component would not be utilized by the optics of the concentrator. Deterioration of the reflecting surfaces exposed to weather, oxidation, and air pollution could be another serious problem.

In general, thermal conversion would require the following elements:

1. radiation concentrator
2. energy receiver
3. heat transfer system
4. energy conversion equipment (e.g., steam turbogenerator)
5. energy storage.

One type of concentrator–receiver that has been proposed is shown in Figure 10.2. A heat transfer fluid flows in the receiver, whose axis will be oriented to track the sun throughout the day, automatically activated, perhaps by a bimetallic strip. Collector efficiencies of 40 to 60 percent have been calculated for this arrangement, with concentration ratios up to 30:1. For low concentrations, special coatings that selectively absorb solar radiation are necessary to develop high temperatures.

The mounting of a solar collector on a tall tower exposed to reflected radiation from a field of tilted flat-plate collectors at ground level appears to be a lower-cost concept (15). Some studies indicate that 50 MWe of electricity could be generated in this way, using an 80-story tower with 0.5 mile^2 (1.5 km^2) of reflectors. The flat reflect-

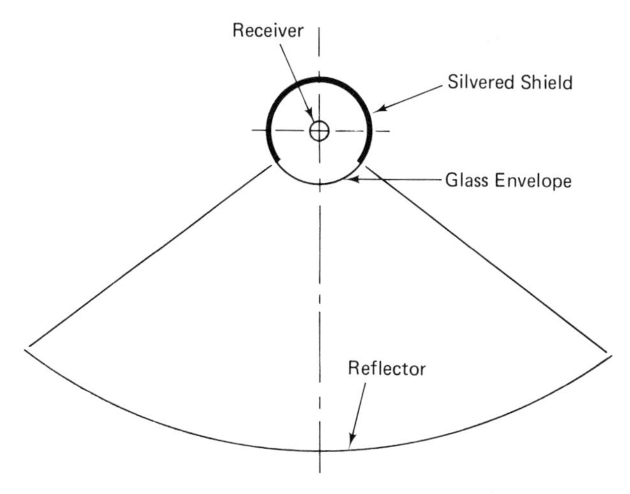

Figure 10.2
Parabolic reflector.

ing surfaces would track the sun during the day and create temperatures as high as 2000°F (1100°C) at the receiver. This could be used in the desert without cooling water in a relatively efficient gas turbine, if a suitable high-temperature material could be found for the receiver.

Another type of concentrator–collector is shown in Figure 10.3. This makes use of a Fresnel lens to concentrate the radiation within an evacuated glass tube. Concentric circular grooves in the lens are successively offset at such an angle that light passes at a different angle through each groove, converging on a point. Selective coatings are indicated, with the hope that working fluid temperatures can be raised to as high as 540°C for efficient steam generation. The selective coatings are thin films, perhaps 0.1μ, designed to absorb sunlight and discourage the radiation of energy in the infrared range, which is characteristic of the thermal energy in the solid material.

Either of these collector–receiver configurations would be expensive. Because the largest part of the plant cost would be associated with the collector, the economically optimum solar–thermal system would likely have a simpler collector, relatively low maximum temperature, and low overall plant efficiency.

Since solar energy is so diffuse, it is desirable in principle to simplify the energy conversion system as much as possible. Elimination of a working fluid, ducting, and moving parts are the potential advantages of thermionic and thermoelectric generators. The difficulty of obtaining high temperatures with low-cost collectors appears to rule out thermionic converters, which require temperatures around 1700°C; at that, the overall efficiency is perhaps only 11.5 percent (17). Thermoelectric converters will operate at much lower tempera-

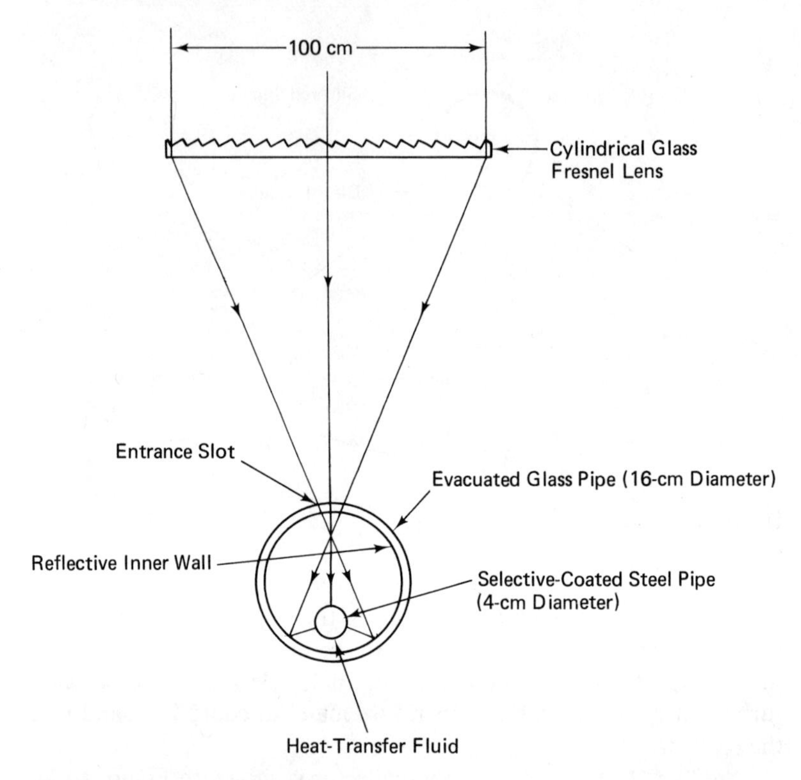

Figure 10.3
Collector proposed by the Meinels (16).

tures, but measured efficiencies are low, with 3.5 percent indicative of best experience (18).

Much thought has been given to Rankine and Brayton cycle plants for manned spacecraft power generation. Table 10.3 summarizes estimated size, cost, and efficiency for four competing systems studied for a 20-kW generation plant. These plants differ fundamentally from terrestrial plants in that, with heat rejection by radiation only, heat rejection temperatures were kept fairly high to minimize overall system mass. Reduction of the heat rejection temperature could improve the efficiency of the H_2O Rankine cycle to at least 20 percent. The efficiencies of the postulated Rankine cycle turbines are low in the small-power range. Further improvement in efficiency would undoubtedly be possible if it were feasible to scale up from 20 kW to many megawatts. However, it seems unlikely that capital cost per kilowatt could be scaled down by two orders of magnitude to compete with fossil fuel or nuclear plants.

It is of interest to consider Table 10.3 in relation to the proposal of P. Glaser that a 10,000-MW power plant could be located in stationary orbit, beaming power to the earth by microwaves (20). Glaser proposed power generation by large arrays of silicon cells.

Table 10.3
Four Spacecraft Systems Considered for 20-kW Power Generation

	Solar Cell	Mercury–Rankine	Water–Rankine	Brayton
Maximum Temperature (°F)	28	1050	582	1490
Efficiency (percent)	12	18.3	14.9	21.6
Cost ($/kWe)	1,000,000	20,000	22,000	25,000
Collector Area (ft²)	6720	1870	2280	1700
Radiator Area (ft²)		244	773	976
Turbine Efficiency		70	50	86
Minimum Temperature (°F)		535	300	75
Collector Concentration Factor	2			
Collector Orientation Accuracy (degrees)		0.25	0.25	0.25
System Mass (kg/kWe)	210	43	62	90

Source: Ref. (19).

Solar–thermal plants have also been considered (21). The cost of placing such a power plant in orbit is itself considerable since (even with very large power plants) the specific mass will likely be at least 5 kg/kWe. The cost of placing mass in orbit via a space shuttle vehicle has been estimated as $200 to $400/kg. Thus the cost of placing the plant in orbit could easily be $2000/kWe. Although the costs of the power plants quoted in Table 10.3 refer to small units, it is difficult to believe that thermal power plants, with their need for expensive collector and radiator surfaces, could be built for less than $2000/kWe. To this must be added the cost of microwave beam antennae in space and a receiver area on earth. Allowing for losses in microwave generation, transmission, and energy conversion, it appears doubtful that a capital cost of less than $10,000/kWe could be realized.

H. Oman and C. J. Bishop have studied the possibility of an 800-kW solar Brayton plant for terrestrial use (13). The plant would have 569 parabolic concentrators, each 13.2 m² in area. A major difficulty of this system is heat loss from the working fluid between collector and engine. The heat transfer problem appears to militate against large-scale power plants. Oman and Bishop consider that a 16.6-kW Stirling engine–generator could be driven with 30 percent efficiency and a maximum temperature of 650°C. For this, the overall capital cost would be about $2700/kW, about 85 percent of this being associated with the collector. Maintenance and other operational costs of the many small units would contribute substantially to overall generation cost.

Solar–thermal power conversion is technically possible. High-temperature concentration of the solar energy requires fairly accurate beam optics and, possibly, selective coatings, as well as elaborate insulation of the heat transfer medium. The design of low-cost tracking collectors of solar energy at the earth's surface is a relatively undeveloped art. Few experimental units have been built, so there may be room for significant technical innovation. Nevertheless, it seems likely that these collector–receivers will be the dominant contributor to the cost of the solar–thermal power plant and will place plant cost in the range \$3000 to \$10,000/kW (at \$200/m^2 and an annual average power level of 0.03 kWe/m^2 of array).

10.4 WIND POWER

Wind power is another diffuse form of solar energy that can be employed for power generation as well as other uses (22–24). In a wind of density ρ and velocity v perpendicular to an area A, there is a kinetic energy flux of $\rho A v^3/2$. If all of this energy were extracted by a wind turbine, the instantaneous power would be

$$P(\text{kW}) = 0.0006\ A v^3,$$

where A is the area in m^2 swept by the turbine blades, and v the air speed in m/sec. A nominal atmospheric density has been assumed.

Even a frictionless turbine could not extract all of this energy since the wind must flow through the turbine at finite speed. The ideal power extracted is the product of the mass flow and the kinetic energy reduction per unit mass, and it may be shown that only about 16/27, or 59 percent, of the wind kinetic energy per unit area is extractable, so the ideal power output is actually

$$P(\text{kW}) = 0.00035\ \eta A v^3,$$

where η is an efficiency in the range 0 to 1.

Owing to frictional effects on the blades and transmission and losses in the generator, the overall efficiency of electrical power generation from the wind turbine is typically in the range 0.5 to 0.8, with 0.7 being considered reasonable.

Since the power density P/A varies with the cube of the wind velocity, the level and steadiness of wind velocity are critically important. Wind velocities are highly dependent on site, even in relatively favorable regions such as the Orkney Islands (Table 10.4).

In North America, average wind intensity is generally much higher in coastal regions than inland. In only a few areas on the eastern and western coasts does average wind velocity exceed 20 mph. Over much of the continent average wind speeds are less than 10 mph.

The average wind velocity is quite sensitive to local terrain. A. H. Stoddart reports that at typical hilltop sites in the United Kingdom average wind velocities are 30 to 35 percent higher than on nearby

Table 10.4
Average Wind Velocities in Various Regions

Region or Site	Average Wind Velocity (mph)	Reference
Orkney Islands		25
Costa Head[a]	19.5 (July–Aug.)	
Vestra Field[a]	28.8 (Dec.–June)	
	15.1 (July–Aug.)	
	24.8 (Dec.–June)	
Kirkwall[a]	12.3 (July–Aug.)	
	15.9 (Dec.–June)	
Amarillo, Texas	13.7	23
Astoria, Oregon Coast	12	23
Grandpa's Knob, Vermont[b]	16.7	22
Mount Washington, N.H.	44	22
Moresly Island, British Columbia	greater than 20	26
Magdalen Islands, Gulf of St. Lawrence	greater than 20	26
Part of Northern British Columbia	15–20	
Northern Ontario and Quebec, Arctic	10–15	
Rest of Canada	less than 10	

[a]Elevation of measurement: 35 ft.
[b]Elevation of measurement: 140 ft.

plains. The measured wind velocity is also quite sensitive to height above the ground.

If one were to take an average velocity of 15 mph (6.7 m/sec) and an efficiency of 70 percent, the power density (in kilowatts per square meter of area within the blade circle) would be about 0.075 kW/m^2. Raising the average velocity to 30 mph would increase this power density to nearly 0.60 kW/m^2, but few sites in the world would have average velocities at that level. If the average wind velocity is 10 mph, the net power output of the turbine is only 0.02 kW/m^2, which is of the same order as the power density that might be expected from solar radiation plants (per unit area of solar collector).

For a particular site, the design of the wind turbine will be a compromise, depending on the annual distribution of wind velocities. For the Costa Head site, E. W. Golding estimates that a wind turbine designed for 30 mph would have an annual load factor of about 50 percent. It would begin to generate electricity at 17 mph, and would be run at constant power in the range 30 to 60 mph; the rotor would be stopped if wind velocities in excess of 60 mph were thought likely. The turbine is not designed for maximum power output, but rather for maximum electrical energy generation per year per kilowatt of installed capacity. The load factor mentioned above must be viewed as exceptionally high. For many of the relatively promising sites for wind power generation, maximum load factor would be in the range 20 to 30 percent.

Stoddart lists the following ranges of possible annual electricity generation from windmills in North America with rated power at 25 mph (23):

Coastal United States:	2250–3750 kWh/kW
Inland United States:	750–2250 kWh/kW
Coastal and Northern Canada:	3750–5000 kWh/kW.

For 5000 kWh/kW the mean wind speed is 20.7 mph; for 3750 kWh/kW it is 18.8 mph.

In visualizing the possibility of a large network of wind turbines, the question of optimum spacing arises. R. J. Templin has estimated that the wind power available to each machine is not appreciably affected by the presence of the others as long as the ratio of disc area to ground surface area does not exceed 0.001 (27). This implies that the windmills must be spaced about 30 diameters apart. Over the plains of Canada this would make possible a wind power potential of 150 to 200 kW/mile^2. (With average wind speed of 15 mph and efficiency of 0.7, the power density is 0.075 kW/m^2 of disc area, or 200 kW/mile^2.) Thus the total Canadian electrical power capacity, which is about 35,000 MW, could be met by spacing windmills uniformly over an area 500 miles × 500 miles.

Wind power generation is very commonly used in remote locations where electrical lines are not available, whereas for quite small power needs, diesel–electric power generation would be relatively inexpensive. Also, for low power levels, lead–acid storage batteries are acceptably low in total cost, so that small variable electrical loads may be met regardless of variability in the wind. In small units, wind power is also useful for water pumping and may even be practical as a heat source.

The use of wind power for supply to national electrical grids has been demonstrated successfully at many sites. In Denmark, for example, machines with tens or hundreds of kilowatts capacity have contributed in total about 100,000 kW to the national electrical supply grid. Used in this way, wind turbines do not require special storage capacity, and serve mainly to reduce the consumption of fuel by the remainder of the system.

The largest wind turbine used for electricity supply appears to be the 1250-kW unit built by Morgan Smith Co. in Vermont in 1941 under the direction of Palmer Putnam (22). The turbine had two blades of stainless steel, with a blade diameter of 187 ft. It developed a maximum power of 1400 kW. Although the machine operated successfully for a brief period, the operation stopped after a blade failure. Repairs were not carried out because by then it had been established that costs of electricity generation with the wind turbine were about 50 percent higher than the corresponding costs of hydroelectric generation at that time in Vermont.

The key factor in assessing the power potential of a single wind turbine for a given site is the optimum size. If a given wind turbine is greatly scaled up in size (preserving geometric similarity), several disadvantages arise. Since centrifugal stresses are proportional to the square of tip speed, the rotor rotational speed ω must vary inversely as the tip diameter D. This will mean either a larger gearbox to drive the generator at moderately high speed, or a generator that runs at relatively low speed and large mass per kilowatt of power.

A more serious problem is that as the size increases, power increases with D^2, but the mass of the rotor and structure varies with D^3. Hence larger structures require more structural mass per kilowatt of capacity simply to hold the structure up. Also the gravitational stresses imposed on the blades are proportional to D; these can be severe for large wind turbines. Still another problem is the effect of wind shear and gusts on blade stresses. Both average and gust velocities will vary substantially from the highest to the lowest point that the blade reaches. The larger the rotor diameter, the greater the amplitude of unsteady pressure loading the blade experiences in traveling from low to high elevation.

A number of ideas have been implemented to alleviate stress problems, including variable-pitch blades, variable flaps, and hinged blades. Variable-pitch blades permit optimum operation over a wide range of wind speeds and prevent rotor overspeed. Variable flaps are effective for controlling maximum rotor speed by destroying lift. Hinges have been used at blade roots to reduce bending stresses, particularly to alleviate the effects of gusts. With hinges at their roots, the blades rotate on a conical surface, the cone angle depending on centrifugal and aerodynamic forces. Other ingenious modifications may yet be developed for control of stresses in very large wind turbines, but there is no apparent reason for supposing that rotors much larger than 200 ft (60 m) in diameter—the size recommended by Putnam—will be feasible. Golding has suggested an optimum rotor tip diameter of 70 ft (21 m) (25). Putnam estimated the economically optimum turbine to have a blade-tip diameter of 175 to 225 ft (53 to 69 m), a tower height of 150 to 175 ft (46 to 53 m), and a maximum power level of 1500 to 2500 kW (20). The total weight of this optimum unit was 230 kg/kWe, or 30 lb/ft^2 (150 kg/m^2) of swept area.

A new type of vertical axis windmill with a high ratio of tip speed to wind speed (5:8) has been developed at the National Research Council of Canada (28). It consists of a long slender airfoil formed in the shape of a hoop designed for minimum stress. A 14-ft (4.3-m) diameter working model of this design is particularly lightweight, with only 2 lb/ft^2 (10 kg/m^2) of swept area. For geometrically similar designs the mass per unit swept area would be proportional to diameter, so that a 175-ft (53-m) diameter version could have a specific mass as high as $2 \times 175/14 = 25$ lb/ft^2 (120 kg/m^2). It is hoped that with due attention to dynamic stresses the specific mass of large models can be held much lower than this value. A 200-kW prototype with 6500 ft^2 (604 m^2) of swept area is now being built for testing in the Gulf of St. Lawrence region.

Capital costs for wind generators can be established only approximately, since so little experience has been obtained with large units (greater than 100 kW). Capital costs per kilowatt will depend strongly on wind speed, since the power per unit size varies with v^3. Aerodynamic drag and required structural strength will increase with v^2, but the structure will no doubt be designed for some very rare high velocity that is perhaps little dependent on local average wind velocity.

The 1975 capital costs of two small windmills, each rated at 6 kW and each developing about 1.5 kW at 15 mph, have been reported by C. K. Brown and D. F. Warne (29). One is of the conventional horizontal axis type; the other is of the new vertical axis type. Capital costs for each of these windmills are quoted as $6000, that is, $1000/kW of rated capacity but $4000/kW of capacity at 15 mph. If 100 kWh of electrical storage capacity are required, an additional capital charge of $10,000 (at $100/kWh) is incurred for each windmill. Though windmill mass per unit power output increases with size, other factors should lead to reduced capital cost per kilowatt for large units.

In 1945 Putnam estimated the capital cost of a 1500-kW wind turbine (rated at 25 mph) at $205/kW. Allowing for an average inflation rate of 4.7 percent, yielding a quadrupling of costs from 1945 to 1975, would bring the cost of such a unit to $800/kW at rated power. At power developed with a 15 mph wind speed the capital cost would be $3700/kW, not including the cost of energy storage. The cost and weight breakdown of the Putnam unit is given in Table 10.5, which shows that the rotor components accounted for 60 percent of the total weight and 40 percent of the total cost. The tower accounted for 24 percent of the total weight and only 7 percent of total cost. Control of blade pitch, rotor speed, and cone angle as a function of wind speed proved to be expensive. This limited experience with a greater-than-1-MWe windmill does not provide a strong base for projection of the future cost of large windmills, but it does indicate that they are quite expensive if one takes availability into account. In 1973 Stoddart implied that for wind power to be economical at current fuel prices, year-round average wind speeds of 25 mph or more would be necessary (23).

Table 10.5
Mass and Cost Distribution of the 1500-kWe Smith-Putnam Windmill

Element	Mass (lb)	Percentage of Total Mass	Percentage of Total Cost
Rotor	422,000	58.6	40
Blades	67,000		
Hub Assembly	155,000		
Pintle Assembly	200,000		
Tower (with elevator hoist)	193,000	26.8	7
Gears and Bearings	40,500	5.6	12
Generator and Transformer	35,250	4.9	9
Miscellaneous (not including foundation)	29,200	4.1	32
Total	719,950	100	100

Source: Design data from Ref. (20).

In summary, electrical power generation by wind turbines is technically feasible but has several serious disadvantages, including the need for large energy storage capacity owing to wind variability. The capital cost per average power output may be two or three times the capital cost of fossil fuel or nuclear plants, even without allowance for the capital cost of energy storage. Maximum power levels per unit appear limited to 1 or 2 MW, and these are likely to be feasible only where average wind velocity exceeds perhaps 20 mph. For a 1000-MWe power level, the use of 1000 individual wind power units rather than one central power station would no doubt result in a severe penalty in operation and maintenance costs.

10.5 ENERGY STORAGE

The intensity of solar energy received by a power plant would be variable not only through the day, but from season to season and with changing weather conditions. Peak radiation intensity would seldom coincide with peak elective power demand, which usually occurs in the early hours of the evening (and either in summer or winter, depending on latitude). Wind energy is even more variable and unpredictable. Thus to make most effective use of either of these energy sources, some form of energy storage would generally be desirable for a large-scale solar energy generation plant. If the solar plant were merely a part of a large system, it could operate without storage merely to save fuel during the hours of sunshine; this could be a benefit to the system only if the capital cost of the plant were small enough, which is unlikely.

10.5.1 Batteries

Electrical storage batteries are limited in energy density to perhaps 20 to 30 Wh/kg (with conventional lead–acid batteries) and to power of 20 to 30 W/kg. Costs are around $100/kWh of capacity. (If the battery can be charged and discharged 1000 times during its lifetime, this would represent a storage charge of $0.1/kWh of electricity stored.) Batteries of potentially higher energy and power densities (around 200 Wh/kg and 200 W/kg, respectively) are under development. Examples of these are the sodium–sulfur and sodium–lithium batteries, which operate at temperatures of 300 to 400°C; it may be a decade or more before these can be developed, and it is not clear that their costs will be lower than present cells, though there is hope that costs could be as low as $10 to $20/kWh of capacity, with a life as long as 10 to 20 years (30). For the present, less costly storage methods are available.

10.5.2 Pumped Storage

One of these is pumped storage, which is particularly useful in conjunction with hydroelectric power generation, where reservoirs are already available (31). In some cases it is economical to build special reservoirs to store hydraulic energy, using reversible pump-turbines.

One recent example is the pumped storage reservoir near Ludington, Michigan. It consists of an 842-acre (3.41-km^2) reservoir, with 100-ft (30-m) maximum depth and elevation 370 ft (113 m) above Lake Michigan. Six pump-turbines, each with 312-MW capacity, provide storage of 16,000 MWh of energy. The cost of the system is equivalent to about $20/kWh of storage capacity and about $180/kW of turbine power capacity. Unfortunately, sites for development of large conventional pumped storage capacity are limited. Possible sites are often in areas of natural beauty, or too far from load centers.

Underground pumped storage is another possibility owing to advances in the art of excavating underground caverns. The system is comprised of a natural or artificial ground-level water reservoir and a lower pool in a cavern deep underground. Reversible pump-turbines can operate with elevation differences of 600 m (or greater with two stages). J. L. Haydock indicates that for a 2000-MW storage plant operating with a 1000-m head and 10-hr operating period, the capital costs could be in the range $150 to $200/kW, not greatly different than conventional pumped storage (32).

10.5.3 Compressed Air

Alternatively, energy may be stored by compressing air and storing it in underground caverns. Energy is recovered by extracting the air, passing it through a combustor, and expanding it in a gas turbine. Since the compressor is driven separately, the turbine will have possibly three times more net power output than if it were operating in a conventional gas turbine. To reduce storage cavern volume and excavation costs, proposed storage pressures are high, in the range 40 to 60 atm. This calls for a fairly elaborate turbomachinery configuration, including two or more compressors, an intercooler, two or more turbines, a reheater, and possibly a regenerator. A compressor aftercooler will also typically be needed since high-temperature air entering and leaving the cavern could cause thermal cracking and disintegration of the walls. With all these complications, direct capital costs can nevertheless be reduced to perhaps $150/kW. Whether this benefit will be worthwhile compared with pumped storage depends on the price of the gas turbine fuel and the operating time of the storage system.

Compressed air storage is a serious possibility. Unlike conventional pumped storage, it can be located almost anywhere. Coupling it with a solar plant would, however, be complex: solar energy would produce electrical power to drive a compressor to store air that would be used in a gas turbine. The gas or oil fuel requirement for the turbine would be considerable in an economically optimum system.

10.5.4 Hydrogen

Still another possibility for energy storage lies in hydrogen generation, which may be done directly from water by electrolysis. This process is expensive; other possibilities, such as thermochemical processes, are being studied but are not yet feasible (34). With electrolysis efficiency of 75 percent (35) and a fuel cell for electricity recovery from the energy of the stored hydrogen and oxygen, at least 40 to 60 percent of the input electrical energy would be lost between the two transformation processes. Capital cost of the fuel cell could be in the range $100 to $200/kW. To this must be added the costs of hydrogen compression, or of liquefaction for low-cost storage in large quantities. Liquefaction would no doubt be preferred, though consuming perhaps 5 kWh per pound of hydrogen. The equivalent capital costs of the liquefaction plant and storage system could easily bring total storage capital costs to the equivalent of $200 to $300/kW output. Consuming a total of 25 kWh per pound of hydrogen for electrolysis and liquefaction, the fuel cell (at 50 percent efficiency) would be able to develop only about 9 kWh of electrical energy. The overall efficiency of the hydrogen storage process would then be only about 35 percent.

Table 10.6 summarizes estimates assembled from various sources of capital costs of various methods of energy storage, including superconducting magnets and flywheels, on which research is under way. For each the capital cost is the sum of two parts—one

Table 10.6
Energy Storage

Mode	Capital Costs[a] A ($/kW)	Capital Costs[a] B ($/kWh)	Energy Density (kWh/ft³)	Expected Life (years)	Input–Output Efficiency (percent)
Pumped Water					
Above Ground	110	1–2	0.001–0.01	50	70–75
Underground	110	9–15	0.07	50	70–75
Compressed Air					
Underground	110	3–5	0.25	30	75–80[b]
Batteries and Converters					
Lead–Acid	80–100	40	1–2	6–8	60–75
Flywheels		35	0.8	30	85
Superconducting Magnets (estimated)	50	25	0.6	30	85
Steam Storage	85	25	0.5	30	75
Electrolysis and Storage					
Gaseous Hydrogen	150–300	2–3		30	35–50
Liquid Hydrogen	200–250	1–2			

[a]Total energy storage capital cost is considered to consist of the sum of the power output capacity multiplied by A plus the energy storage capacity multiplied by B.
[b]The overall efficiency quoted for compressed air storage is an equivalent figure obtained by dividing the electricity output by the sum of the electricity input and 0.33 times the heating value of the gas turbine fuel input. The factor 0.33 would stand for the average efficiency of electricity generation from fossil fuel. A compressed air storage plant might have a heat rate of 4700 Btu/kWh and an electricity input rate of 0.8 kWh(in)/kWh(out). The equivalent overall efficiency would then be $1/(0.8 + 0.33 \times 4700/3413) = 0.8$.
Sources: Refs. (30) (32) (36) (37) (38) (39).

proportional to the power in kilowatts, the other proportional to the energy storage in kilowatt-hours. For pumped storage, for example, the first will depend on the cost of the pump-turbines, motors, and generators; the second will depend on the size of the storage reservoir. The total cost of storage is therefore linearly dependent on the length of the generation period. As indicated previously, pumped underground storage is relatively expensive and has fairly high input–output efficiency. So does compressed air storage, though one must allow for the cost of gas turbine fuel consumption. Batteries are relatively expensive, and are not usually as durable as Table 10.6 would indicate. Hydrogen storage will suffer from low overall efficiency (which may be closer to 30 percent than to 50 percent) as well as high capital cost. Magnetic and flywheel storage look very expensive; both, however, are still in the laboratory stage.

The methods of energy storage considered up to this point all deal with the electricity produced by the solar- or wind-powered plant. An alternative possibility for a solar–thermal plant only would be a thermal energy storage unit between the collector and the heat engine; a variety of organic fluids and molten salts could store thermal energy up to 400°C. Liquid metals may be useful at still higher temperatures. An important problem with such a scheme would be the heat loss from ducts and tanks used for the thermal energy collection and storage. The capital costs of adequate insulation would need to be carefully evaluated in any feasibility analysis.

10.6 COSTS OF A SOLAR POWER PLANT

As already shown, a principal disadvantage of solar energy is that it is so diffuse. Power per unit area of solar collector for photovoltaic and solar thermal plants, and power per unit flow area of wind turbines, are so low that these plants are expensive to build.

A second major disadvantage of these solar plants is power variability, since even if substantial storage capacity is provided, the plant would be idle for a large fraction of the day or year. The costs of storage, as indicated in Table 10.6, are considerable. Suppose a solar plant were able to function 8 hr/day, and storage were required for 4 days output. The total capital cost per kilowatt could then be $\$110 + 4 \times 8 \times \$5 = \$270$/kWe for compressed air storage.

Two methods of solar power production that do not require such storage systems would be ocean thermal plants and photosynthesis plants.

10.6.1 Ocean Thermal Plants

The ocean thermal plant is a Rankine cycle designed to use temperature differences around 20 to 25°C between surface and deep layers of the ocean, as in the Gulf Stream (39, 40). The idea has been of interest for many years; a demonstration plant producing 22 kW was built by George Claude and operated in the Caribbean. In the Straits of Florida, 30 miles from Miami, measurements indicate a surface temperature of 25°C and steady temperatures of 7 or 8°C at a depth of 360 m. For an 18°C temperature differential the efficiency is only 6 percent. Taking into account the temperature differences necessary to transfer heat into and out of the plant at reasonable rates, the efficiency drops to perhaps 4 percent. Then allowing for the thermodynamic limitations of the working fluid and various system losses, including the power required to circulate vast volumes of water into and out of the plant to provide the thermal energy for boiling and condensation, the overall efficiency obtainable may be only 1 or 2 percent.

Possible working fluids are propane and ammonia, which have

reasonably large vapor pressures at the low operating temperatures. Because of the low efficiency, the required heat transfer and condenser area (per kW of output) would likely be an order of magnitude larger than for a conventional Rankine cycle plant. This and the peculiar problems of the marine environment suggest that overall plant size and cost may be an order or magnitude larger than a fossil fuel Rankine cycle plant.

10.6.2 Photosynthesis

Another mode of continuous use of solar radiation is photosynthesis, which could be the source of fuel for a conventional type of steam plant. Hans Thirring concludes that the fuel energy available from growth of a variety of algae called chlorella would be about 3.7×10^9 Btu/acre/yr (0.030 kW/m^2) (3). Using this fuel in a thermal plant with an overall efficiency of 40 percent would give a plant power density of about 0.012 kWe/m^2; G. C. Szego, in a discussion of the energy plantation concept, suggests that since forests may capture 1 to 3 percent of the incident solar energy, the fuel for a 1000-MWe electrical plant could be grown continuously in a forest of area 200 to 600 miles^2 (42). This is equivalent to 0.001 to 0.003 kWe/m^2. The overall efficiency of using solar radiation energy in this way would be the product of the power plant efficiency and the incident solar energy capture fraction, or less than 1 percent. The cost of power from a continuous forest fuel source does not appear to be unreasonably high, but the total land requirement may well be prohibitive.

10.6.3 Plant Availability and Power Density

With the exception of ocean thermal plants and "photosynthesis fuel" plants, power variability of solar plants is a serious problem. One way to describe this variability is to define a plant availability factor as the ratio of average annual power available to maximum power output. The availability factor does not tell the whole story, of course; the power available is useful only if it is in phase with power demand or if it can be coupled to storage of acceptably high overall efficiency. The ratio of average power used to plant capacity may be much lower than the availability factor for a solar-powered plant. Typical availability factors for various kinds of plant are shown in Table 10.7.

Except for maintenance and overhaul, fossil fuel plants are available all the time. For some types of nuclear plants the time required for refueling during shutdown may be significant. Plants that burn photosynthesis fuel (produced continuously in fuel farms) would also operate with high availability. So would ocean thermal plants since ocean thermal gradients are relatively steady. Solar radiation plants would have very low availability, perhaps 15 percent

Table 10.7
Characteristic Availabilities and Efficiencies of Power Plants

Power Source	Plant Availability Factor	Characteristic Efficiency
Fossil Fuel	0.9	0.4
Nuclear	0.8	0.3–0.4
Photovoltaic	0.15	0.1
Solar Thermal	0.15	0.1–0.2
Wind	0.2–0.4	0.7
Photosynthesis	0.9	0.004–0.01
Ocean Thermal	0.9	0.01–0.02

for stationary collectors, although possibly up to 30 percent for collectors oriented toward the sun. These low availabilities, and the fact that electricity load demand will not typically peak at noontime, mean that solar radiation collectors will generally require some form of energy storage. Wind power plants, as Table 10.7 suggests, have typically higher availability than solar radiation plants, but are restricted to sites where winds are strong and relatively steady.

Power density is another important criterion, for it affects the land area required and costs of plant and structure. To compare various types of power plants, we select for each a cross-sectional area for energy flow that fundamentally limits the size of the plant per kilowatt, and we call this the critical area. The areas selected for various plants are shown in Table 10.8.

For the fossil fuel–steam plant, the largest single element is the furnace; the gas-side heat transfer rates in the boiler are typically limited to 8000 to 10,000 Btu/ft^2/hr (25 to 31.5 kW/m^2). With a plant efficiency of 0.4 this is equivalent to 10 to 13 kWe/m^2 of furnace heat transfer area. In a different way, nuclear plants are also heat transfer limited, with heat transfer rates being quite dependent on the coolant used:

Water (H_2O or D_2O): 100,000–200,000 Btu/ft^2/hr (300 to 600 kW/m^2)
Gas (CO_2 or Helium): 10,000–50,000 Btu/ft^2/hr (30 to 150 kW/m^2)
Liquid Sodium: 500,000–800,000 Btu/ft^2/hr (1500 to 2400 kW/m^2).

Table 10.8 summarizes these power densities and other factors; Figure 10.4 indicates the ranges of capital costs per average kilowatt of power that can be generated for various power plants. The separation of solar from nuclear plants is striking. All solar plants have power densities less than 0.1 kWe/m^2 of primary heat transfer area; fossil and nuclear plants all have power densities 10 to 100 kWe/m^2. In general, this order-of-magnitude difference of two or three in power density is reflected in capital costs, with wind turbines and fuel farms being possible exceptions.

The main difficulty with wind turbines is the lack of suitable sites with average winds above 20 mph and the limitation of unit

Table 10.8
Typical Power Plant Characteristics

Power Source	Thermal Efficiency	Availability Factor	Typical 1976 Capital Cost[a] ($/kW)	Critical Power Density		
				Area	Capacity (kWe/m²)	Average (kWe/m²)
Nuclear						
Water	0.3	0.8	600	Reactor	130–250	100–200
CO_2	0.4		800	heat		
He	0.4		800	transfer	10–50	8–40
Na	0.4		800	area	40–60	35–50
Fossil Fuel	0.4	0.9	400	Combustor heat transfer area	10–13	9–12
Solar						
Photovoltaic	0.1	0.15	10^4–10^5	Solar collector	0.1	0.01–0.015
Thermal	0.1–0.2	0.15	10^3–10^4	area	0.1–0.2	0.02–0.03
Wind	0.5–0.8	0.2–0.4	1000–2000	Turbine area	0.2–0.3	0.04–0.1
Ocean Thermal	0.01–0.02	0.9	10^3–10^4	Evaporator area	0.025–0.06	0.02–0.06
Photosynthesis (forest)	0.004–0.01	0.9	800	Fuel farm area	0.001–0.004	0.001–0.004

[a]Capital costs of storage not included.

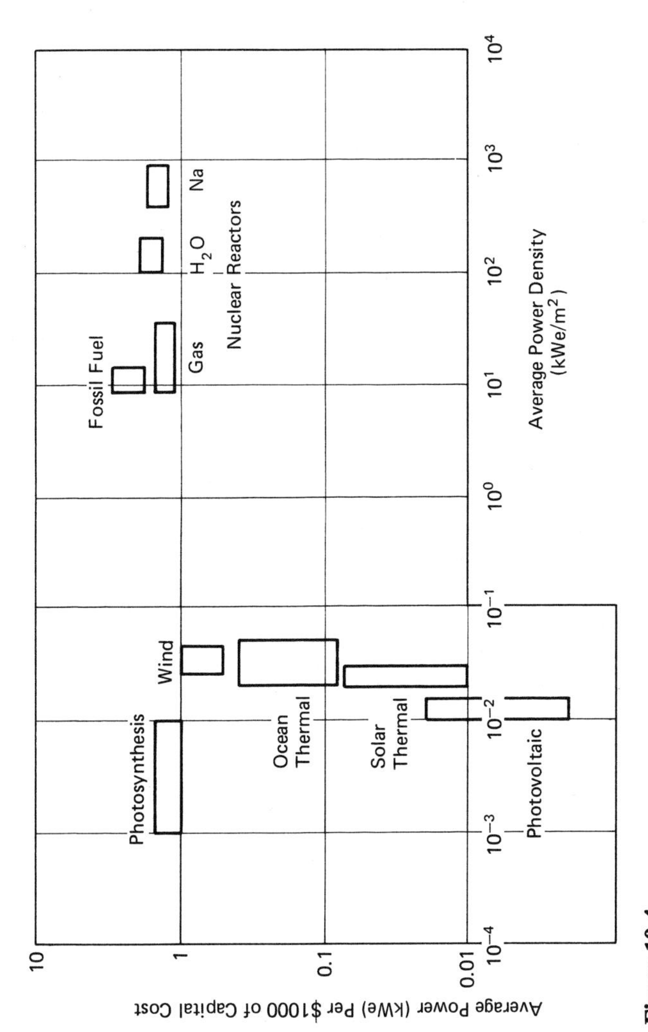

Figure 10.4
Energy conversion capital costs and power densities. No allowance is made for energy storage. Average power = Maximum power × Availability.

size to 1 or 2 MW, for which labor and maintenance costs would be much higher per kilowatt than for a 1000 to 2000-MW plant. Also, turbines in large numbers might be an unacceptable blight on the landscape, as well as a hazard to people nearby. Wind turbines therefore appear unlikely to generate any large fraction of future electricity demand, however useful they may be in remote areas and for special tasks.

Solar photovoltaic and thermal plants could be located in areas unsuitable for agriculture, but they too are unlikely to be developed soon. A reduction of two or three orders of magnitude in unit costs might be needed before photovoltaic cells could become competitive with nuclear power.

Table 10.9 shows approximate relative costs of solar–thermal, coal, and nuclear power for the 1980s. A solar plant capital cost of \$1500/kWe would have to be regarded as an optimistic figure for the 1980s, corresponding to installed capacity of 0.1 kWe/m^2 of collector area and a plant cost of only \$100/m^2, or \$10/ft^2. Allowing for storage of 50 kWh/kW of capacity at \$10/kWh of storage capacity, the required storage plant would have a capital cost of \$500/kW. The availability factor could be higher than 0.15 with tracking collectors, but the estimate does not allow for losses in energy storage. A wind power plant could cost less, although operation and maintenance costs (for many small units) could be very much higher than shown in Table 10.9.

Great inventions are always possible. Who would have foreseen nuclear power in 1900, or even in 1935? But until the invention arrives that makes solar power economical, we can hardly be sure that it is coming. We can say that if fossil and nuclear fuels could no

Table 10.9
Approximate Relative Costs of Solar, Coal, and Nuclear Power in the 1980s

	Coal	Nuclear	Solar
Capital Cost ($/kW)	700	850	1500 (plant) 500 (storage)
Load Factor	0.8	0.8	0.15
Fixed Charges[a] (mills/kWh)	16.0	19.4	243
Fuel Charges[b] (mills/kWh)	9.0	2.4	
Operation and Maintenance	2.0	1.5	2.0
Total Costs at Switchboard (mills/kWh)	27.0	23.3	245

[a]At 16 percent.
[b]Coal price \$25/ton or about \$1/10^6 Btu.

longer be used for electricity generation, solar energy *could* meet world power needs, but with the possible penalty of an order-of-magnitude increase in cost.

References

1. Farrington Daniels. *Direct Use of the Sun's Energy*. New Haven: Yale University Press, 1964.

2. John I. Yellott. "Solar energy progress—A world picture." *Mechanical Engineering*, July 28, 1970.

3. Hans Thirring. *Energy for Man*. New York: Greenwood Press, 1968.

4. R. A. Tybout and G. O. G. Löf. "Solar house heating." *Natural Resources Journal*, vol. 10, no. 2 (April 1970), pp. 268–326. See also G. O. G. Löf and R. A. Tybout. "Cost of house heating with solar energy." *Solar Energy*, vol. 14 (1973), p.253.

5. H. C. Hottel and J. B. Howard. *New Energy Technology: Some Facts and Assessments*. Cambridge, Mass.: MIT Press, 1971.

6. Dwain F. Spencer. "Solar energy: A view from an electric utility standpoint." *Proceedings of the American Power Conference*, vol. 37 (1975).

7. E. J. Truhlar, J. R. Latimer, G. L. Mateer, and W. L. Godson. "Solar radiation measurements in Canada." Paper 5/18, *Proceedings of the United Nations Conference on New Sources of Energy and Development*, New York, 1961.

8. S. Angrist. *Direct Energy Conversion*. Boston: Allyn and Bacon, 1965.

9. *Workshop Proceedings: Photovoltaic Conversion of Solar Energy for Terrestrial Applications*. Vols. I and II, NSF-RA-N-74-013. Organized by Jet Propulsion Laboratory, California Institute of Technology, October 23–25, 1973.

10. W. R. Cherry. "The generation of pollution-free electrical power from solar energy." *ASME Transactions: Journal of Engineering for Power*, April 1972, pp. 78–82.

11. W. R. Cherry and J. A. Zoutendyk. "State of the art in solar cell array for space electrical power." In *Progress in Astronautics and Aeronautics*, vol. 16: *Space Power Systems and Engineering*. New York: Academic Press, 1966.

12. W. R. Cherry and F. H. Morse. "Conclusions and recommendations of the Solar Energy Panel." Paper 72–WA/Sol-5 presented at the Annual Winter Meeting of the American Society of Mechanical Engineers, New York, New York, November 26–30, 1972.

13. H. Oman and C. J. Bishop. "A look at solar power for Seattle." *Proceedings. Intersocieties Energy Conversion Engineering Conference*, Philadelphia, August 1973.

14. John A. Duffie and William Beckman. *Solar Energy Thermal Processes*. New York: John Wiley & Sons, 1974.

15. Piet B. Bos. "Solar realities." *EPRI Journal*, February 1976, pp. 7–13.

16. Aden B. Meinel and Marjorie P. Meinel. "Solar energy: Is it a feasible option?" Paper presented at the Symposium of the Forum on Physics and

Society, Albuquerque, New Mexico, June 5, 1972. See also A. B. Meinel. "Physics looks at solar power." *Physics Today*, vol. 25 (1972), p. 44.

17. Peter Rouklove. "Status report on solar thermionic power stations." In *Progress in Astronautics and Aeronautics*, vol. 16: *Space Power Systems and Engineering*. New York: Academic Press, 1966.

18. D. A. Warner and J. L. McCabris. "Solar thermoelectric power conversion coupled with thermal storage for orbital space applications." In *Progress in Astronautics and Aeronautics*, vol. 16: *Space Power Systems and Engineering*. New York: Academic Press, 1966.

19. Donald L. Southam. "Power systems comparison for manned space station applications." In *Progress in Astronautics and Aeronautics*, vol. 16: *Space Power Systems and Engineering*. New York: Academic Press, 1966.

20. *Satellite Solar Power Station: An Option for Power Generation.* Report prepared by A. D. Little, Inc., Grumman Aerospace Corp., Raytheon Co., and Textron, Inc., for the Task Force on Energy of the Subcommittee on Science Research and Development of the Committee on Science and Astronautics, U.S. House of Representatives, 92nd Congress, vol. 2, p. 146, 1972.

21. J. Patha and G. R. Woodcock. "Feasibility of large-scale orbital solar/thermal power generation." *Proceedings. Intersocieties Energy Conversion Engineering Conference*, Philadelphia, August 1973.

22. Palmer C. Putnam. *Power from the Wind.* New York: Van Nostrand, 1948.

23. *Proceedings of the Wind Energy Conversion Systems Workshop*, Washington, D.C., June 11–13, 1973. NTIS Publication PB–231–341 prepared for the National Science Foundation, December 1973.

24. G. Gimpel. "The windmill today." *Engineering*, May 30, 1958.

25. E. W. Golding. *Large-Scale Generation of Electricity by Wind Power, Preliminary Report.* Technical Report C/T/01, British Electrical and Allied Research Association, Leatherhead, Surrey, 1949.

26. Gordon N. Patterson. *Canada's Energy Corridors to the Future: A Background Study for the Science Council of Canada*, 1974.

27. R. J. Templin. *An Estimate of the Interaction of Windmills in Widespread Arrays.* Report LTR–LA–171, National Research Council of Canada, December 1974.

28. P. South and R. S. Rangi. *A Wind Tunnel Investigation of a 14-ft Diameter Vertical Axis Windmill.* Report LTR-LA-105, National Research Council of Canada, September 1972.

29. C. K. Brown and D. F. Warne. *An Analysis for the Potential for Wind Energy Production in Northwestern Ontario.* Sheridan Park, Ontario: Ontario Research Foundation, November 1975.

30. R. A. Fernandez, O. D. Gildersleeve, and R. T. Schneider. "Assessment of advanced concepts in energy storage and their application on electric utility systems." Paper 6.1–17, *Transactions of the 9th World Energy Conference*, Detroit, Michigan, September 22–27, 1974.

31. S. R. Knapp. "Pumped storage: The handmaiden of nuclear power." *IEEE Spectrum*, April 1969.

32. James L. Haydock. "Energy storage and its role in electric power systems." Paper 6.1–21, *Transactions of the 9th World Energy Conference*, Detroit, Michigan, September 22–27, 1974.

33. R. Decker. "Compressed air energy storage system characteristics." *Proceedings. Intersocieties Energy Conversion Engineering Conference*, Philadelphia, August 1973.

34. G. DeBeni and C. Marchetti. "Mark I: A chemical process to decompose water using nuclear heat." Paper presented at the Symposium on Non-Fossil Chemical Fuels, 163rd National Meeting, American Chemical Society, Division of Fuel Chemistry, Boston, April 10–14, 1972.

35. John W. Michel. "Hydrogen and synthetic fuels for the future." Paper presented at the American Chemical Society Symposium on Chemical Aspects of Hydrogen as a Fuel, Chicago.

36. F. R. Kalahammer and P. S. Zygelbaum. "Potential for large-scale storage in electric utility systems." Paper 74-WA/Ener-9 presented at the Annual Winter Meeting of the American Society of Mechanical Engineers, New York, New York, November 17–22, 1974.

37. R. F. Post and S. F. Post. "Flywheels." *Scientific American*, December 1973, pp. 17–23.

38. P. V. Gilli and G. Beckman. "The nuclear steam storage plant: An economic method of peak power generation." *Transactions of the 9th World Energy Conference*, Detroit, Michigan, September 22–27, 1974.

39. E. Neuman. "Thermodynamic and economic analysis of compressed air energy storage power plants." M.Sc. thesis, Queen's University, Kingston, Ontario, 1976.

40. J. G. McGowan, W. E. Heronemus, J. W. Connell, and P. D. Clouter. "Ocean thermal difference power plant design." Paper 73-WA/Oct-S presented at the Annual Winter Meeting of the American Society of Mechanical Engineers, Detroit, Michigan, November 11–15, 1973.

41. J. H. Anderson, Jr. "Economic power and water from solar energy." Paper 72-WA/Sol-2 presented at the Annual Winter Meeting of the American Society of Mechanical Engineers, New York, New York, November 26–30, 1972.

42. G. C. Szego. "The energy plantation." Hearings before the Committee on Science and Astronautics, Subcommittee on Science Research and Development, U.S. House of Representatives, May 9–30, 1972, pp. 693–705.

Supplementary Reference

G. Gimpel and A. H. Stoddart. *Windmills for Electricity Supply in Remote Areas*. Technical Report C/T120, Electrical Research Association, Leatherhead, Surrey, 1958.

Problems

10.1 Show, via the following reasoning due to A. Betz, that the maximum efficiency of a horizontal-axis windmill is $\eta_{\max} = 16/27 = 0.593$.

Figure 10.A represents the windmill as an "actuator disc" and shows a streamtube of air flowing through the disc. Show, from the Bernoulli equation, that the pressure difference across the disc is

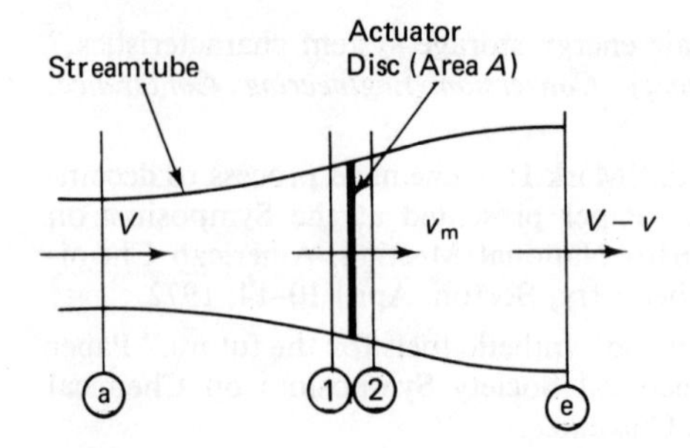

Figure 10.A
The windmill as an "actuator disc."

$$\frac{P_1 - P_2}{\rho} = \frac{V^2 - (V - v)^2}{2},$$

where P denotes pressure, ρ density, and V velocity, with v_m the mean velocity in the actuator disc plane and v the decrease in streamtube velocity due to the actuator disc. Then show that the thrust T exerted on the disc is

$$T = A\,(P_1 - P_2) = \rho\,A\,v_m\,v$$

and hence that

$$v_m = V - \frac{1}{2}v.$$

That is, the velocity of flow through the disc is the average of upstream and downstream velocities.

Neglecting the effects of friction, and of swirl in the downstream flow, show that the power developed is

$$P = \rho\,A\left(V - \frac{1}{2}v\right)\left(\frac{V^2}{2} - \frac{(V - v)^2}{2}\right),$$

which has a maximum of

$$P = \frac{16}{27}\,\frac{\rho\,A V^3}{2}.$$

Thus

$$\eta_{max} = \frac{16}{27}.$$

10.2 A windmill is to be sited in a region where the number of hours per year that the wind speed equals or exceeds v is as shown in Table 10.A. Consider two designs with the characteristics shown in Table 10.B. In Design 1, for example, no power is produced until the wind speed is 15 mph, and the blades are designed so that at all speeds above 30 mph the power output is the same as at 30 mph. Find the

Table 10.A
Wind Speeds at a Hypothetical Windmill Site

v (mph)	No. of Hours per Year That v Is Equalled or Exceeded
50	500
40	1300
30	3000
25	4200
20	5600
15	6900
10	7850

Table 10.B
Two Windmill Designs

	Wind Speed at Maximum Power (mph)	Wind Speed at Minimum Power (mph)
Design 1	30	17
Design 2	40	21.5

number of kilowatt-hours generated per year per kilowatt of installed capacity (ratio at wind speed for maximum velocity) for each of the designs.

10.3 Estimate the optimum size of a "large" windmill using the following simplified economic assumptions.

Let the capital cost of the wind turbine, generator, and electrical and mechanical transmission be proportional to power output. These items taken together account for $250/kWe.

Suppose the capital cost of the tower is proportional to the tower mass, which varies with D^3, while the power output varies with D^2, D being the rotor diameter. It is known that the tower for a 100-kWe windmill costs approximately $75,000.

Assume that the optimum windmill is large enough that it requires a man on duty at all times, so that total operation and maintenance costs depend little on size and are about $60,000/yr.

Suppose the total annual charge on invested capital is 15 percent per year (interest, depreciation, and taxes). The windmill produces 4000 kWh/yr per kilowatt capacity.

a. What is the optimum windmill size?

b. What is the minimum generation cost (mills/kWh)?

c. What would be the generation cost if the windmill were twice the optimum size?

10.4 R. J. Templin has estimated that the power absorbed by windmills will not significantly reduce local wind velocities, provided the spacing between them is 30 rotor-diameters or greater (27).

Suppose a grid of large-size windmills extends over a territory where windmills can be designed for 3500 kWh/yr per kilowatt of capacity at rated wind speed of 25 mph and 70 percent efficiency.

a. What are the maximum and annual-average power levels (kWe/km^2)?

b. What fraction of annual average solar radiation intensity could be transformed into wind-generated electrical power, taking solar energy as $0.17\ kW/m^2$?

c. M. K. Hubbert states that over the earth's surface 370×10^{12} W of solar energy are used to drive winds and waves. If this expenditure were uniformly distributed over the earth's surface, what fraction of it could be absorbed in wind power generation?

10.5 H. C. Hottel and J. B. Howard have estimated that a flat solar collector located at El Paso, Texas, and tilted toward the equator at an angle equal to the latitude, could collect 61 percent of the impinging solar radiation (5). Under best operating conditions, the maximum temperature would vary from 50 to 200°C through the day. The average Carnot efficiency corresponding to these numbers would be 0.18. The year-round solar radiation intensity at El Paso is 3.72 $kWh/m^2/day$.

Estimate the maximum yearly average electrical power density (kWe/km^2) if the tilted collectors occupy half the land area on the solar collector site and the thermal efficiency of the power plant which is supplied by the solar collector is 10 percent. Assume also that two-thirds of the electricity generated is stored with a storage efficiency of 75 percent, while the remainder is supplied directly after generation to users.

10.6 The Meinels (15) have been much more optimistic than Hottel and Howard (5) regarding possible solar–electric power density. They suggest the possibilities shown in Table 10.C. Calculate the corresponding average annual power density (kWe/km^2 of land area) allowing for energy storage losses. Identify the main factors that lead to the difference between these estimates and those of Problem 10.5.

10.7 Suppose that 1 to 3 percent of solar energy is captured by photosynthesis. What size of forest would be required for continuous supply of wood fuel to a 1000-MWe power plant?

Given the following data, estimate the power generation cost for

Table 10.C
Possible Characteristics of Solar–Electric Power

Parameter	Early Plant	Highly Developed Plant
Average Incident Energy (8-hr day) (W/m^2)	700	700
Input Energy to Fluid (W/m^2)	360	540
Maximum Fluid Temperature (°C)	450	600
Efficiency Ratio η/η_{Carnot}	0.37	0.48
Thermal Efficiency	0.28	0.36
8-hr Average Power Output (W/m^2)	100	194

a 1000-MWe plant operating at load factor 0.75 under the following assumptions:

Solar intensity:	1000 Btu/ft^2-day
Solar energy capture rate:	1 percent
Land cost:	$250/acre
Planting, harvesting, and transport cost:	$800/acre
Harvesting period:	30 years
Plant thermal efficiency:	0.35
Plant capital cost:	$500/kWe
Plant operation and maintenance cost:	2 mills/kWh
Annual capital charge rate:	15 percent.

10.8 Energy storage is to be provided for a solar-powered thermal plant of 2000-MWe capacity. The storage unit is to be large enough to store 4 days' energy output of the solar power generator, which can operate at full capacity for 6 hr/day. Over the course of a year the storage unit will probably store 80 percent of the output of the plant (equivalent to operating at full load 6 hr/day for 250 days/yr). The maximum power level for the storage unit is 1000 MWe.

Using data from Table 10.6 for

a. underground pumped storage,

b. lead–acid batteries,

estimate total capital costs and total costs per kilowatt-hour of electricity supplied from the storage unit.

Assume annual capital charge rates of 15 percent for pumped storage and 30 percent for lead–acid batteries (to account for their relatively short operating life). For both cases, take operating and maintenance costs as 2 mills/kWh. Assume that the efficiency of the storage unit is 75 percent in both cases, and that the "lost electricity" has a value of 20 mills/kWh.

10.9 Make an approximate check of certain of the energy storage density figures provided in Table 10.6 under the following assumptions:

a. Pumped water: Storage elevation is 100 to 1000 ft for above-ground storage and 3000 ft for underground storage.

b. Compressed air: Air is stored in caverns at 40 atm and 100°F. The work done by the air in a turbine (after it has been taken from the storage cavern and heated by combustion) is 300 Btu/lb.

c. Lead–acid batteries: Energy storage density is the same as for an automobile battery.

d. Flywheel: The energy storage corresponds to the kinetic energy of a rotating circular rim of material whose tangential speed is 2500 ft/sec (a material of unusually high strength-to-mass ratio). The material density is 100 lb/ft^3, and the storaged unit volume is four times the volume of the rim.

e. Steam storage: The heat transfer to the stored steam is 1000 Btu/lb, and 35 percent of this can be converted into electricity. The steam is stored at 1000 psia and 1000°F.

10.10 It has been suggested that energy produced by solar power plants could be stored in hydrogen produced by electrolysis. Electrolysis efficiency in current practice is not greater than 75 percent, and energy stored in hydrogen could be converted into electrical energy at maximum practical efficiencies of about 60 percent via a fuel cell.

Assume the following:

1. For hydrogen electrolysis: capital cost is \$75/kWe of input power; efficiency is 75 percent; load factor is 0.6; capital charge rate is 17 percent.

2. For hydrogen storage: storage for 48 hr of output from electrolyzer is required; capital cost is \$2/kWh electrical output capacity; capital charge rate is 17 percent.

3. For fuel cell: capital cost is \$50/kWe output capacity; fuel cell capacity/electrolyzer capacity = 2; efficiency is 61 percent; annual capital charge rate is 25 percent.

Suppose that, for the entire storage and regeneration unit, the operation and maintenance costs are 3 mills/kWh. Assume also that the value of the electrical energy lost in storage and regeneration is 30 mills/kWh.

Determine the storage cost (mills/kWh) of electrical energy supplied from the storage unit.

Appendix A
Conversion of Units

Length

1 m = 3.281 ft
= 6.214×10^{-4} miles

Area

1 m^2 = 10.76 ft^2
= 3.861×10^{-7} $miles^2$
= 2.47×10^{-4} acres

Volume

1 m^3 = 10^3 liters
= 35.31 ft^3
= 1.308 yd^3

Mass

1 kg = 2.205 lb
= 1.103×10^{-3} tons
= 10^{-3} tonnes

1 tonne = 1.102 tons

Pressure

1 Pa = 1 N/m^2

1 atm = 101.325 kPa

1 bar = 10^5 Pa

1 psia = 6.895 kPa

Temperature

$T(°C) = (T(°F) - 32)/1.8$
$T(°K) = T(°C) + 273.1$
$T(°R) = T(°F) + 459.6$

Energy

1 J = 9.478×10^{-4} Btu
= 2.778×10^{-7} kWh
= 0.2388 cal.
= 6.242×10^{12} MeV
= 1 W-sec
= 0.7375 ft-lb

1 Btu = 1055 J
= 2.931×10^{-4} kWh
= 778.2 ft-lb
= 252 cal.
= 6.586×10^{15} MeV

1 kWh = 3.6 MJ
= 3412 Btu

1 MeV = 1.602×10^{-13} J

Energy Equivalents (approximate)

10^6 bbl/day of crude oil = 2.1×10^{15} Btu/yr = 2.2×10^{18} J/yr
1 bbl of crude oil = 5.8×10^6 Btu = 6.12 GJ
10^3 ft^3 of natural gas = 1.03×10^6 Btu = 1.09 GJ
1 ton of coal = 22 to 28 $\times 10^6$ Btu
1 ton of crude oil = 40 to 45 $\times 10^6$ Btu

Power

1 kW = 3412 Btu/hr
= 1.341 horsepower

1 horsepower = 550 ft-lb/sec

Power Density or Heat Flux Density

1 W/m^2 = 0.3172 $Btu/hr/ft^2$
1 $Btu/hr/ft^2$ = 3.152 W/m^2
1 $Btu/hr/ft^2/°F$ = 5.678 $W/m^2/°C$

Prefixes

k	kilo	10^3
M	mega	10^6
G	giga	10^9
T	tera	10^{12}

Appendix B
Resource Depletion with Exponential Growth

Exponential growth may be described by the equation

$$\dot{Q} = \dot{Q}_0 \exp(\mathrm{kt}),$$

where
$\dot{Q}$ is the consumption rate,
$\dot{Q}_0$ is the consumption rate at time $t = 0$.
t is the time in years, and
k is the annual growth rate of consumption, a constant.
Alternatively, this may be written

$$\dot{Q} = \dot{Q}_0 \exp[(\ln 2)\, t/t_{\mathrm{d}}],$$

where $t_{\mathrm{d}} = (\ln 2)/k$ is the *doubling time* in years. The relationship between the growth rate k and the doubling time is shown in Figure B.1.

It is convenient to introduce R, the total quantity of reserves initially in place (prior to any consumption), and r_0, the fraction of R that has been used at a particular time $t = 0$. The total amount consumed from $t = -\infty$ until the point of exhaustion of all resources at $t = t_{\mathrm{e}}$ is then

$$R = \int_{-\infty}^{t_e} Q\, dt = Q_0 (t_{\mathrm{d}}/\ln 2) \exp[(\ln 2)\, t_{\mathrm{e}}/t_{\mathrm{d}}]. \qquad (1)$$

The quantity consumed up to $t = 0$ is

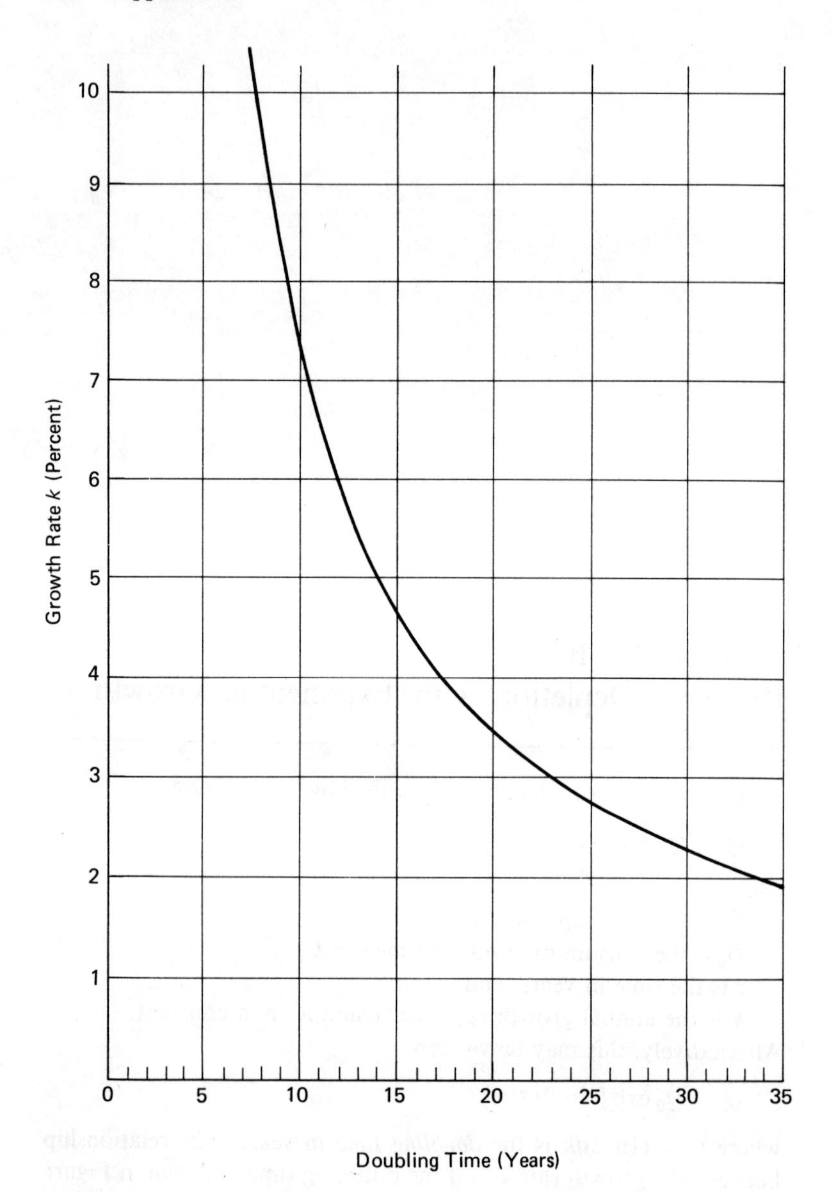

10
9
8
7
6
5
4
3
2
1
0
Growth Rate k (Percent)
5
10
15
20
25
30
35
Doubling Time (Years)

Figure B.1

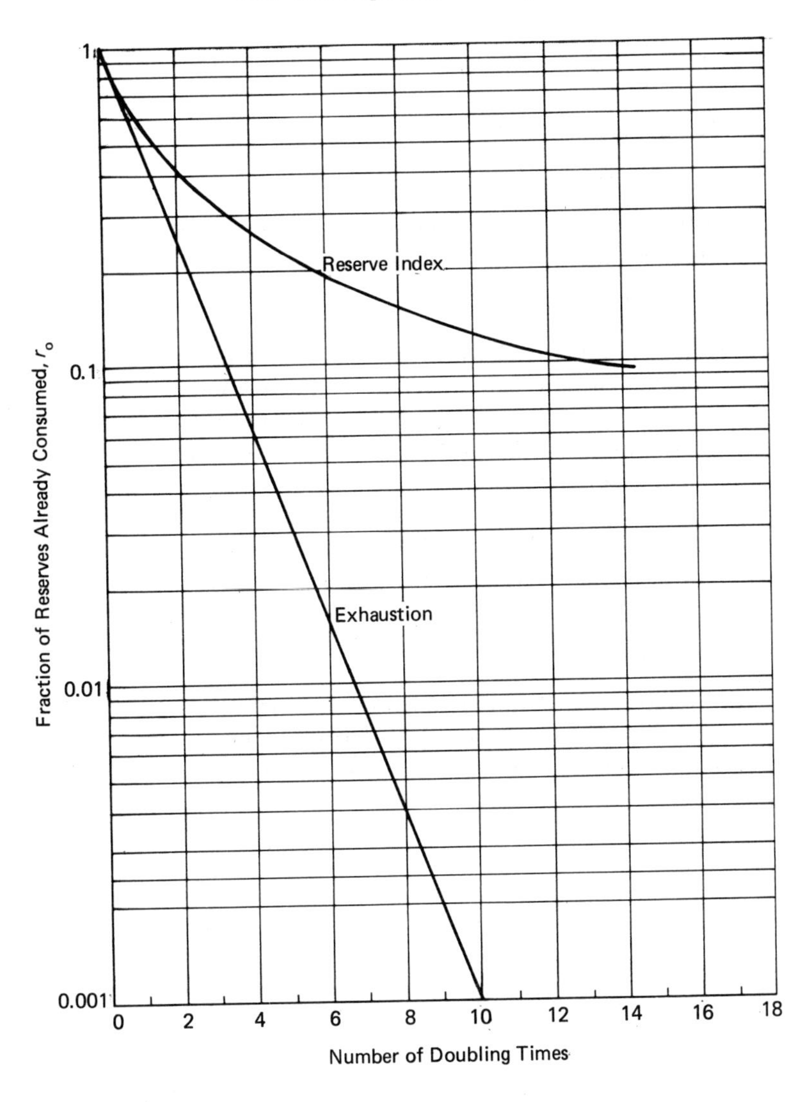
1
0.1
0.01
0.001
Reserve Index
Exhaustion
Fraction of Reserves Already Consumed, r_o
0
2
4
6
8
10
12
14
16
18
Number of Doubling Times

Figure B.2

$$r_0 R = \int_{-\infty}^{0} \dot{Q}\, dt = Q_0\, t_{\mathrm{d}}/\ln 2. \tag{2}$$

substituting (2) into (1) yields

$$1 = r_0 \exp\,[(\ln 2)\, t_{\mathrm{e}}/t_{\mathrm{d}}],$$

so that the number of years remaining until all reserves have been exhausted (assuming the continuation of exponential growth) is

$$t_{\mathrm{e}} = t_{\mathrm{d}} \ln\,(1/r_0)/\ln 2, \tag{3}$$

as illustrated in Figure B.2, which shows how the number of doubling times until exhaustion, $t_{\mathrm{e}}/t_{\mathrm{d}}$, is related to the fraction r_0 of reserves consumed by the start of the counting period. The formula and curve assume, of course, that the growth rate k is constant throughout the period of consumption.

The reserve life t_{ri} may be defined as the ratio of remaining reserves at $t = 0$ to the initial consumption rate $\dot{Q}_0$. Thus $t_{\mathrm{ri}} = R(1 - \dot{r}_0)/Q_0$ and, using (2), we can write the reserve index, $t_{\mathrm{ri}}/t_{\mathrm{d}}$, as

$$\frac{t_{\mathrm{ri}}}{t_{\mathrm{d}}} = \frac{1 - r_0}{r_0 \ln 2}. \tag{4}$$

If, for example, four doubling times are required till exhaustion, the life index will be 3.5 times the actual exhaustion life.

For oil, with $t_{\mathrm{d}} = 10.3$ and $r_0 = 0.136$ in 1970, $t_{\mathrm{e}} = 30$ and $t_{\mathrm{ri}} = 94$.

Appendix C
Cost of Useful Heat Generation at a Nuclear Plant

Retaining the definition of plant efficiency η as the ratio of plant electrical output power $\boldsymbol{P}_{\mathrm{e}}$ kWe to thermal input rate $\dot{Q}_1$ (kW thermal), we have

$$\eta = \frac{\boldsymbol{P}_{\mathrm{e}}}{\dot{Q}_1}.$$

For a given plant output, with variable temperature extraction from the steam turbine,

$$\frac{\boldsymbol{P}_{\mathrm{e}}}{\boldsymbol{P}_{\mathrm{e}0}} = \frac{\eta}{\eta_0},$$

where the subscript 0 denotes no useful heat output, i.e., the turbine exit temperature is the same as for electricity generation only, $T = T_0$.

With $T > T_0$ the loss of electrical power is

$$\Delta \boldsymbol{P}_{\mathrm{e}} = \boldsymbol{P}_{\mathrm{e}0} - \boldsymbol{P}_{\mathrm{e}} = \boldsymbol{P}_{\mathrm{e}0}\left[1 - \frac{\eta}{\eta_0}\right].$$

The hourly cost of this power loss is

$$\dot{C} = r_{\mathrm{e}}\, \boldsymbol{P}_{\mathrm{e}0}\left[1 - \frac{\eta}{\eta_0}\right],$$

where r_{e} is the unit value of electricity at the plant (mills/kWh).

Assuming (for the nuclear plant) that essentially all the heat rejected from the plant at $T > T_0$ is useful heat, the hourly quantity of useful heat is

$$\dot{Q}_h = (1 - \eta)\,\dot{Q}_1 = (1 - \eta)\,\frac{P_{e0}}{\eta_0}.$$

For a fossil fuel plant (allowing for stack gas and other losses) the useful heat can be expressed as

$$\dot{Q}_h = (1 - \eta)\,\frac{P_{e0}}{\eta_0} - K\dot{Q}_1,$$

with K approximately equal to 0.12.

The unit cost of this useful heat (in mills/kW thermal) is then

$$r_h = \frac{\dot{C}}{\dot{Q}_h} = \frac{r_e\,(\eta_0 - \eta)}{1 - \eta - K}.$$

Using the approximate result shown in Figure 3.7 for the variation of nuclear and fossil fuel plant efficiency with condensation temperature, we have the parameters shown in Table C.1. This result is shown in Figure 3.9.

Table C.1

	Nuclear Plant ($K = 0$)		Fossil Fuel Plant ($K = 0.12$)	
T(°F)	η	r_h/r_e	η	r_h/r_e
100	0.31	0	0.40	0
200	0.24	0.092	0.33	0.13
300	0.167	0.172	0.255	0.23

Index